CHEMICAL NANOSCIENCE AND NANOTECHNOLOGY

New Materials and Modern Techniques

AAP Research Notes on Nanoscience and Nanotechnology

CHEMICAL NANOSCIENCE AND NANOTECHNOLOGY

New Materials and Modern Techniques

Edited by
Francisco Torrens, PhD
A. K. Haghi, PhD
Tanmoy Chakraborty, PhD

Apple Academic Press Inc.
3333 Mistwell Crescent
Oakville, ON L6L 0A2
Canada USA

Apple Academic Press Inc.
1265 Goldenrod Circle NE
Palm Bay, Florida 32905
USA

ISBN 13: 978-1-77463-448-6 (pbk)
ISBN 13: 978-1-77188-774-8 (hbk)

Library and Archives Canada Cataloguing in Publication

Title: Chemical nanoscience and nanotechnology : new materials and modern techniques / edited by Francisco Torrens, PhD, A.K. Haghi, PhD, Tanmoy Chakraborty, PhD.

Names: Torrens, Francisco (Torrens Zaragoza), editor. | Haghi, A. K., editor. | Chakraborty, Tanmoy, editor.

Series: AAP research notes on nanoscience & nanotechnology.

Description: Series statement: AAP research notes on nanoscience & nanotechnology | Includes bibliographical references and index.

Identifiers: Canadiana (print) 20190142081 | Canadiana (ebook) 20190142103 | ISBN 9781771887748 (hardcover) | ISBN 9780429398254 (ebook)

Subjects: LCSH: Nanotechnology. | LCSH: Nanostructured materials. | LCSH: Nanoscience. | LCSH: Chemistry.

Classification: LCC T174.7 .C54 2019 | DDC 620/.5—dc23

CIP data on file with US Library of Congress

ABOUT THE AAP RESEARCH NOTES ON NANOSCIENCE & NANOTECHNOLOGY BOOK SERIES:

AAP Research Notes on Nanoscience & Nanotechnology reports on research development in the field of nanoscience and nanotechnology for academic institutes and industrial sectors interested in advanced research.

Editor-in-Chief: A. K. Haghi, PhD

Associate Member of University of Ottawa, Canada;
Member of Canadian Research and Development Center of Sciences
and Cultures Email: akhaghi@yahoo.com

BOOKS IN THE AAP RESEARCH NOTES ON NANOSCIENCE & NANOTECHNOLOGY BOOK SERIES:

- **Nanostructure, Nanosystems and Nanostructured Materials: Theory, Production, and Development**
  Editors: P. M. Sivakumar, PhD, Vladimir I. Kodolov, DSc,
  Gennady E. Zaikov, DSc, A. K. Haghi, PhD
- **Nanostructures, Nanomaterials, and Nanotechnologies to Nanoindustry**
  Editors: Vladimir I. Kodolov, DSc, Gennady E. Zaikov, DSc,
  and A. K. Haghi, PhD
- **Foundations of Nanotechnology:**
  Volume 1: Pore Size in Carbon-Based Nano-Adsorbents
  A. K. Haghi, PhD, Sabu Thomas, PhD, and Moein MehdiPour MirMahaleh
- **Foundations of Nanotechnology: Volume 2: Nanoelements Formation and Interaction**
  Sabu Thomas, PhD, Saeedeh Rafiei, Shima Maghsoodlou, and Arezo Afzali
- **Foundations of Nanotechnology: Volume 3: Mechanics of Carbon Nanotubes**
  Saeedeh Rafiei
- **Engineered Carbon Nanotubes and Nanofibrous Material: Integrating Theory and Technique**
  Editors: A. K. Haghi, PhD, Praveen K. M., and Sabu Thomas, PhD
- **Carbon Nanotubes and Nanoparticles: Current and Potential Applications**
  Editors: Alexander V. Vakhrushev, DSc, V. I. Kodolov, DSc,
  A. K. Haghi, PhD, and Suresh C. Ameta, PhD
- **Advances in Nanotechnology and the Environmental Sciences: Applications, Innovations, and Visions for the Future**
  Editors: Alexander V. Vakhrushev, DSc, Suresh C. Ameta, PhD,
  Heru Susanto, PhD, and A. K. Haghi, PhD
- **Chemical Nanoscience and Nanotechnology: New Materials and Modern Techniques**
  Editors: Francisco Torrens, PhD, A. K. Haghi, PhD,
  and Tanmoy Chakraborty, PhD
- **Nanomechanics and Micromechanics**
  Editors: Satya Bir Singh, PhD, Alexander V. Vakhrushev, DSc,
  and A. K. Haghi, PhD

ABOUT THE EDITORS

Francisco Torrens, PhD
Lecturer, Physical Chemistry, Universitat de València, València, Spain

Francisco Torrens, PhD, is a lecturer in physical chemistry at the Universitat de València in Spain. His scientific accomplishments include the first implementation at a Spanish university of a program for the elucidation of crystallographic structures and the construction of the first computational chemistry program adapted to a vector-facility supercomputer. He has written many articles published in professional journals and has acted as a reviewer as well. He has handled 26 research projects, has published two books and over 350 articles, and has made numerous presentations.

A. K. Haghi, PhD
Professor Emeritus of Engineering Sciences, Formerly Editor-in-Chief, International Journal of Chemoinformatics and Chemical Engineering and Polymers Research Journal; Member, Canadian Research and Development Center of Sciences and Cultures (CRDCSC), Canada

A. K. Haghi, PhD, is the author and editor of 165 books, as well as 1000 published papers in various journals and conference proceedings. Dr. Haghi has received several grants, consulted for a number of major corporations, and is a frequent speaker to national and international audiences. Since 1983, he served as professor at several universities. He is the former currently Editor-in-Chief of the *International Journal of Chemoinformatics and Chemical Engineering* and *Polymers Research Journal* and on the editorial boards of many international journals. He is also a member of the Canadian Research and Development Center of Sciences and Cultures (CRDCSC), Montreal, Quebec, Canada.

Tanmoy Chakraborty, PhD
Associate Professor, Department of Chemistry,
Manipal University Jaipur, Jaipur, India

Tanmoy Chakraborty, PhD, is now working as Associate Professor in the Department of Chemistry at Manipal University Jaipur, India. He has been working in the challenging field of computational and theoretical chemistry for the last 11 years. He has completed his PhD from the University of Kalyani, West Bengal, India, in the field of application of QSAR/QSPR methodology in the bioactive molecules. He has published many international research papers in peer-reviewed international journals with high impact factors. Dr. Chakraborty is serving as an international editorial board member of the *International Journal of Chemoinformatics and Chemical Engineering*. He is also reviewer of the *World Journal of Condensed Matter Physics* (WJCMP). Dr. Tanmoy Chakraborty is the recipient of prestigious Parameswar Mallik Smawarak Padak, from Hooghly Mohsin College, Chinsurah (University of Burdwan), India, in 2002.

CONTENTS

CONTRIBUTORS

Sanjay Kumar Bharti
Institute of Pharmaceutical Sciences, Guru Ghasidas Vishwavidyalaya (A Central University), Bilaspur 495009, Chhattisgarh, India

Gloria Castellano
Departamento de Ciencias Experimentales y Matemáticas, Facultad de Veterinaria y Ciencias Experimentales, Universidad Católica de Valencia San Vicente Mártir, Guillem de Castro-94, E-46001 València, Spain

Tanmoy Chakraborty
Department of Chemistry, Manipal University Jaipur, Dehmi Kalan, Jaipur 303007, India. E-mail: tanmoychem@gmail.com, tanmoy.chakraborty@jaipur.manipal.edu

A. K. Haghi
Textile Engineering Department, University of Guilan, Rasht, Iran. E-mail: AKHaghi@yahoo.com

Shohreh Kasaei
Department of Computer Engineering, Sharif University of Technology, Tehran, Iran

Rajpreet Kaur
Department of Chemistry, BBK DAV College for Women, Amritsar 143001, Punjab, India

Ajay Kumar
Department of Mechatronics Engineering, Manipal University Jaipur, Dehmi Kalan, Jaipur 303007, India

Poonam Khullar
Department of Chemistry, BBK DAV College for Women, Amritsar 143001, Punjab, India. E-mail: virgo16sep2005@gmail.com

Debarshi Kar Mahapatra
Department of Pharmaceutical Chemistry, Dadasaheb Balpande College of Pharmacy, Nagpur 440037, Maharashtra, India. E-mail: mahapatradebarshi@gmail.com

Bentolhoda Hadavi Moghadam
Textile Engineering Department, University of Guilan, Rasht, Iran

Sukanchan Palit
43, Judges Bagan, Post Office-Haridevpur, Kolkata 700082, India. E-mail: sukanchan68@gmail.com, sukanchan92@gmail.com

Prabhat Ranjan
Department of Mechatronics Engineering, Manipal University Jaipur, Dehmi Kalan, Jaipur 303007, India

P. L. Reshma
Laboratory of Biopharmaceuticals and Nanomedicine, Division of Cancer Research, Regional Cancer Centre, Thiruvananthapuram 695011, Kerala, India

T. T. Sreelekha
Laboratory of Biopharmaceuticals and Nanomedicine, Division of Cancer Research,
Regional Cancer Centre, Thiruvananthapuram 695011, Kerala, India.
E-mail: ttsreelekha@gmail.com, ttsreelekha@rcc.gov.in

Heru Susanto
Department of Computer Science and Information Management, Tunghai University,
Taichung, Taiwan
Computational Science, Indonesian Institute of Sciences, Serpong, Indonesia
School of Business and Economics, University of Brunei, Bandar Seri Begawan,
Brunei Darussalam

H. P. Syama
Laboratory of Biopharmaceuticals and Nanomedicine, Division of Cancer Research,
Regional Cancer Centre, Thiruvananthapuram 695011, Kerala, India

Lavanya Tandon
Department of Chemistry, BBK DAV College for Women, Amritsar 143001, Punjab, India.
E-mail: Lavanyatandon@yahoo.in

Francisco Torrens
Institut Universitari de Ciència Molecular, Universitat de València, Edifici d'Instituts de Paterna,
P. O. Box 22085, E-46071 València, Spain. E-mail: torrens@uv.es

B. S. Unnikrishnan
Laboratory of Biopharmaceuticals and Nanomedicine, Division of Cancer Research,
Regional Cancer Centre, Thiruvananthapuram 695011, Kerala, India

ABBREVIATIONS

2D	two-dimensional
3D	three-dimensional
AgNPs	silver nanoparticles
APIs	application programming interfaces
B3LYP	Lee–Yang–Parr exchange correlation functional
BSA	bovine serum albumin
CAD	computer-aided design
CDFT	conceptual density functional theory
Ce 6	Chlorin e6
CGAL	Computational Geometry Algorithms Library
CT	computed tomography
DFT	density functional theory
DTX	docetaxel
DVI	digital volumetric imaging
EBSD	electron back-scatter diffraction
EGFR	epidermal growth factor receptor
EPR	enhanced permeation and retention
FFT	fast Fourier transform
FIB-SEM	focused ion beam-scanning electron microscope
GGA	generalized gradient approximation
GPU	graphics processing unit
GSH	glutathione
HA	hypocrellin A
HOMO–LUMO	highest occupied molecular orbital–lowest unoccupied molecular orbital
HRP	horse radish peroxidase
INVITE	inulin-d-α-tocopherol succinate
IT	information technology
KNIME	Konstanz Information Miner
LbL	layer-by-layer
LSCM	laser scanning confocal microscope
LSPR	localized surface plasmon resonance
MMP	matrix metalloproteinase
MNPs	magnetic nanoparticles

MNPs	metal nanoparticles
MRI	magnetic resonance imaging
NF	nanofiltration
NGO	nanographene oxide
NIR	near infrared
NMs	nanomaterials
NPs	nanoparticles
NSs	neutron stars
PAH-MNPs	PAH-stabilized magnetic nanoparticles
PCL	polycaprolactone
PCR	polymerase chain reaction
PDT	photodynamic therapy
PEG	polyethylene glycol
PS	Photosensitizer
PSA	prostate-specific antigen
PTE	periodic table of the elements
PTT	photothermal therapy
PTX	paclitaxel
PVP	polyvinylpyrolidine
RGD peptide	arginylglycylaspartic acid peptide
RISC	RNA-induced silencing complex
RMSE	root mean square error
ROS	reactive oxygen species
SA 2	Screen Assistant 2
SCCA	squamous cell carcinoma antigen
SEM	scanning electron microscope
SERS	surface enhanced Raman spectroscopy
SIFT	scale-invariant feature transform
siRNA	small interfering RNA
SPION	super paramagnetic iron oxide nanoparticle
STL	stereo lithography
TEM	transmission electron microscopy
TGS	transcriptional gene silencing
UCNPs	upconversion nanoparticles
UV	ultraviolet
VEGF	vascular endothelial growth factor
VTK	Visualization Toolkit
WPN	Working Party of Nanotechnology

PREFACE

The fast growth of nanotechnology with modern methods gives scientists the possibility to design multifunctional materials and provide products in human surroundings. Understanding the nanoworld makes up one of the frontiers of modern science. One reason for this is that technology based on nanostructures promises to be hugely important economically.

The scope of this book is to provide practical experience of commercially available systems. It includes small-scale nanotechnology-related projects for academics, researchers, and scientists to present their research and development work that have potential for applications in several disciplines of chemical nanotechnology.

The evolution of chemical nanotechnology has brought chemists working in close collaboration with other scientific and engineering disciplines, with physics and biochemistry, as well as with materials scientists and industrialists. In order for this area of chemistry to progress rapidly, it is crucial that these diverse specialists understand the requirements of each other and work together efficiently.

In this book, contributions range from new methods to novel applications of existing methods to gain understanding of the material and/or structural behavior of new and advanced systems. This book provides innovative chapters on the growth of educational, scientific, and industrial research activities among chemical scientists. It also provides a medium for mutual communication between international academia and the industry.

The chapters provide comprehensive coverage on the latest developments of research in the ever-expanding area of theoretical and experimental chemistry and their applications to broad scientific fields spanning physics, chemistry, biology, materials, and so on.

This book will serve as a major single source of information on the latest research that can be broadly defined to be in the general area of theoretical and experimental nanochemicals. This book includes original contributions on broad aspects: from both the development of fundamental theoretical methodology, to extensive specific scientific problems.

This volume brings new knowledge to the attention of our readers. As such, this volume not only reports results but also draws conclusions and explores implications of the work presented.

The collection of topics in this book reflects the diversity of recent advances in nanotechnology with a broad perspective that may be useful for graduate students and industrial sectors. Reporting new methodologies and important applications in nanochemistry, this book aims to present leading-edge research from around the world in the dynamic field of nanotechnology.

CHAPTER 1

ADVANCING COMPUTATIONAL METHODS IN CHEMICAL ENGINEERING AND CHEMOINFORMATICS

HERU SUSANTO*

Department of Computer Science and Information Management, Tunghai University, Taichung, Taiwan

Research Center for Informatics, Indonesian Institute of Sciences, Cibinong, Indonesia

School of Business, University Technology of Brunei, Bandar Seri Begawan, Brunei Darussalam

*E-mail: heru.susanto@lipi.go.id

ABSTRACT

Cheminformatics, which is the combination of chemistry and informatics, this field mixing of those information resources to transform data into information and knowledge for the intended purpose of making better decisions faster in the area of drug lead identification and optimization. Here, cheminformatics refers to the application of computer technology in anything that is associated before the development of computerized data storage for chemical compounds. An effective data mining system helps to create and study new chemical objects in which it allows authentication and checking of physical and chemical characteristics among a large collection of described compounds. It is without doubt that cheminformatics helps humanity in many ways especially in the drug-making activity and in discovery of new drugs. Furthermore, it also opens up new area for research and development and also increases opportunities for education. Information system and technology plays a major and crucial part in cheminformatics, especially to store the tremendous amount of data relating to chemical compounds and other chemical information. In this study was highlighted the impact of information system and technology to enabling accessibility of these data since it can be easily obtainable for further stages and processes. Without information system and technology, chemists would not be able to create new drugs easily, resulting in a huge setback on the drug discovery process. Cheminformatics engineering has found a particular application especially in the pharmaceutical industry, but it is now beginning to penetrate into other areas of chemistry.

1.1   INTRODUCTION

As time passed and technology improved, it was discovered that it was more efficient to store information on a computer desktop than just the desk. Hence, the birth of cheminformatics, which is the combination of chemistry and informatics. Initially, this field of chemistry did not have a name until 1998, when cheminformatics was first coined by Brown.[23] His definition is the "mixing of those information resources to transform data into information and information into knowledge for the intended purpose of making better decisions faster in the area of drug lead identification and optimization." Generally, cheminformatics refers to the application of computer technology in anything that is associated before the development of computerized data storage for chemical compounds; we have the traditional documentation system using books and documents. One of the earliest encyclopedias on chemical compounds is Beilstein's Handbuch der Organischen Chemie or *Beilstein's Handbook of Organic Chemistry* whose first edition was published in 18th century, consisting of two volumes, with more than 2000 pages and registering 1500 compounds. This comprehensive encyclopedia of organic structures covers chemical literature from 1771 to date. Information system concept is an integral part of cheminformatics. In this report, we will look at the history of cheminformatics and its applications and usefulness in the field of chemistry and beyond.[1–3,25]

1.2 MAIN IDEAS

1.2.1 HISTORY OF CHEMINFORMATICS

An effective data mining system helps to create and study new chemical objects in which it allows authentication and checking of physical and chemical characteristics among a large collection of described compounds. Getting all this information had always been a problem in chemistry. Chemists should be able to get access to accurate data whenever it is needed. Therefore, from the beginning, chemists have been developing documentation systems on chemical compounds. The earliest and oldest journal on chemistry is Chemisches Zentralblatt, which appeared as early as 1830; another example is as mentioned above, the Beilstein's Handbuch der Organischen Chemie, Handbook of Organic Chemistry and Chemical Abstracts which has been published since 1907. One more notable example is March's Advanced Chemistry. The problem is these data storage systems are studied in basic chemical handbooks.[9,10]

Chemical information branches wanted to benefit from computers as it is much more effective and efficient to save information or data on the computer than just on the desk. Therefore, computations and chemical information are two important components that made up cheminformatics. Polanski[25] stated that Johann Gasteiger, a famous German chemist, supported this statement with the fact that in 1975, *Journal of Chemical Documentation* (a journal specializing in chemical information compiled by American Chemical Society, one of the most authoritative providers of chemistry-related information) changed its name to *Journal of Chemistry and Computer Science*. Similarly, we can use the same title to show recent developments in this field of chemistry, since the journal had just changed the name to *Journal of Chemical Information and Modeling* in 2004. Thus, the journal's history briefly illustrates the scope of the discipline of cheminformatics.

1.2.2 USEFULNESS OF CHEMINFORMATICS AND ITS APPLICATIONS

The main purpose of cheminformatics is to preserve and allow access to tons of data (big data) and information that are related to chemistry;

moreover integrating informations needed on specific tasks or studies. Another purpose is to aid in the discovery of new drugs.

Possible use of information technology (IT) is to plan intelligently and to automate the processes associated with chemical synthesis of components of the treatment is a very exciting prospect for chemists and biochemists. One example of the most successful drug discoveries is penicillin. Penicillin is a group of β-lactam antibiotic used in the treatment of infectious diseases caused by bacteria, usually Gram-positive manifold. The way to discover and develop drugs is the result of chance, observation, and many intensive and slow chemical processes. Until some time ago, drug design is considered labor intensive, and the test process always failed.

Nowadays, drug abuse is becoming a major issue around the world. Cheminformatics contributes to this issue by predicting or designing drugs. Immunoassays are commonly used to detect or screening test of drug abuse on individuals through their urine or other body fluids. Immunoassays screening commonly points toward classes of drugs such as amphetamines, cannabinoids, cocaine, methadone, and opiates.

Moreover, drugs identification can be identified in specific details by applying mass spectrometry methods such as liquid and gas chromatography. Chromatography is the process which suggested that cheminformatics is just a new name for an old problem (Hann and Green, 1999). While some current interest in chemoinformatics can come as natural enthusiasm for new things, of separating constituents of a solution by exploiting different bonding properties of different molecules.

In cheminformatics, there is a good relationship between chemistry and technology. The development of IT has evolved considerably over time.[1] The development of IT developed in conjunction with a variety of disciplines and applied in various fields. Advances in IT can be accessed quickly and precisely. IT has changed the way we look at science. Surrounded by a sea of data and phenomenal computing capacity, methodologies and approaches to scientific issues developed into a better relationship between theory, experiment, and data analysis).[5,7,8]

Consequently, cheminformatics is related to the application of computational methods to solve chemical problems, with special emphasis on manipulation of chemical structural information. As said before, the term was introduced by Brown[23] but there has been no universal agreement about the correct term for this field. Cheminformatics is also known as chemical informatics. Many of the techniques used in cheminformatics are actually rather well established, be the result of years if not decades of research in academic, government, and industrial laboratories. The main reason for its inception can be traced to the need for dealing with large amounts of data generated by the new approach to drug discovery, such as high-througpout screening and combinatorial chemistry. Increase in computer power, especially for desktop engine, has provided the resources to handle this flood. Many other aspects of drug discovery make use of the techniques of cheminformatics, from design of new synthetic route by searching a database of reactions known through development of computational models such as quantitative structure–activity relationship (QSAR), which associates something that is observed through the biological activity of chemical structures through the use of molecular docking program to predict the three-dimensional structure protein–ligand complexes and then chosen from a set of compounds for screening. One common characteristic is that this method cheminformatics must apply to a large number of molecules. Cheminformatics, which seams together chemistry and informatics, is obviously linked to computer applications. But, not all chemical branches that relied on computers should automatically be included in the field. Although this term was introduced in the 1980s, it has a long history with its roots going back more than 40 years. The principles of cheminformatics are used in chemical representation and search structure, quantitative structure–activity relationships, chemometrics, molecular modeling, and structural elucidation of computer aid design and synthesis. Each area of chemistry from analytical chemistry to drug design can benefit from the methods of cheminformatics. This report will briefly discuss the accomplishments in chemistry that information system and technology had helped to support.[11,12]

1.2.3   INFORMATION TECHNOLOGY: BIOTECHNOLOGY

IT applications in the field of molecular science has spawned the field of biotechnology. Biotechnology is a branch of science that studies the use of living organisms (bacteria, fungi, viruses, etc.) as well as products from living organisms (enzyme, alcohol) in the production process to produce goods and services. This study is increasingly important because the development has been encouraging and impactful on the field of medicine, pharmacy, environmental, and others. These fields include the application of methods of mathematics, statistics, and informatics to answer biological problems, especially with the use of DNA and amino acid sequences as well as the information related to it. Insistence of the need to collect, store, and analyze biological data from a database of DNA, RNA, or protein acts as a spur to the development of bioinformatics. There are nine branches of biotechnology, and one of them is cheminformatics. Cheminformatics is one of biotechnology's disciplines that is a combination of chemical synthesis, biological filtering, and data mining is used for drug discovery and development.

1.2.4 DRUG DISCOVERY

Cheminformatics has developed ever since the time of its establishment although throughout the decades when computers and technologies had also developed. Researchers and scientists have been developing a way of assembling information and data within computers. These innovations on computers and technologies have the ability of storing and obtaining chemical informations.

The meaning of these disciplines mentioned above is the identification of one of the most popular activities compared with various fields of study that may exist below this field. Drug discovery requires some experts to bring in technology from other fields, and a great time to be provocative and give a long overdue paradigm shift.

The way to discover and develop drugs is the result of the agreement, observation, and chemical processes, many intensive and slow. Until some time ago, considered drug design should always use a labor-intensive process and the test fails (the process of trial-and-error). The possibility of the use of IT to plan intelligently and to automate processes associated with chemical synthesis components of the treatment is a very exciting prospect for chemists and biochemists. Award to produce a drug that can be marketed faster is huge, so the target is at the core of cheminformatics.

The scope is very broad chemical from academic. Examples of areas of interest include: planning synthesis, reactions and structure retrieval, modeling, 3D structure retrieval, computational chemistry, visualization tools, and utilities.

1.2.5 CHEMINFORMATICS: EDUCATION

In today's society, technology plays an important role, not only on unlimited access for communicating purposes but also educational. Cheminformatics opens up a new field of education, although there are countless educational areas that exist today. Cheminformatics has numerous and distinct applications and database. Cheminformatics can be learned and taught online. This could give the society easy access about learning the disciplines and methods of cheminformatics. Universities have been partnering up with companies that offer higher education learnings such as Coursera, Udacity, and edX where world experts can connect up with society with an internet connection and a computer for the purpose of tutoring or other educational purposes.

The use of information system is important to science educators; it can help them to do their task by exploring the technology that already exist. It can also make science educations that use information system in their learning become a higher education. It is important to note that the insertion of information system and technology into education or other courses will make a difference in higher education as it has in the simple insertion into public school. Information system and technology is not a vehicle for change, but technology is simply a tool used by its user to do certain objectives.

1.2.6 HEALTH INFORMATION SYSTEM

Health information system manages information related to health. Everything related to the health is managed by the information system in order to make it easy to be used. It also can help any activities of an organizations or individuals to access some information about their health easily. Some examples about it are hospital patient administration system (PAS) and human resource management information system (HRMIS). Overall, it used by the organizations to integrate effort to collect, process, report, and use health information and knowledge to impact the policy and decision making, program action, individuals and public health outcomes, and research.[14,15,21,22]

The main function of the system is to set up some guidelines to help health workers, health information managers, and administrators at all levels to focus on increasing the effectiveness of time, the accuracy and reliability of health data. Although the emphasis might seem to be on hospitals medical records, these guidelines have been designed to address all areas in healthcare where data are collected and information generated. Guidelines describe activities that must be taken into consideration when responding to the question of data quality in health care, regardless of the setting. Readers are guided to access and enhance the quality of the data generated in the environment in which they function, regardless of size, isolation, or sophistication. These guidelines are intended for government policy makers and healthcare administrators in the primary level, secondary, and tertiary healthcare. There are doctors, nurses, other healthcare providers, and health information managers as well. All these share a responsibility for documentation, implementation, development, and management of health information services.[4,13,18]

1.2.7   REDUCTION OF RISKS IN DRUG DEVELOPMENT

Cheminformatics can help chemists and other scientists to produce and manage information. In silico analysis using cheminformatics techniques can actually reduce the risks in drug development. Techniques such as virtual screening, library design, and docking figures go into the analysis. Physical properties that may have an impact on whether a substance has the potential to be developed as a drug are often examined by one of the cheminformatics features which makes comparison among a large number of substances. Examples are clogP, a measure of the amount of molecular obesity in the system. Sometimes, conclusions can be drawn about a set of associated properties, such as when Chris Lipinski, an experienced medicinal chemist, formulated the famous Rule of Five which is saying that compounds such as drugs tend to have five hydrogen bond donors or fewer, 10 maximum hydrogen bond acceptors, the calculated logP should be less than or equal to 5, and having a molecular mass of up to 500 Da. The compounds which showed greater values than these criteria tend to have poor absorption or permeation, meaning the drug is not orally active drug in humans.

1.2.8 FORECASTING COMPOUND AGAINST NEUROLOGICAL DISEASES

Many compounds kept in databases have already been explored for multiple aims as portion of drug-discovery programs. Excavating this information can provide experimental evidence useful for structuring pharmacophores to determine the main pharmacological groups of the compound. Predictor model and DrugBank predictor model dataset was built using data collected from the ChEMBL database by means of a probabilistic method. The model can be used to forecast both the primary target and off-targets of a compound based on the circular fingerprint methodology, one of the technology developed by cheminformatic method. The study of off-target connections is now known to be as important as to recognize both drug action and toxicology. These molecular structures are drug targets in the treatment neurological diseases such as Alzheimer's disease, obsessive disorders, and Parkinson's disease and depression. In future, developing these multitargeted compounds with selection and chosen ranges of cross-reactivity can report disease in a more subtle and effective ways and will be a key pharmacological concept in future.

1.2.9 HELPS IN OPTIMIZING ANTIBACTERIAL PEPTIDES WHEN COMBINED WITH GENETIC ALGORITHM

Research done by University of Columbia, Department of Medicine, had correctly identified additional activity of peptides with 94% accuracy among the top ranked 50 peptides chosen from an in silico library of approximately 100,000 sequences by adopting genetic algorithm. These methods lets a radically increased capability to recognize antimicrobial peptide candidates. A genetic algorithm is well suited for difficulties involving string-like data method for search-and-approximation problems. Implementing an iterative method whereby computational and experimental methodologies were used to find new improved starting point for beginning of genetic algorithms are more effective tactic. Training the machine to learn algorithms using the new data improve the ability to predict peptide activity. Based on the research, it has been reported that there is several peptides that are active against pathogens of clinical importance, despite these limitations on overview of prediction.

1.2.10 THE POTENTIAL OF CHEMINFORMATICS DATABASE IN FOOD SCIENCE

Computer databases nowadays have become important tools in biological sciences. Bioinformatics tools support the identification of chemical compounds, for example, peptides and proteins by mass spectrometry which is the most reliable tool for application s of the type. Cheminformatics databases are generally used in biological and medical sciences and play an increasingly significant role in modern science. Emphasizing the developing character of data-mining and management techniques in animal breeding and food technology made computer databases the most extensive resource for finding and processing such information. One of the main goals of cheminformatics is to clarify life from the chemical outlook. The biological activity of chemical compounds thus falls into both bioinformatics and cheminformatics, for example, food scientists and researcher will need to access databases to find information about the biological activity and behavior of several food components.

1.2.11 INCREASED EDUCATIONAL OPPORTUNITIES

People sometimes have difficulties to obtain formal qualification but yet are interested to learn about the methods of cheminformatics due to limitations such as time, finances, and commitment. Nowadays with modern technology, online learning by video and web conferencing as well as free online learning resources are available for cheminformatics. Online learning is another option to be considered in learning cheminformatics and enabling the wisdom impact to a wide range of practitioners and disciplines and in time will increase rapidly growing communities around the globe. Online learning also increase expertise of the number of people with capability in its many techniques and applications in chemistry field. Online courses have been offered in universities together with the collaboration with expertise in chemistry.

The expanding significance of health, science, data, and informatics is evidence of the critical importance of cheminformatics in future. Using computer and connection, people easily have accessibility to cheminformatics education. Live discussion also plays a major role on how to share idea and thought in cheminformatics education in the globally connected world we live in.

1.3   APPLICATION AND SOFTWARE

In this chapter, applications of cheminformatics will be briefly highlighted. The applications included are the most commonly used and contribute on the field of cheminformatics.

1.3.1   SCREEN ASSISTANT 2

Screen Assistant 2 (SA 2), an open-source JAVA software dedicated to the storage and analysis of small to grand size chemical libraries. SA 2 contains information and data on molecules in a MySQL database. SQL, which stands for structured query language, is the common language that is used for the purpose of adding, collecting, regulating, and operating the content in a database. It is much preferred by a whole lot because of its quick response and processing that satisfy a lot of users.

1.3.2   BIOCLIPSE

In cheminformatics, there is a need on applications which can help users with an extensible tool in the obtaining and calculating process of what cheminformatics has to offer. Bioclipse contains 2D editing, 3D visualization, converting files into various formats, and calculation of chemical properties. All these combined into a user friendly application, where preparing and editing are easy such as copy and paste, dragging and drop, and to redo and undo process. Bioclipse is in the form of Java and based on Eclipse Rich Client Platform. Bioclipse has the advantages on other systems as it can be used in anyway based on the field of cheminformatics.

1.3.3   CINFONY

Toolkits such as Rational Discovery Kit for building predictive models, an Open-Source Cheminformatics Software (RDKit), Cyclin-dependent kinases, which drive the cell cycle and therefore cell proliferation, have been a focus for drug developers since the 1990s (CDK), and a chemical toolbox designed to speak the many languages of chemical data. It's an open, collaborative project allowing anyone to search, convert, analyze, or store data from molecular modeling, chemistry, solid-state materials, biochemistry, or related areas (OpenBabel) function are very similar with each other but the differences are each support various sets of data formats. Although they have complementary features, operating these toolkits on similar programs is challenging because they run on different languages, different chemical models, and have different application programming interfaces (APIs).

Cinfony, a Python module that introduce all three toolkits in an interface which make it easier for users to integrate methods and outcomes from any of those three toolkits. Cinfony makes it easier to perform common tasks in cheminformatics tasks that include calculating and reading.

1.3.4   KNIME-CDK: WORKFLOW-DRIVEN CHEMINFORMATICS

Konstanz Information Miner (KNIME): Workflow-driven cheminformatics is one of modern data analytics platform, open-source library fully open, and it is shared and exposed to the commodity.[19,20] One of the features available are plug in features which allows for efficient and easy to use and better suited, whereby it enable researcher to automate the routine task and data analysis and also enables to build additional nodes; data analysis pipelines from defined components that work well combine with the existing molecule presentation.[6,16,17] KNIME allows you to execute complex statistics and data mining by using tools like clustering and machine learning, even plotting and chart tools on the data to examine trends and forecast possible results.

One of the standard roles includes data manipulation tools to manage data in tables, for example, joining, filtering, and partitioning as well as execute these molecule transformation according to common formats. Other tools to manage data use are substructure searching, signatures generating, and fingerprints for a molecular properties. KNIME also use target prediction tool which can predict the effects of existing drug in term of their toxicity by giving suggestion on what molecular mechanism is observed behind the undesirable side effects and repurposing by exposing the new uses of the current existing drugs.

1.3.5   CHEMOZART: VISUALIZER PLATFORM AND A WEB-BASED 3D MOLECULAR STRUCTURE EDITOR

Chemozart provides the ability to create 3D molecular structure. This application tool which is web based also use for viewing and editing of these molecular structures. As modern technology evolved, Chemozart which have flexible core technologies can be accessed easily via a UR. The platform is independent and compatible and has been deliberately created in a way that it is compatible with the latest devices, that is, mobile. This application also enables the process of teaching given that it works on mobile devices. It gives benefit to students as it has user-friendly interface, as they easily understand the concept of stereochemical of molecules when constructing, drawing, and viewing 3D structures; therefore, it is used for educational purpose. With the help of this web-based platform, user can simply create as well as modify or edit or just viewing the structures of the molecular compounds with just rearranging the position of atoms as simple as dragging them around or use keyboard or now we can use any touch screens devices.

1.3.6   Open3DALIGN: SOFTWARE FOCUSES ON UNSUPERVISED LIGAND ALIGNMENT

One of the classical tasks of cheminformatics is unsupervised alignment of a structurally varied series of biologically active ligands which leads to various ligand-based drug design methodologies. The most important ligand-based drug design methods are pharmacophore elucidation and 3D quantitative structure–activity relationship studies. Open3DALIGN together with its scriptable interface had the capability of carrying out both conformational searches and multi-unsupervised conformational alignment of 3D molecular structures rigid body which makes automated cheminformatics workflows an ideal component of high throughput. Now different algorithms have been applied to perform single and multiconformation superimpositions on one or more templates. Alignments which contains two operations; feature matching and conformational search can be achieved by corresponding pharmacophores and heavy atoms or any mixed of the two. Feature matching can be accomplished through field-based, pharmacophore-based, and atom-based methods approaches whether to find a same matching molecular interaction fields or searching a collection of pharmacophoric points or heavy atom pairs. Finding the best suited conformer for each ligand which may be mined from pre-built libraries are the strategy use in conformational search and can be achieved by following rigid alignment on the template and candidate ligands which may also be easily adaptable and aligned on the template. Regardless of the methods and approaches, great computational performance has been achieved through well-organized parallelization of the code features.

1.3.7   CDK-TAVERNA: OPEN WORKFLOW ENVIRONMENT SOLUTION FOR THE BIOSCIENCES

Computational process and analysis of small molecule are among the essential for both cheminformatics and structural bioinformatics, for example, in drug discovery application. CDK-Taverna 2.0 through combination of unsimilar open-source projects whose goals are structuring a freely available open-source cheminformatics pipelining solution and become a progressively influential tool for the biosciences. CDK-Taverna 2.0 was effectively applied and tested and verified in academic and industrial environments with sea of data of small molecules combined with workflows from bioinformatics statistics and images as well as analysis being made. CDK-Taverna supports the process of varied sets of biological data by constructing a complex systems biology oriented workflows. In old days, insufficiencies like workarounds for iterative data reading are removed by sharing the previously accessible workflows developed by a lively community of and available online which enables molecular scientists to quickly compute, process, and analyze molecular data as typically found in today's systems biology scenarios. Graphical workflow editor is currently maintained and are being supported by design and manipulation of workflows. The features are considerably enhanced by the combinatorial chemistry related reaction list. Implementing the identification of likely drug candidates are one of the additional functionality for calculating score for a natural product similarity for small molecules. The CDK-Taverna project was recent and constantly up-to-date and multiusage by paralleled threads are now enabled to carry out analysis of large sets of molecules as well as faster in memory processing.

1.3.8   ODDT: OPEN DRUG DISCOVERY TOOLKIT

Drug discovery has become a significant element supplementing classical medicinal chemistry and high-throughput screening these results to many computational chemistry methods were developed to aid them in learning capable drug candidates. ODDT open-source player in the drug discovery field which aims to fulfill the need for comprehensive and open-source drug discovery software because there has been enormous progress in the open cheminformatics field in both methods and software development. Sadly, there has only been little effort to combine them in only one package.

Structure based are the most general and successful methods in drug discovery in which the methods are commonly actively used to screen large small molecule datasets, that is online databanks or smaller sets, that is, tailored combinatorial chemistry libraries. These methods are crucial for decision making. Today, much effort is focused toward machine learning which is the most valuable in clarifying both nonlinear and trivial correlations in data, respectively.

1.3.9   INDIGO: UNIVERSAL CINFORMATICS API

Indigo is an open-source library which allows developers and chemists to solve many cheminformatics tasks. During the past years, a collection of more specific tools have enormous development by the universal portable library. Tools suitable for scientists are popular programing languages as well as some GUI and command line. Performance and important chemical features are the core of this C++ cheminformatics library. Among the chemical features of Indigo are it supports popular chemistry formats and cis–trans stereochemistry.

1.4   POSSIBLE FUTURES FOR CHEMINFORMATICS

Information system has changed the way we do things nowadays. The process of storing, finding the exact molecule by indexing, searching, retrieving information, and applying information about chemical compounds are made easy with modern technology. Cheminformatics uses computer and informational techniques and applied it to a range of problems in the field of chemistry including chemical problems from making chemical analysis and biochemistry to pharmacology and drug discovery. Cheminformatics contributions to decision making are known to help certain scientists and chemist, for example, to extract right combination of density and structures from a database containing thousands of molecular structures data that are most likely to provide a specific function or healing effect. These are made available in the database to support research for better chemical decision-making by storing and integrating data in maintainable ways.

Improvement in technology assist cheminformatics in a way that now is potential to simulate protein complexes in solution, for example, pharmacophore analysis/visualization/pattern recognition and also the most complex biological networks that were not possible with the use of pen and paper. Faster working capability of computer and high speed networks connection had develop the quality of algorithms and eventually data sources will significantly increase as well. D. E. Shaw group recent publication shows that now millisecond simulations of drug docking using molecular dynamics are possible. This information can enhance knowledge worldwide. According to Glen,[24] there are altogether three areas that had to advance and develop in order to realize the possible growth of computational chemistry.

The first essentially area is finding the most accurate and relevant data available. Collaboration among researcher, organization that involved in chemical research, and relevant parties who are interested in discovery of new development in chemistry would make it possible that the relevant as well as credible and supported data and information are more open and available for anyone. These could be made possible with crowdsourcing and group of community sharing forum network as well as developing social system for sharing specific area of interest. Too much information could eventually become one of the main challenge face. Today, in order to face the main challenge, language processing had been developed even the most complex complicated documents made by researcher can be processed by robots to extract and find the most useful information. These robot will assist in finding the best suited data, then filter and organize them in a way they can cater our filed of interest. Another possible solution will be developing network that are self-organized and able to search and navigate the pool of data faster and relevant to our area of interest.

The second area of concern is how cheminformatics will be presented in computers whether the highly professional researcher in chemistry is willing to share information or they were bind by the confidentiality and secrecy of their work. These researcher tend to manipulate data by using complex language or symbols to present their research and chemistry concepts although the changes of name and description detail and easier term had been come up by three most influential in chemistry field namely Berzelius, Archibald Scott-Couper, and Frederick Beilstein almost 200 years ago. Now the trend is slowly change the name form complex to much simpler and the complex description become more easily to understand, the complicated detail attached to its function are made more simpler term yet there are present in the accurate and relevant information with much more detail information. The description of the information is not easily accessible where we can easily target the most relevant data that we are mostly curious in. More so, the symbols used in presenting molecules and structures are not easily made and understood. Proper indexing had also been made although until now it had not been fully incorporated yet and much improvement are still needed.

Third area is to replicate reality in the simulation. It is one of the main aims in computational chemistry. With modern technology nowadays petabyte computing are offered. Their capacity can simulations of real biological systems within millisecond with hardware precisely designed for simulation. With vastly increase development of hardware, the complexity of simulation would be made easier and user friendly. There will be much improvement in accuracy also. It is also dependent on the capability of the hardware design in future to cater the need for the development in simulation. When computing capacity are no longer limited in term of its capacity, extracting, filtering, and evaluating as well as recalculating could be done in more straightforward form and the complexity ill no longer be exists. IBM now is developing chips known as neural chips and with the emergence of cloud computing can improve further the area of cheminformatics. Super high speed connection, vast memory capability, and now environmental virtualization had made it possible in future to change the current cheminformatics. With the modern technology where everything is done by technology, it is possible that according to Glen[24] these machines will ask and think on their own to assist human in future.

Compounds will not be classified at molecular level anymore but rather in genetic and clinical effects term. Modern technology has been one of the major driving forces of the development of cheminformatics. It is hard to predict the future of cheminformatics but rather these report will cater for the possible future for cheminformatics and how the information system can support them.

1.4.1 FREE TEXT MACHINE LEARNING MODELS

Currently, the difficulty to process large information of data and extract information that is relevant to molecular discovery that in future automation are expected to enable to extract relevant information and conclusions rather than relying on human curation.

The free text machine learning models are likely to enable machine learning process and statistical model generation and would combine multiple controlled vocabularies and ontologies, for example, chemical–biological and biological assays interactions in an appropriate manner.

1.4.2 AN INCREASE IN ENVIRONMENTAL CHEMINFORMATICS

Currently, the awareness for the society to study the relationship between environmental exposure and human health had increased from time to time. Nowadays, human being are exposed to a wide range of environmental chemicals includes pollution present in our air and water, integrally exist in our food, medicines, cosmetics, and many others.

Regardless these chemicals still exist at a low concentration of risk below the toxicological concern but these matter should be taken into serious attention. These combination of chemical can potentially affect human health as fast as lightning. Integration of assessable data monitored studies should be cautiously taking into account not just the concentration of chemicals but also exposure of time and life period through a collaborative framework. The expectation of chemical–chemical collaboration tools (here referring to their effects in living organisms, not to chemistry and physics) would be tremendously valued in risk assessment.

1.4.3 SIGNIFICANT DECISION-MAKING SUPPORT

In order to support decision makers to minimize risk of exposure, support and develop legislative and societal demands a forecast support systems could be used in evidence-based medicine to assess environmental threat and chemical warfare agents. The future system may assist with optimizing the selection of active medical ingredients during the process of formulating new drug mixtures.

The system will have the capability to handle combinations of drug mixtures of cheminformatics at the "fixed" and "unspecified" mixture level. These technologies of computer-assisted chemical production might become universal in our determination to lessen the synthesis cost of making drugs that are approved and to reduce the impact on environmental chemical reagents.

1.4.4   DRUG DISCOVERY CONDUCTED BY AUTOMATED ROBOT

In future, robot scientists would perform procedures on the medicinal chemistry and toxicology works in search of additional proof to highlight the best likely chemicals and suggest these proof with supporting evidence to conduct cheminformatics experiments. These robots will identify safe chemicals that are supported by systems chemical biology within the fundamental human context by using computer-aided molecular and synthesis proposal, for example, conducting biomolecular screening and toxicity experiments.

1.4.5   EXPANSION OF KNOWLEDGE AND OPEN-ENDED SCIENCE

Cheminformatics can assist the expansion of knowledge. The problem faced currently are the vast amounts of molecular property and sea of data often lead to difficulty in making choices and assumptions about the data essentially related to the quality and type of data input that are no longer relevant. Scientists with lack of experience are less likely to investigate multiple diverging options thoroughly while conducting their research. These situation are called epistemological and they are likely to occur when mining difficult data. This epistemological situation are proof and supported by anecdotal behavior, for example, most internet users prefer to take top 10 hits offered by search engines and do not bother to scroll down below 10 hits.

By developing integrative tools and approach based on knowledge within the Cytoscape framework will enable users to visualize molecular interaction networks which provides a basic features for data integration and analysis and share them via a public setting. The example of public setting are crowdsourcing. Crowdsourcing can help knowledge sharing as it is easier for information to be gathered. This tools will make possible in future to share and explore alternative hypotheses and multiple situations given the available limitations as well as building high confidence in prediction data and models. Thus, unlikely for the occurrence of epistemological situation.

1.4.6 INNOVATION AND THE IMPACT OF CHEMINFORMATICS ON SOCIETY

By combining cheminformatics with bioinformatics and other computational systems are expected to give benefit to the developments in proteomics, metabolomics, and metagenomics as well as other sciences. Now in future, we will not need to estimate the bioactivity profile of a chemical at the molecular level but rather we will study biomedical information with the addition of inherent polymorphisms and clinical effects.

These development in proteomics, metabolomics, and metagenomics as well as other sciences can be supported by developing tools for integrated chemical–biological data acquirement, filtering, and processing by taking into account relevant information related to collaborations between proteins and small molecules as well as possible metabolic alterations. These tools will be integrated into the virtual physiological human.

1.4.7 INCREASE IN COMPUTATIONAL SPEED

Graphics processing unit (GPU) technologies are likely to result in tremendous improvements over existing pharmacophore and fingerprint technologies when the cheminformatics software is being transfer on GPU platform and the ability to do research on highly accurate electronic densities will improve dramatically.

1.5 OPINION

Cheminformatics contributes to the community in many different ways. One of the ways is that it helps in discovering new drugs and also to predict drugs. Cheminformatics helps researchers to experiment on chemical compounds and molecules through computers rather than to practically conduct the experiments. This can reduce costs on the process of the experiment itself.

Although cheminformatics contributes have strong impacts on modern society and science, there are also many obstacles faced by eager and anxious users on learning cheminformatics. This followings could easily be a disadvantages for those keen people in learning cheminformatics. The most common difficulties encountered are time limitation. Cheminformatics has various types of database and software that were installed in the computer. Amateurs will need to spend time on the way of how cheminformatic works, their database and software. Thus, most of people that learn cheminformatics has a shallow knowledge about it due to shortage of time.

Financial limitations, although experimenting on new drugs and discovering new drugs do not requires high expenses but taking the course in learning cheminformatics requires high price of admission among learners on the field of cheminformatics.

There are also challenges faced by the implementation of cheminformatics. The world nowadays promotes the practice of going green and environmental friendly. It is crucial to search for the right chemicals which are the ones that has low toxicity and low environmental threat properties. Although cheminformatics have the respect of working highly reliable through computers on prediction of drugs on chemical combinations.

Other challenges faced by cheminformatics are that it should not be focused on chemistry alone, this is because chemicals have also influence and impacts on cellular functions which trickles a way to biological field.

1.6 CONCLUSION

Cheminformatics has a history almost as long as the computer itself. It is the application of computer technology and methods for the chemical related field is molecular modeling and computational chemistry. It is without doubt that chemoinformatics help humanity in many ways especially in the drug-making activity and in discovery of new drugs. Furthermore, it also opens up new area for research development and also increases opportunities for education.

Information system and technology plays a major and crucial part in cheminformatics especially to store the tremendous amount of data relating to chemical compounds and other chemical information. Information system and technology also enable easy accessibility of these data because they can be easily obtainable whereas in the old days, all these data and information were only kept in handbooks which made it hard to get the data when needed. Without information system and technology, chemists would not be able to create new drugs easily resulting in a huge setback on the drug discovery process. In the beginning, cheminformatics engineering has found particular application especially in the pharmaceutical industry, but it is now beginning to penetrate into other areas of chemistry.

KEYWORDS

- cheminformatics
- health information system
- drug development
- drug discovery
- open source
- data mining
- neurological diseases

REFERENCES

1. Susanto, H; Almunawar, M. N. Information Security Management Systems: A Novel Framework and Software as a Tool for Compliance with Information Security Standard; CRC Press, USA, 2018.
2. Susanto, H.; Chen, C. K. Macromolecules Visualization Through Bioinformatics: An Emerging Tool of Informatics. *Appl. Phys. Chem. Multidiscip. Approaches* **2018,** 383–406.
3. Susanto, H.; Chen, C. K. Informatics Approach and Its Impact for Bioscience: Making Sense of Innovation. *Appl. Phys. Chem. Multidiscip. Approaches* **2018,** 407–426.

4. Susanto, H. Smart Mobile Device Emerging Technologies: An Enabler To Health Monitoring System. In *Kalman Filtering Techniques for Radar Tracking*; CRC Press, USA, 2018; p 241.

5. Liu, J. C.; Leu, F. Y.; Lin, G. L.; Susanto, H. An MFCC-based Text-independent Speaker Identification System for Access Control. In *Concurrency and Computation: Practice and Experience*; Wiley, USA, 2018; Vol. 30, p e4255.

6. Almunawar, M. N.; Anshari, M.; Susanto, H.; Chen, C. K. How People Choose and Use Their Smartphones. In *Management Strategies and Technology Fluidity in the Asian Business Sector;* IGI Global, 2018; pp 235–252.

7. Susanto, H.; Chen, C. K.; Almunawar, M. N. Revealing Big Data Emerging Technology as Enabler of LMS Technologies Transferability. In *Internet of Things and Big Data Analytics Toward Next-Generation Intelligence*; Springer: Cham, 2018; pp 123–145.

8. Almunawar, M. N.; Anshari, M.; Susanto, H. Adopting Open Source Software in Smartphone Manufacturers' Open Innovation Strategy. In *Encyclopedia of Information Science and Technology, 4th ed.*; IGI Global, USA, 2018b; pp 7369–7381.

9. Susanto, H. Cheminformatics: The Promising Future: Managing Change of Approach Through ICT Emerging Technology. In *Applied Chemistry and Chemical Engineering, Volume 2: Principles, Methodology, and Evaluation Methods*, CRC Press, USA,; 2017; p 313.

10. Susanto, H. Biochemistry Apps as Enabler of Compound and Dna Computational: Next-generation Computing Technology. *Applied Chemistry and Chemical Engineering, Volume 4: Experimental Techniques and Methodical Developments*; CRC Press, USA, 2017; p 181.

11. Susanto, H. Electronic Health System: Sensors Emerging and Intelligent Technology Approach. In *Smart Sensors Networks*; 2017; pp 189–203.

12. Leu, F. Y.; Ko, C. Y.; Lin, Y. C.; Susanto, H.; Yu, H. C. Fall Detection and Motion Classification by Using Decision Tree on Mobile Phone. In *Smart Sensors Networks*; 2017; pp 205–237.

13. Susanto, H.; Chen, C. K. Information and Communication Emerging Technology: Making Sense of Healthcare Innovation. In *Internet of Things and Big Data Technologies for Next Generation Healthcare;* Springer: Cham, 2017.

14. Susanto, H.; Almunawar, M. N.; Leu, F. Y.; hen, C. K. Android vs iOS or Others? SMD-OS Security Issues: Generation Y Perception. *Int. J. Technol. Diffu.* **2016,** *7* (2), 1–18.

15. Susanto, H. *Managing the Role of IT and IS for Supporting Business Process Reengineering*, SSRN Elsevier, 2016; p 16.

16. Susanto, H.; Kang, C.; Leu, F. *Revealing the Role of ICT for Business Core Redesign*, SSRN Elsevier; 2016.

17. Susanto, H.; Almunawar, M. N. Security and Privacy Issues in Cloud-based E-Government. In *Cloud Computing Technologies for Connected Government* IGI Global, USA; 2016; pp 292–321.

18. Leu, F. Y.; Liu, C. Y.; Liu, J. C.; Jiang, F. C.; Susanto, H. S-PMIPv6: An Intra-LMA Model for IPv6 Mobility. *J. Netw. Comput. Appl.* **2015,** *58,* 180–191.

19. Susanto, H.; Almunawar, M. N. Managing Compliance with an Information Security Management Standard. In *Encyclopedia of Information Science and Technology,* 3rd ed.; IGI Global, 2015; pp 1452–1463.
20. Almunawar, M. N.; Susanto, H.; Anshari, M. The Impact of Open Source Software on Smartphones Industry. In *Encyclopedia of Information Science and Technology,* 3rd ed.; IGI Global, 2015; pp 5767–5776.
21. Almunawar, M. N.; Anshari, M.; Susanto, H. Crafting Strategies for Sustainability: How Travel Agents Should React in Facing a Disintermediation. *Operational Res.* **2013,** *13* (3), 317–342.
22. Almunawar, M.; Susanto, H.; Anshari, M. A Cultural Transferability on IT Business Application: iReservation System. *J. Hospitality Tourism Technol.* **2013,** *4* (2), 155–176.
23. Brown, F. Editorial Opinion: Chemoinformatics: A Ten Year Update. *Curr. Opin. Drug Discov. Dev.* **2005,** 8 (3), 296–230.
24. Glen, R. Computational Chemistry and Cheminformatics: an Essay on the Future. *J. Cheminformatics; Springer Science-Business Media B.V.* Switzerland **2011,** *26,* 47–49.

CHAPTER 2

PERIODIC TABLE, QUANTUM BITING ITS TAIL, AND SUSTAINABLE CHEMISTRY

FRANCISCO TORRENS[1,*] and GLORIA CASTELLANO[2]

[1]*Institut Universitari de Ciència Molecular, Universitat de València, Edifici d'Instituts de Paterna, P.O. Box 22085, E-46071 València, Spain*

[2]*Departamento de Ciencias Experimentales y Matemàticas, Facultad de Veterinaria y Ciencias Experimentales, Universidad Catòlica de Valencia San Vicente Màrtir, Guillem de Castro-94, E-46001 València, Spain*

Corresponding author. E-mail: torrens@uv.es

ABSTRACT

A relation exists between biological evolution and the periodic table of the elements: all organisms, since their first evolutionary stages, made use of the chemical properties of many metal ions for the development of their essential biochemical functions. Cadmium is found in the periodic table in the same column as Hg and Zn, but its chemical properties are closer to Zn, whose similarity affects Cd distribution and toxic properties. Arsenic and P, also found in the same column of the periodic table, present a similar chemical behavior; however, oxidation-state reduction from five to three results easier in As. Throughout biological evolution, some elements substituted others, for example, Fe substituted Cu in many redox biosystems and processes. Heavy elements such as Nb and the like were formed in merging neutron stars. The key experiment in interpreting quantum-systems behavior was *double-slit experiment*. The thought experiment *Schrödinger's cat* illustrates the concept of *superposition of states*, which is basis of *quantum computation*. Thought experiments were important in understanding complex concepts (e.g., Einstein's special and general theories of relativity). In sustainable chemistry and photocatalysis, metal nanoparticles are used as fotocatalysts for the synthesis of organic compounds.

2.1 INTRODUCTION

A relationship exists between the biological evolution and the periodic table of the elements (PTE): all organisms, since their first evolutionary stages, made use of the chemical properties of many metal ions for the development of their essential biochemical functions. Cadmium is found in PTE in the same column as Hg and Zn, but its chemical properties are closer to Zn, whose similarity affects Cd distribution and its toxic properties. Arsenic and P, also found in the same PTE column, present a similar chemical behavior; however, oxistation-state (OS) reduction from five to three results easier in As. Throughout biological evolution, some elements substituted others, for example, under the anoxic atmosphere, Cu was important, but it was substituted by Fe under the oxic atmosphere in many biosystems and processes. Heavy PTE elements such as Nb and the like were formed in merging neutron stars (NSs).

The key experiment in interpreting quantum-systems behavior was the *double-slit experiment*. The thought experiment *Schrödinger's cat* illustrates the concept of *superposition of states*, which is the basis of *quantum computation*. Throughout science history, thought experiments were important in understanding complex concepts (e.g., Einstein's special (STR, 1905) and general (STR, 1915) theories of relativity). Venturi group reviewed molecular devices, machines, concepts, and perspectives for the nanoworld.[1] They revised the electrochemistry of functional supramolecular systems.[2] They developed sustainability via scientific and ethical issues.[3] Hazard screening methods for nanomaterials (NMs) were comparatively studied.[4] In sustainable chemistry and photocatalysis, metal nanoparticles (MNPs) are used as fotocatalysts for the synthesis of organic compounds.

In earlier publications, it was reported PTE,[5-7] quantum simulators,[8-16] science and ethics of developing sustainability via nanosystems and devices,[17] *green nanotechnology* as an approach toward environment safety,[18] and molecular devices and machines as hybrid organic–inorganic structures.[19] Back to PTE? The present study discusses toxic metals, the emergence of the heavy elements, key experiment in interpreting quantum-systems behavior, thought experiment illustrating the concept *superposition of states* (which is the basis of *quantum computation*), sustainable chemistry, photocatalysis, and MNPs. There is a general interest in photocatalysis, in the field of sustainable chemistry, aiming to transform solar into chemical energy, in order to obtain products or processes that are thermally forbidden or take place with low yield.

2.2 TOXIC METALS

A narrow relationship exists between the biosphere evolution and PTE: all organisms, since their first evolutionary stages, made use of the chemical properties of many metal ions for the development of their essential biochemical functions.[20] As a consequence, the metallic elements, even in low doses, are essential for development of their vital functions. Their insufficient contribution leads to organisms-developmental anomalies; however, when their concentration is higher to optimum, they exert toxic effects on organisms, thereby limiting their development. Cadmium is found in PTE in the same column as Hg and Zn, but its chemical properties are closer to Zn, whose similarity affects Cd distribution and its toxic properties. Arsenic and P, also found in the same PTE column, present a similar chemical behavior; however, OS reduction from five to three results easier in As.

Along the evolutionary history of the biosphere, some elements substituted others, for example, under the anoxic atmosphere, Cu was important, but it was substituted by Fe under the oxic atmosphere in many redox biosystems and bioprocesses.

2.3 THE EMERGENCE OF THE HEAVY ELEMENTS

Heavy PTEs such as Nb, Mo, Ru–Xe, Cs–Nd, Sm–Rn, and Fr–U (OSs 3, 4, 2, 1, 5, 6, −1, 0) were formed in merging NSs (Fig. 2.1).

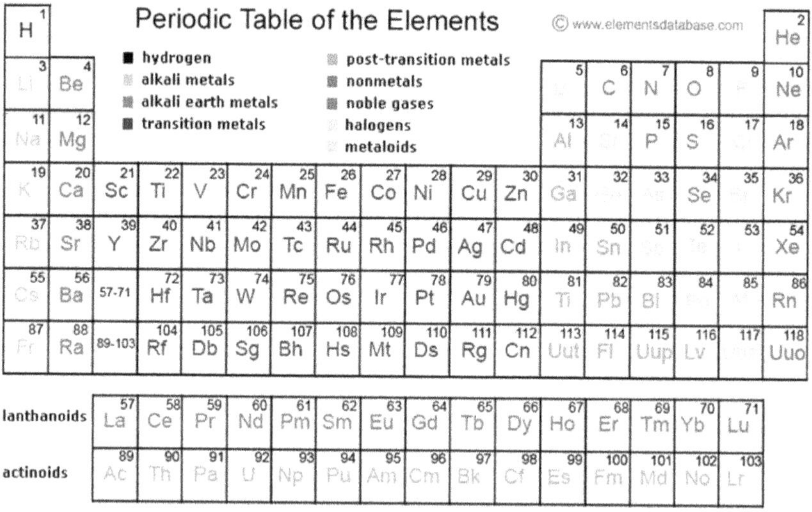

Periodic Table of the Elements

© www.elementsdatabase.com

- ■ hydrogen
- ■ alkali metals
- ■ alkali earth metals
- ■ transition metals
- ■ post-transition metals
- ■ nonmetals
- ■ noble gases
- ■ halogens
- ■ metaloids

1 H																	2 He
3 Li	4 Be											5	6 C	7 N	8 O	9	10 Ne
11 Na	12 Mg											13 Al	14	15 P	16 S	17	18 Ar
19 K	20 Ca	21 Sc	22 Ti	23 V	24 Cr	25 Mn	26 Fe	27 Co	28 Ni	29 Cu	30 Zn	31 Ga	32	33	34 Se	35	36 Kr
37 Rb	38 Sr	39 Y	40 Zr	41 Nb	42 Mo	43 Tc	44 Ru	45 Rh	46 Pd	47 Ag	48 Cd	49 In	50 Sn	51	52	53	54 Xe
55 Cs	56 Ba	57-71	72 Hf	73 Ta	74 W	75 Re	76 Os	77 Ir	78 Pt	79 Au	80 Hg	81 Tl	82 Pb	83 Bi	84	85	86 Rn
87 Fr	88 Ra	89-103	104 Rf	105 Db	106 Sg	107 Bh	108 Hs	109 Mt	110 Ds	111 Rg	112 Cn	113 Uut	114 Fl	115 Uup	116 Lv	117	118 Uuo

lanthanoids	57 La	58 Ce	59 Pr	60 Nd	61 Pm	62 Sm	63 Eu	64 Gd	65 Tb	66 Dy	67 Ho	68 Er	69 Tm	70 Yb	71 Lu
actinoids	89 Ac	90 Th	91 Pa	92 U	93 Np	94 Pu	95 Am	96 Cm	97 Bk	98 Cf	99 Es	100 Fm	101 Md	102 No	103 Lr

FIGURE 2.1 **(See color insert.)** In emergence of some heavy PTEs, most were formed merging NSs: Pt/Au factory in sky.

2.4 THE QUANTUM THAT BITES ITS TAIL

Nebot Gil described the key experiment in interpreting quantum-systems behavior: *double-slit experiment.*[21] He used thought experiment (*Gedankenexperiment*) *Schrödinger's cat* to illustrate the concept *superposition of states*, which is the basis of *quantum computation*. He described the calculations that were published on molecular systems performed with *quantum computers.*[22]

2.5 SUSTAINABLE CHEMISTRY AND PHOTOCATALYSIS: METAL NANOPARTICLES

González-Béjar discussed the general interest of photocatalysis, in the field of sustainable chemistry, aiming to transform solar energy into chemical energy, in order to obtain products or processes that are thermally forbidden or take place with low yield.[23] She included an introduction of general concepts in photocatalysis, and a brief description of the main characteristics of MNPs and their capabilities as photocatalysts on irradiation. She incorporated representative examples of syntheses of organic compounds to illustrate the capabilities, with special emphasis on supported MNPs for heterogeneous photocatalysis.

2.6 DISCUSSION

A relationship exists between the biological evolution and PTE: really, all organisms, since their first evolutionary stages, made use of the chemical properties of many metal ions for the development of their essential biochemical functions. Cadmium is found in PTE in the same column as Hg and Zn, but its chemical properties are closer to Zn, whose similarity affects Cd distribution and its toxic properties. Arsenic and P, also found in the same PTE column, present a similar chemical behavior; however, OS reduction from five to three results easier in As. Along biological evolution, some elements substituted others, for example, under the anoxic atmosphere, Cu was important, but it was substituted by Fe under the oxic atmosphere in many redox biosystems and processes. Heavy PTE elements such as Nb and the like (OSs 3, 4, 2, 1, 5, 6, etc.) were formed in merging NSs.

The key experiment in interpreting quantum-systems behavior was *double-slit experiment*. Thought experiment *Schrödinger's cat* illustrates the concept *superposition of states*, which is basis of *quantum computation*. Throughout science history, thought experiments were important in understanding complex concepts (e.g., Einstein's special (STR, 1905) and general (STR, 1915) theories of relativity).

In sustainable chemistry and photocatalysis, MNPs are used as fotocatalysts for the synthesis of organic compounds.

2.7 FINAL REMARKS

From the results and discussion of this study, the following final remarks can be drawn:

(1) A relation exists between bioevolution and PTEs: really, all organisms, since their first evolutionary stages, made use of chemical properties of many metal ions for development of their essential biochemical functions. Cadmium is found in the periodic table in the same column as Hg and Zn, but its chemical properties are

closer to Zn, whose similarity affects Cd distribution and its toxic properties. Arsenic and P, found in the same column of the periodic table, present a similar chemical behavior but oxidation state reduction from five to three is easier in As.

(2) Along biological evolution, some elements substituted others, for example, under the anoxic atmosphere, Cu was important, but it was substituted by Fe under the oxic atmosphere in many redox biosystems and processes.

(3) Heavy elements such as Nb and the like (oxidation states 3, 4, 2, 1, etc.) of the periodic table were formed in merging NSs.

(4) The key experiment in interpreting quantum-systems behavior was *double-slit experiment*. Thought experiment *Schrödinger's cat* illustrates the concept of *superposition of states*, which is the basis of *quantum computation*. Thought experiments were important in understanding complex concepts (e.g., Einstein's special and general theories of relativity).

(5) In sustainable chemistry and photocatalysis, MNPs are used as fotocatalysts for the synthesis of organic compounds.

(6) The codes of ethics are important but, in the end, they are a personal theme. The actions must be carried out in an ambit of respect and without eagerness to intimidation. The principle is noble, and the way to transmit the message and obtain its acceptance must also be dignified.

(7) Further work will deal with conductive two-dimensional metal–organic frameworks as multifunctional materials.

KEYWORDS

- sustainable chemistry
- photocatalysis
- metal nanoparticle
- periodic table of the elements
- superposition of states
- quantum simulation
- quantum computation

REFERENCES

1. Balzani, V.; Credi, A.; Venturi, M. *Molecular Devices and Machines: Concepts and Perspectives for the Nanoworld;* Wiley-VCH: Weinheim, Germany, 2008.

2. Ceroni, P.; Credi, A.; Venturi, M., Eds. *Electrochemistry of Functional Supramolecular Systems;* Wiley: New York, NY, 2010.

3. Venturi, M. Developing Sustainability: Some Scientific and Ethical Issues. In *Sustainable Nanosystems Development, Properties, and Applications;* Putz, M. V., Mirica M. C., Eds.; IGI Global: Hershey, PA, 2017; pp 657–680.

4. Sheehan, B.; Murphy, F.; Mullins, M.; Furxhi, I.; Costa, A. L.; Simeone, F. C.; Mantecca, P. Hazard Screening Methods for Nanomaterials: A Comparative Study. *Int. J. Mol. Sci.* **2018**, *19*, 1–22.

5. Torrens, F.; Castellano, G. Reflections on the Nature of the Periodic Table of the Elements: Implications in Chemical Education. In *Synthetic Organic Chemistry;* Seijas, J. A., Vázquez Tato, M. P., Lin, S. K., Eds.; MDPI: Basel, Switherland, 2015; Vol. 18, pp 1–15.

6. Torrens, F.; Castellano, G. Nanoscience: From a Two-dimensional to a Three-dimensional Periodic Table of the Elements. In *Methodologies and Applications for Analytical and Physical Chemistry;* Haghi, A. K., Thomas, S., Palit, S., Main, P., Eds.; Apple Academic–CRC: Waretown, NJ, 2018; pp 3–26.

7. Torrens, F.; Castellano, G. Periodic Table. In *The Explicative Handbook of Nanochemistry;* Putz, M. V., Ed.; Apple Academic–CRC: Waretown, NJ, in press.

8. Torrens, F.; Castellano, G. Ideas in the History of Nano/Miniaturization and (Quantum) Simulators: Feynman, Education and Research Reorientation in Translational Science. In *Synthetic Organic Chemistry;* Seijas, J. A., Vázquez Tato, M. P., Lin, S. K., Eds.; MDPI: Basel, Switzerland, 2015; Vol. 19, pp 1–16.

9. Torrens, F.; Castellano, G. Reflections on the Cultural History of Nanominiaturization and Quantum Simulators (Computers). In *Sensors and Molecular Recognition;* Laguarda Miró, N., Masot Peris, R., Brun Sánchez, E., Eds.; Universidad Politécnica de Valencia: València, Spain, 2015; Vol. 9, pp 1–7.

10. Torrens, F.; Castellano, G. Nanominiaturization and Quantum Computing. In *Sensors and Molecular Recognition;* Costero Nieto, A. M., Parra Álvarez, M., Gaviña Costero, P., Gil Grau, S., Eds.; Universitat de València: València, Spain, 2016; Vol. 10, pp 1–5.

11. Torrens, F.; Castellano, G. Nanominiaturization, Classical/Quantum Computers/ Simulators, Superconductivity and Universe. In *Methodologies and Applications for Analytical and Physical Chemistry;* Haghi, A. K., Thomas, S., Palit, S., Main, P., Eds.; Apple Academic–CRC: Waretown, NJ, 2018; pp 27–44.

12. Torrens, F.; Castellano, G. Superconductors, Superconductivity, BCS Theory and Entangled Photons for Quantum Computing. In *Physical Chemistry for Engineering and Applied Sciences: Theoretical and Methodological Implication;* Haghi, A. K., Aguilar, C. N., Thomas, S., Praveen, K. M., Eds.; Apple Academic–CRC: Waretown, NJ, 2018; pp 379–387.

13. Torrens, F.; Castellano, G. EPR Paradox, Quantum Decoherence, Qubits, Goals and Opportunities in Quantum Simulation. In *Theoretical Models and Experimental*

Approaches in Physical Chemistry: Research Methodology and Practical Methods; Haghi, A. K., Ed.; Apple Academic–CRC: Waretown, NJ, 2018; Vol. 5, pp 317–334.

14. Torrens, F.; Castellano, G. Nanomaterials, Molecular Ion Magnets, Ultrastrong and Spin–Orbit Couplings in Quantum Materials. In *Physical Chemistry for Chemists and Chemical Engineers: Multidisciplinary Research Perspectives;* Vakhrushev, A. V., Haghi, R., de Julián-Ortiz, J. V., Allahyari, E., Eds.; Apple Academic–CRC: Waretown, NJ, in press.

15. Torrens, F.; Castellano, G. Nanodevices and Organization of Single Ion Magnets and Spin Qubits. In *Chemical Science and Engineering Technology: Perspectives on Interdisciplinary Research;* Balköse, D., Ribeiro, A. C. F., Haghi, A. K., Ameta, S. C., Chakraborty, T., Eds.; Apple Academic–CRC: Waretown, NJ, in press.

16. Torrens, F.; Castellano, G. Superconductivity and Quantum Computing via Magnetic Molecules. In *New Insights in Chemical Engineering and Computational Chemistry;* Haghi, A. K., Ed.; Apple Academic–CRC: Waretown, NJ, in press.

17. Torrens, F.; Castellano, G. Developing Sustainability via Nanosystems and Devices: Science–Ethics. In *Chemical Science and Engineering Technology: Perspectives on Interdisciplinary Research;* Balköse, D., Ribeiro, A. C. F., Haghi, A. K., Ameta, S. C., Chakraborty, T., Eds.; Apple Academic–CRC: Waretown, NJ, in press.

18. Torrens, F.; Castellano, G. Green Nanotechnology: An Approach towards Environment Safety. In *Advances in Nanotechnology and the Environmental Sciences: Applications, Innovations, and Visions for the Future;* Vakhrushev, A. V., Ameta, S. C., Susanto, H., Haghi, A. K., Eds.; Apple Academic–CRC: Waretown, NJ, in press.

19. Torrens, F.; Castellano, G. Molecular Devices/Machines: Hybrid Organic–Inorganic Structures. In *Research Methods and Applications in Chemical and Biological Engineering;* Pourhashemi, A., Deka, S.C., Haghi, A.K., Eds.; Apple Academic–CRC: Waretown, NJ, in press.

20. Spiro, T. G.; Stigliani, W. M. *Chemistry of the Environment;* Prentice Hall: Upper Saddle River, NJ, 2003.

21. Nebot Gil, I., Personal Communication.

22. Aromí, G.; Gaita-Ariño, A.; Luis, F. Computación Cuántica con Moléculas Magnéticas. *Revista Española de Física* **2016,** *30* (3), 21–24.

23. González-Béjar, M. Química Sostenible y Fotocatálisis: Nanopartículas Metálicas Como Fotocatalizadores Para la Síntesis de Compuestos Orgánicos. *An. Quím.* **2018,** *114,* 31–38.

CHAPTER 3

THE STUDY OF PHYSICOCHEMICAL PROPERTIES OF BIMETALLIC CuAU$_n$ (n = 1–8) NANOALLOY CLUSTERS

PRABHAT RANJAN[1], TANMOY CHAKRABORTY[2,*], and AJAY KUMAR[1]

[1]*Department of Mechatronics Engineering, Manipal University Jaipur, Dehmi Kalan, Jaipur 303007, India*

[2]*Department of Chemistry, Manipal University Jaipur, Dehmi Kalan, Jaipur 303007, India*

Corresponding author. E-mail: tanmoychem@gmail.com; tanmoy.chakraborty@jaipur.manipal.edu

ABSTRACT

The study of bimetallic clusters has become an important topic of research due to its ability to connect with the vital applications of science and technology. In recent years, bimetallic clusters have received large attention because of its large applications in fabrications, biology, nanoscience, catalysis, and medicine. Among these bimetallic clusters, the compound formed between Cu and Au is of immense importance due to its interesting electronic and optical properties. Density functional theory (DFT) is the most efficacious approach of quantum mechanics to study the electronic properties of matter. In this chapter, we have studied the physico-chemical properties of CuAu$_n$ (n = 1–8) invoking conceptual density functional theory (CDFT)-based descriptors. Our results reveal that Cu–Au has highest HOMO–LUMO energy gap and CuAu$_8$ has least energy gap in this series. The linear correlation between CDFT-based descriptors and HOMO–LUMO energy gap supports our analysis.

3.1 INTRODUCTION

Nanoclusters consist of millions of nanoparticles or molecules; the integral atoms may be same or different. The size of nanoparticles varies between 1 and 100 nm.[1-5] A number of researchers have reported that due to variation in composition and size, nonlinear transition of physical and chemical properties of molecules may also vary. It is also reported that physicochemical properties of clusters vary interestingly as a function of size, which are different from those of discrete molecules or bulk materials.[6] A detailed analysis about nanoclusters with well-defined dimension and configuration may lead to some better alternatives.[7] Because of the potential application of nanoclusters in fabrication, nanoscience, biology, and catalysis, they are of huge importance.[1,3,8-10]

From experimental and theoretical point of view, study of bi-metallic clusters have received a lot of attention, since they reveal new kind of properties that are distinct from the pure and bulk metal clusters.[11] Noble-metal clusters can be extensively applied in potential technological areas due to their interesting electronic and optical properties.[12-18] The positive conjoint effects of two or more noble metals on the abovementioned properties have been lucidly explained by researchers.[13,19-20] In recent years, different compositions of nanoalloys are being utilized for advancement of methodologies and characterization techniques.[13,19,21] The study related to core-shell structure of nanoclusters is in demand because their properties can be tuned by proper control of other structural and chemical parameters. Group 11 metal (Cu, Ag, and Au) clusters demonstrate the filled inner d-orbitals with one unpaired electron in the valence s shell.[22] This electronic arrangement is responsible for reproduction of exactly similar shell effects,[23-27] which are practically observed in the alkali metal clusters.[28-30] In the periodic table, among the nanoalloy clusters of group 11 elements, the compound formed between Cu and Au is of immense importance due to its large-scale technological applications in fabrication, material science, catalysis, biology, and medicine.[6,31-42] It has been already established that s–d hybridization is strong in gold as compared to copper and silver, due to relativistic contraction of the d-orbitals.[43-46] In recent years, many experimental and theoretical studies have been performed to improve the stability and physico-chemical properties of Au clusters.[6,11,13,47-57] Recently, it has also been observed that copper–gold bi-metallic nanoalloy clusters, as catalysts, can enhance the reaction efficiency and selectivity.[11,47,49,58,59]

Though a number of studies have been performed on this particular type of compounds, a theoretical analysis invoking conceptual density functional based descriptors is yet to explore in this domain.

Density functional theory (DFT) is the most popular approach of quantum mechanics to study the electronic properties of matter.[7,60] Due to its computational friendly behavior, DFT is a widely accepted method to study the many-body systems. In the domain of material science research, particularly in super conductivity of metal-based alloys,[61] magnetic properties of nanoalloy clusters,[62] quantum fluid dynamics, molecular dynamics,[63,64] and nuclear physics,[65,66] DFT has gained a huge importance. The study of DFT covers three major domains, that is, theoretical, conceptual, and computational.[67–70] Conceptual density functional theory (CDFT) is established as an important approach to study the chemical reactivity of materials.[71–73] The CDFT is highlighted following Parr's dictum "Accurate calculation is not synonymous with useful interpretation. To calculate a molecule is not to understand it."[74] We have rigorously applied conceptual density functional based global and local descriptors for studying of physicochemical properties of nano-engineering materials and drug designing process.[75–89]

In this report, we have studied the physico-chemical properties of CuAu$_n$ ($n = 1$–8) invoking CDFT-based descriptors, that is, highest occupied molecular orbital–lowest unoccupied molecular orbital (HOMO–LUMO) energy gap, molecular hardness, softness, electrophilicity index, electronegativity, and dipole moment.

3.2 COMPUTATIONAL DETAILS

In this report, we have studied physico-chemical properties of bi-metallic CuAu$_n$ ($n = 1$–8) nanoalloy clusters. Three-dimensional (3D) modeling and structural optimization of all the compounds have been performed using Gaussian 03[90] within density functional theory framework. For optimization purpose, generalized gradient approximation (GGA)–B3PW91 exchange correlation functional with basis set LanL2dz has been adopted. Invoking Koopmans' approximation,[78] ionization energy (I) and electron affinity (A) of all the nanoalloys have been computed using the following ansatz:

$$I = -\varepsilon_{HOMO} \qquad (3.1)$$

$$A = -\varepsilon_{LUMO} \qquad (3.2)$$

Thereafter, using I and A, the conceptual DFT-based descriptors, that is, electronegativity (χ), global hardness (η), molecular softness (S), and electrophilicity index (ω) have been computed. The equations used for these calculations are as follows:

$$\chi = -\mu = \frac{I + A}{2}, \qquad (3.3)$$

where, μ represents the chemical potential of the system.

$$\eta = \frac{I - A}{2} \qquad (3.4)$$

$$S = \frac{1}{2\eta} \qquad (3.5)$$

$$\omega = \frac{\mu^2}{2\eta} \qquad (3.6)$$

3.3 RESULTS AND DISCUSSION

The physico-chemical properties of Cu–Au nanoalloy clusters have been performed invoking density functional theory methodology. The orbital energies in form of HOMO–LUMO energy gap along with computed CDFT-based descriptors, that is, electronegativity (χ), hardness (η), softness (S), and electrophilicity index (ω) have been reported in Table 3.1. The molecular dipole moment in Debye unit is also shown in this table. The result reveals that HOMO–LUMO energy gap of the nanoclusters have linear relationship with the computed hardness values. It has been already observed that as the frontier orbital energy gap increases, their hardness value also increases. The molecule having the utmost value of HOMO–LUMO energy gap will be highly stable which means it will not give any response against any external perturbation. The result from Table 3.1 shows that CuAu cluster with $C_{\infty v}$ symmetry group has maximum HOMO–LUMO energy gap of 3.483 eV, whereas CuAu$_8$

cluster with 1.442 eV has least energy gap. Because optical properties of nanoalloy clusters depend on the flow of electrons within the molecular system, which depends on the energy required for an electron to move from valence band to conduction band. Xiao and colleagues[91] have already reported that there is a linear relationship between HOMO and LUMO gaps with the difference in the energy levels of valence-conduction band. Based on this fact, we can assume that optical properties of CuAu$_n$ clusters are having direct qualitative correlation with their evaluated HOMO–LUMO energy gap. Keeping this in view, we may conclude that optical properties of bi-metallic Cu–Au nanoalloy clusters increases with increase in their molecular hardness values. Similarly, the softness data displays an inverse relationship with the experimental optical properties. Similar relationships are also observed between other DFT-based descriptors and HOMO–LUMO energy gap. The linear correlation between HOMO and LUMO energy gap along with their computed softness and electrophilicity index are lucidly plotted in Figures 3.1 and 3.2, respectively. The results from Figures 3.1 and 3.2 reveal that softness and HOMO–LUMO energy gap have regression coefficient of 0.940, whereas regression coefficient between electrophilicity index and HOMO–LUMO energy gap is 0.749. The maximum regression coefficient, that is, $R^2 = 1$ is obtained between hardness and HOMO–LUMO energy gap, whereas the least regression coefficient, that is, $R^2 = 0.020$ is found between electronegativity and HOMO–LUMO energy gap.

TABLE 3.1 Computed DFT-based Descriptors of CuAu$_n$ $(n = 1–8)$ Nanoalloy Clusters.

Species	Symmetry group	HUMO-LUMO gap (eV)	χ (eV)	η (eV)	S (eV)	ω (eV)	Dipole moment (Debye)
CuAu	C$_{\infty v}$	3.483	4.626	1.741	0.287	6.144	2.275
CuAu$_2$	C$_s$	2.068	4.109	1.034	0.484	8.163	2.626
CuAu$_3$	C$_s$	1.850	4.735	0.925	0.540	12.115	0.802
CuAu$_4$	D$_{2h}$	1.551	4.612	0.775	0.645	13.715	1.795
CuAu$_5$	C$_{2v}$	3.320	4.816	1.660	0.301	6.987	1.525
CuAu$_6$	C$_1$	2.367	4.449	1.184	0.422	8.361	3.163
CuAu$_7$	C$_{2v}$	1.850	5.115	0.925	0.540	14.143	1.453
CuAu$_8$	C$_{2v}$	1.442	4.993	0.721	0.693	17.287	1.246

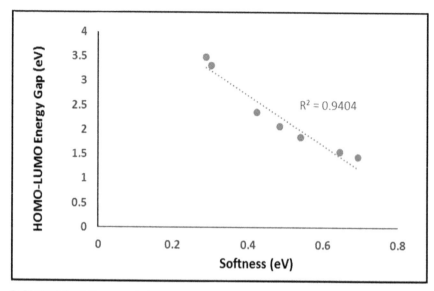

FIGURE 3.1   A linear correlation plot between softness versus HOMO–LUMO energy gap.

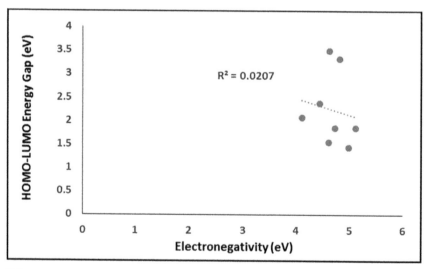

FIGURE 3.2   A linear correlation plot between electronegativity versus HOMO–LUMO energy gap.

3.4 CONCLUSION

In recent years, bi-metallic nanoalloy clusters have received a lot of attention. The study of group 11 metals, namely Cu and Au is of considerable interest due to their unique electronic and optical properties. In this chapter, we have studied the physico-chemical properties of CuAu$_n$ (n = 1–8) nanoalloy clusters invoking CDFT-based descriptors. We have used GGA–B3PW91 exchange correlation with basis set LanL2dz basis set for geometry optimization. The result reveals that HOMO–LUMO energy gap maintains direct relationship with evaluated hardness and inverse relationship with softness values. The data also reveals the equivalency between optical properties of Cu–Au nanoalloy cluster with their computed HOMO–LUMO energy gap. It shows that optical property of these clusters has direct relationship with computed hardness and inverse relationship with computed softness. This trend is expected from experimental point of view also. The linear correlation between HOMO and LUMO energy gaps along with their computed CDFT-based descriptors is also observed. The maximum regression coefficient, that is, R^2 = 1 is obtained between hardness and HOMO–LUMO energy gap, whereas the least regression coefficient, that is, R^2 = 0.020 is observed between electronegativity and HOMO–LUMO energy gap.

KEYWORDS

- bi-metallic cluster
- copper–gold
- conceptual density functional theory
- HOMO–LUMO gap
- molecular softness
- dipole moment

REFERENCES

1. Khosousi, A. Z.; Dhirani, A. A. *Chem. Rev.* **2008,** *108*, 4072.
2. Daniel, M. C.; Astruc, D. *Chem. Rev.* **2004,** *104*, 293.
3. Ghosh, S. K.; Pal, T. *Chem. Rev.* **2007,** *107*, 4797.

4. Chaudhuri, R. G.; Paria, S. *Chem. Rev.* **2012,** *112*, 2373.

5. Alivisatos, A. P. *Science* **1996,** *271*, 933.

6. Wang, H. Q.; Kuang, X. Y.; Li, H. F. *Phys. Chem. Chem. Phys.* **2010,** *12*, 5156.

7. Ismail, R. Theoretical Studies of Free and Supported Nanoalloy Clusters. Ph.D. Thesis, University of Birmingham, UK, 2012.

8. Roucoux, A.; Schulz, J.; Patin, H. *Chem. Rev.* **2002,** *102*, 3757.

9. Munoz-Flores, B. M.; Kharisov, B. I.; Jimenez-Perez, V. M.; Martinez, P. E.; Lopez, S. T. *Ind. Eng. Chem. Res.* **2011,** *50*, 7705.

10. Murray, R. W. *Chem. Rev.* **2008,** *108*, 2688.

11. Yin, F.; Wang, Z. W.; Palmer, R. E. *J. Nanopart. Res.* **2012,** *14*, 1124.

12. Teng, X.; Wang, Q.; Liu, P.; Han, W.; Frenkel, A. I.; Wen, M. N.; Hanson, J. C.; Rodriguez, J. A. *J. Am. Chem. Soc.* **2008,** *130*, 1093.

13. Ferrando, R.; Jellinek, J.; Johnston, R. L. *Chem. Rev.* **2008,** *108*, 845.

14. Henglein, A. *J. Phys. Chem.* **1993,** *97*, 5457.

15. Davis, S. C.; Klabunde, K. J. *Chem. Rev.* **1982,** *82*, 153.

16. Lewis, L. N. *Chem. Rev.* **1993,** *93*, 2693.

17. Schmid, G. *Chem. Rev.* **1992,** *92*, 1709.

18. Schon, G.; Simon, U. *Colloid. Polym. Sci.* **1995,** *273*, 101.

19. Oderji, H. Y.; Ding, H. *Chem. Phys.* **2011,** *388*, 23.

20. Liu, H. B.; Pal, U.; Medina, A.; Maldonado, C.; Ascencio, J. A. *Phys. Rev. B* **2005,** *71*, 075403.

21. Baletto, F.; Ferrando, R. *Rev. Mod. Phys.* **2005,** *77*, 371.

22. Alonso, J. A. *Chem. Rev.* **2000,** *100*, 637.

23. Katakuse, I.; Ichihara, T.; Fujita, Y.; Matsuo, T.; Sakurai, T.; Matsuda, H. Int. *J. Mass Spectrom. Ion Processes* **1985,** *67*, 229.

24. Katakuse, I.; Ichihara, T.; Fujita, Y.; Matsuo, T.; Sakurai, T.; Matsuda, H. *Int. J. Mass Spectrom. Ion Processes* **1986,** *74*, 33.

25. Heer, W. A. D. *Rev. Mod. Phys.* **1993,** *65*, 611.

26. Gantefor, G.; Gausa, M.; Meiwes-Broer, K. H.; Lutz, H. O. J. *Chem. Soc. Faraday Trans.* **1990,** *86*, 2483.

27. Leopold, D. G.; Ho, J.; Lineberger, W. C. *J. Chem. Phys.* **1987,** *86*, 1715.

28. Lattes, A.; Rico, I.; Savignac, A. D.; Samii, A. A. Z. *Tetrahedron* **1987,** *43*, 1725.

29. Chen, F.; Xu, G. Q.; Hor, T. S. A. *Mater. Lett.* **2003,** *57*, 3282.

30. Taleb, A.; Petit, C.; Pileni, M. P. *J. Phys. Chem. B* **1998,** *102*, 2214.

31. Liu, H. Q.; Tian, Y.; Xia, P. P. *Langmuir* **2008,** *24*, 6359.

32. Noonan, K. J. T.; Gillon, B. H.; Cappello, V.; Gates, D. P. *J. Am. Chem. Soc.* **2008,** *130*, 12876.

33. Wang, T.; Hu, X. G.; Dong, S. J. *J. Phys. Chem. B* **2006,** *110*, 16930.

34. Majumder, C.; Kandalam, A. K.; Jena, P. *Phys. Rev.* **2006,** *74*, 205437.

35. Li, X.; Kiran, B.; Cui, L. F.; Wan, L. S. *Phys. Rev. Lett.* **2005,** *95*, 253401.

36. Torres, M. B.; Fernandez, E. M.; Balbas, L. C. *J. Phys. Chem. A* **2008,** *112*, 6678.

37. Teles, J. H.; Brode, S.; Chabanas, M.; *Angew. Chem.* **1998,** *110*, 1475.

38. Valden, M.; Lai, X.; Goodman, D. W. *Science* **1998,** *281*, 1647.

39. Yoon, B.; Häkkinen, H.; Landman, U.; Wörz, A. S.; Antonietti, J. –M.; Abbet, S.; Judai, K.; Heiz, U. *Science* **2005,** *307*, 403.

40. McRae, R.; Lai, B.; Vogt, S.; Fahrni, C. J. *J. Struct. Biol.* **2006,** *155*, 22.

41. Ackerson, C. J.; Jadzinsky, P. D.; Jensen, G. J.; Kornberg, R. D. *J. Am. Chem. Soc.* **2006,** *128,* 2635.
42. Shaw III, C. F. *Chem. Rev.* **1999,** *99,* 2589.
43. Kabir, M.; Mookerjee, A.; Bhattacharya, A. K. *Eur. Phys. J. D* **2004,** *31,* 477.
44. Heard, C. J.; Johnston, R. L. *Eur. Phys. J. D* **2013,** *67,* 34.
45. Hakkinen, H.; Moseler, M.; Landman, U. *Phys. Rev. Lett.* **2002,** *89,* 176103.
46. Massobrio, C.; Pasquarello, A.; Car, R. *Chem. Phys. Lett.* **1995,** *238,* 215.
47. Bauer, J. C.; Mullins, D.; Li, M.; Wu, Z., Payzant, E. A.; Overbury, S. H.; Dai, S. *Phys. Chem. Chem. Phys.* **2011,** *13,* 2571.
48. Bouwen, W.; Vanhoutte, F.; Despa, F.; Bouckaert, S.; Neukermans, S.; Theil, K. L.; Weidele, H.; Lievens, P.; Silverans, R. E. *Chem. Phys. Lett.* **1999,** *314,* 227.
49. Bracey, C. L.; Ellis, P. R.; Hutchings G. J. *Chem. Soc. Rev.* **2009,** *38,* 2231.
50. Chen, W.; Yu, R.; Li, L.; Wang, A.; Peng, Q.; Li, Y. *Angew. Chem. Int. Ed.* **2010,** *49,* 2917.
51. Liu, X.; Wang, A.; Li, L.; Zhang, T.; Mou, C.; Lee, J. *J. Catal.* **2011,** *278,* 288.
52. Rao, C. N. R.; Kulkarni, G. U.; Thomas, P. J.; Edwards, P. P. *Chem. Soc. Rev.* **2000,** *29,* 27.
53. Pauwels, B.; Tendeloo, G. V.; Zhurkin, E.; Hou, M.; Verschoren, G.; Kuhn, L. T.; Bouwen, W.; Lievens, P. *Phys. Rev. B* **2001,** *63,* 165406.
54. Sra, A. K.; Schaak, R. E. *J. Am. Chem. Soc.* **2004,** *126,* 6667.
55. Toai, T. J.; Rossi, G.; Ferrando, R. *Faraday Discuss.* **2008,** *138,* 49.
56. Yasuda, H.; Mori, H. *Z. Phys. D* **1994,** *31,*131.
57. Yin, F.; Wang, Z. W.; Palmer, R. E.; *J. Am. Chem. Soc.* **2011,** *133,* 10325.
58. Rapallo, A.; Rossi, G.; Ferrando, R.; Fortunelli, A.; Curley, B. C.; Lloyd, L. D., Tarbuck, G. M.; Johnston, R. L. *J. Chem. Phys.* **2005,** *122,* 194308.
59. Molenbroek, A. M.; Norskov, J. K.; Clausen, B. S. *J. Phys. Chem. B* **2001,** *105,* 5450.
60. Cramer, C. J.; Truhlar, D. G. *Phys. Chem. Chem. Phys.* **2009,** *11,* 10757.
61. Wacker, O. J.; Kümmel, R.; Gross, E. K. U. *Phys. Rev. Lett.* **1994,** *73,* 2915.
62. Illas, F.; Martin, R. L. *J. Chem. Phys.* **1998,** *108,* 2519.
63. Kümmel, S.; Brack, M. *Phys. Rev. A* **2001,** *64,* 022506.
64. Car, R.; Parrinello, M. *Phys. Rev. Lett.* **1985,** *55,* 2471.
65. Koskinen, M.; Lipas, P.; Manninen, M. *Nucl. Phys. A* **1995,** *591,* 421.
66. Schmid, R. N.; Engel, E.; Dreizler, R. M. *Phys. Rev. C* **1995,** *52,* 164.
67. Parr, R. G.; Yang, W. *Ann. Rev. Phy. Chem.* **1995,** *46,* 701.
68. Kohn, W.; Becke, A. D.; Parr, R. G. *J. Phys. Chem.* **1996,** *100,* 12974.
69. Liu, S.; Parr, R. G. *J. Chem. Phys.* **1997,** *106,* 5578.
70. Ziegler, T. *Chem. Rev.* **1991,** *91,* 651.
71. Parr, R. G.; Yang, W., Ed. *Density Functional Theory of Atoms and Molecules;* Oxford University Press, Oxford, 1989.
72. Chermette, H. *J. Comput. Chem.* **1999,** *20,* 129.
73. Geerlings, P.; Proft, F. D.; Langenaeker, W. *Chem. Rev.* **2003,** *103,* 1793.
74. Geerlings, P.; Proft, F. D. *Int. J. Mol. Sci.* **2002,** *3,* 276.
75. Ranjan, P.; Dhail, S.; Venigalla, S.; Kumar, A.; Ledwani, L.; Chakraborty, T. *Mater. Sci.-Pol.* **2015,** *33,* 719.
76. Ranjan, P.; Venigalla, S.; Kumar, A.; Chakraborty, T. *New Front. Chem.* **2014,** *23,* 111.

77. Venigalla, S.; Dhail, S.; Ranjan, P.; Jain, S.; Chakraborty, T. *New Front. Chem.* **2014,** *23,* 123.
78. Ranjan, P.; Kumar, A.; Chakraborty, T. *AIP Conf. Proc.* **2016,** *1724,* 020072.
79. Ranjan, P.; Kumar, A.; Chakraborty, T. *Mat. Today Proc.* **2016,** *3,* 1563.
80. Ranjan, P.; Kumar, A.; Chakraborty, T. *Environmental Sustainability: Concepts, Principles, Evidences and Innovations;* Mishra, G. C., Ed.; Excellent Publishing House: New Delhi, 2014; pp 239–242.
81. Ranjan, P.; Venigalla, S.; Kumar, A.; Chakraborty, T. *Recent Methodology in Chemical Sciences: Experimental and Theoretical Approaches;* Chakraborty, T., Ledwani, L., Eds.; Apple Academic Press and CRC Press: USA, 2015; pp 337–346.
82. Ranjan, P.; Kumar, A.; Chakraborty, T. *J. Phys.* **2016,** *759,* 012045.
83. Dhail, S.; Ranjan, P.; Chakraborty, T. *Crystallizing Ideas: The Role of Chemistry;* Ramasami, P., Bhowon, M. G., Laulloo, S. J., Li, H., Wah, K., Eds.; Springer International Publishing: Switzerland, 2016; pp 97.
84. Ranjan, P.; Kumar, A.; Chakraborty, T. *Computational Chemistry Methodology in Structural Biology & Material Sciences;* Chakraborty, T., Ranjan, P., Pandey, A., Eds.; CRC & Apple Academic Press: USA, 2016; ISBN-9781771885683.
85. Ranjan, P.; Chakraborty, T.; Kumar, A. *Applied Chemistry and Chemical Engineering;* Haghi, A. K., Pogliani, L., Castro, E. A., Balköse, D., Mukbaniani, O. V., Chia, C. H., Eds.; CRC & Apple Academic Press; USA, 2016; Vol. 4, ISBN-9781771885874.
86. Ranjan, P.; Kumar, A.; Chakraborty, T. *Mater. Sci. Eng.* **2016,** *149,* 012172.
87. Ranjan, P.; Chakraborty, T.; Kumar, A. *Nano Hybrids and Composites* **2017,** *17,* 62.
88. Ranjan, P.; Chakraborty, T.; Kumar, A. *Phys. Sci. Rev.* **2017,** *2,* DOI: 10.1515/psr-2016-0112.
89. Ranjan, P.; Chakraborty, T. J. *Inter. Acad. Phys. Sci.* **2013,** *17,* 1.
90. Gaussian 03, Revision C.02, Frisch, M. J.; Trucks, G. W.; Schlegel, H. B.; Scuseria, G. E.; Robb, M. A.; Cheeseman, J. R.; Montgomery, Jr., J. A.; Vreven, T.; Kudin, K. N.; Burant, J. C.; Millam, J. M.; Iyengar, S. S.; Tomasi, J.; Barone, V.; Mennucci, B.; Cossi, M.; Scalmani, G.; Rega, N.; Petersson, G. A.; Nakatsuji, H.; Hada, M.; Ehara, M.; Toyota, K.; Fukuda, R.; Hasegawa, J.; Ishida, M.; Nakajima, T.; Honda, Y.; Kitao, O.; Nakai, H.; Klene, M.; Li, X.; Knox, J. E.; Hratchian, H. P.; Cross, J. B.; Bakken, V.; Adamo, C.; Jaramillo, J.; Gomperts, R.; Stratmann, R. E.; Yazyev, O.; Austin, A. J.; Cammi, R.; Pomelli, C.; Ochterski, J. W.; Ayala, P. Y.; Morokuma, K.; Voth, G. A.; Salvador, P.; Dannenberg, J. J.; Zakrzewski, V. G.; Dapprich, S.; Daniels, A. D.; Strain, M. C.; Farkas, O.; Malick, D. K.; Rabuck, A. D.; Raghavachari, K.; Foresman, J. B.; Ortiz, J. V.; Cui, Q.; Baboul, A. G.; Clifford, S.; Cioslowski, J.; Stefanov, B. B.; Liu, G.; Liashenko, A.; Piskorz, P.; Komaromi, I.; Martin, R. L.; Fox, D. J.; Keith, T.; Al-Laham, M. A.; Peng, C. Y.; Nanayakkara, A.; Challacombe, M.; Gill, P. M. W.; Johnson, B.; Chen, W.; Wong, M. W.; Gonzalez, C.; and Pople, J. A.; Gaussian, Inc., Wallingford CT. The bibliographic detail is complete. This is the reference of the computational software used in this study, 2004.
91. Xiao, H.; Kheli, J. T.; Goddard III, W. A. The bibliographic detail is complete. This is the reference of the computational software used in this study. *J. Phys. Chem. Lett.* **2011,** *2,* 212.

CHAPTER 4

STRUCTURAL, ELECTRONIC, AND OPTICAL PROPERTIES OF $AuCu_n^\lambda$ ($\lambda = 0, \pm 1; n = 1$–$8$) NANOALLOY CLUSTERS: A DENSITY FUNCTIONAL THEORY STUDY

PRABHAT RANJAN[1], AJAY KUMAR[1], and
TANMOY CHAKRABORTY[2,*]

[1]*Department of Mechatronics Engineering, Manipal University Jaipur, Dehmi Kalan, Jaipur 303007, Rajasthan, India*

[2]*Department of Chemistry, Manipal University Jaipur, Dehmi Kalan, Jaipur 303007, Rajasthan, India*

Corresponding author. E-mail: tanmoychem@gmail.com; tanmoy.chakraborty@jaipur.manipal.edu

ABSTRACT

The study of bimetallic nanoalloy clusters is of high importance nowadays due to its unique electronic, optical, and magnetic properties. Among such clusters, the compound formed between copper (Cu) and gold (Au) has received considerable interest because of its potential applications in nanoscience, medicine, biology, and catalysis. In recent times, density functional theory (DFT) is one of the most used methodology of quantum mechanics to study the electronic structure of matter. Conceptual DFT-based descriptors have been used to reveal experimental properties qualitatively. In this report, structural, electronic, and optical properties of bimetallic Cu–Au nanoalloy clusters with one Au atom $AuCu_n^\lambda$ ($\lambda = 0, \pm 1$; $n = 1$–8) have been studied invoking conceptual DFT-based descriptors. The results display that the ground state, configurations of Cu–Au clusters look like pure copper clusters in shape. The ground state configurations of clusters have planar geometry till $n = 5$ and three-dimensional structure for $n = 6, 7,$ and 8. The computed highest occupied molecular orbital–lowest unoccupied molecular orbital (HOMO–LUMO) energy gap as a function of cluster size reveal pronounced odd–even oscillation behavior for neutral and charged Cu–Au clusters. The DFT-based descriptors, namely, electronegativity, hardness, softness, electrophilicity index are also discussed. The high value of linear correlation between HOMO–LUMO gaps along with computed softness validates our analysis.

4.1 INTRODUCTION

During recent years, gold (Au) nanoclusters have been very active field of research from experimental and theoretical point of view due to their potential applications in numerous areas like material science, catalysis, medicine.[1-6] Apart from bare Au nanoclusters, there are various studies available in which doped clusters containing Au atoms have been well documented.[7-17] It has been observed that with appropriate impurity, physicochemical properties of Au clusters can be altered in desired manner.[6] Neukermans et al.[9] have reported mass abundance spectra upon photofragmentation of Au clusters doped with 3d-shell metallic atoms (Sc, Ti, V, Cr, Mn, Fe, Co, Ni). They established that size and dopant-dependent magic numbers are related to shell structure which helps in the enhancement of stability of Au clusters. Wang et al.[17] have investigated single copper atom-doped anionic gold clusters [Cu–Au$_{16}$]$^-$ and [Cu–Au$_{17}$]$^-$ using photoelectron spectroscopy technique. Electronic and structural properties of copper-doped gold clusters [Au$_{n-1}$–Cu]$^-$ ($n = 13$–19) have been studied by using density functional theory (DFT). The authors established that addition of single Cu atom enhanced the binding energy and stability of anionic Au clusters.

In order to enhance the physicochemical properties of Ag and Au clusters, we have systematically investigated the structural, electronic, and optical properties of Au-doped copper $[Au–Cu_n]^\lambda$ clusters invoking DFT-based descriptors. The DFT-based descriptors, namely, electronegativity, hardness, softness, electrophilicity index, dipole moment along with HOMO–LUMO energy gap of Cu–Au clusters have been computed. Our computed results are in well agreement with the experimental data.

4.2 COMPUTATIONAL DETAILS

DFT has evolved as an important theoretical approach for bimetallic and multimetallic clusters, since its computational friendly methodology. DFT methods are open to many new innovative fields in material science, physics, chemistry, surface science, nanotechnology, biology, and earth sciences.[18] Among all the DFT approximations, the hybrid functional Becke's three parameter Lee–Yang–Parr exchange correlation functional (B3LYP) has been proven very efficient and used successfully for mixed and impurity-doped clusters.[19–21] The basis set LanL2dz has high accuracy for metallic clusters which has been recently reported.[22,23] All the modeling and structural optimization of compounds have been performed using Gaussian 03 software package[24] within DFT framework. For optimization purpose, B3LYP exchange correlation with basis set LanL2dz has been adopted. The used computation methodology in this paper is based on the molecular orbital approach, using linear combination of atomic orbitals. We have chosen Z-axis as spin polarization axis. The symmetrized fragment orbitals (SFOs) are combined with auxiliary core functions (CFs) to ensure orthogonalization on the (frozen) core orbitals (COs).

Invoking Koopmans' approximation,[25] we have calculated ionization energy (I) and electron affinity (A) of all the nanoalloys using the following ansatz:

$$I = -\varepsilon_{HOMO} \qquad (4.1)$$

$$A = -\varepsilon_{LUMO} \qquad (4.2)$$

Thereafter, using I and A, the conceptual DFT-based descriptors, namely, electronegativity (χ), global hardness (η), molecular softness (S), and electrophilicity index (ω) have been computed. The equations used for such calculations are as follows-

$$\chi = -\mu = \frac{I + A}{2} \tag{4.3}$$

where, μ represents the chemical potential of the system.

$$\eta = \frac{I - A}{2} \tag{4.4}$$

$$S = \frac{1}{2\eta} \tag{4.5}$$

$$\omega = \frac{\mu^2}{2\eta} \tag{4.6}$$

4.3   RESULTS AND DISCUSSION

4.3.1   EQUILIBRIUM GEOMETRIES OF AuCu$_n^\lambda$ (λ = 0, ± 1; n = 1–8) NANOALLOY CLUSTERS

4.3.1.1   PURE COPPER Cu$_{n+1}^\lambda$ (λ = 0, ±1; n = 1–8) CLUSTERS

In order to study the effect of Au doping on pure Ag clusters and obtain the ground state configurations and low lying isomers of bimetallic Cu–Au clusters, we have optimized the lowest energy structure of neutral, cationic, and anionic Ag clusters. There are a number of reports available in which structures and properties of pure Ag clusters have been well documented.[26–31] We have extensively explored the lowest energy structures for neutral, cationic, and anionic Cu clusters which are shown in Figures 4.1–4.3. Our optimized ground state configurations of neutral, cationic, and anionic Cu clusters are in agreement with the previous reported data.[26] The ground state configurations of neutral, anionic, and cationic copper clusters favor 3D structures at $n = 6$, which is also in line with the previous DFT studies.[26]

4.3.1.2 BIMETALLIC NEUTRAL, CATIONIC, AND ANIONIC $AuCu_n^\lambda$ ($\lambda = 0, \pm1$; $n = 1$–8) NANOALLOY CLUSTERS

In this section, we have discussed the ground state configurations and low lying isomers of $AuCu_n^\lambda$ ($\lambda = 0, \pm1$; $n = 1$–8) clusters. The optimized ground state configurations and low lying isomers of Au–Cu_n clusters are shown in Figures 4.1–4.3. The clusters are arranged in ascending energy order from low to high, the low lying isomers are selected as N-n-1, N-n-2, N-n-3, N-n-4, N-n-5; C-n-1, C-n-2, C-n-3, C-n-4, C-n-5; A-n-1, A-n-2, A-n-3, A-n-4, and A-n-5 (where, n represents total atoms, N-neutral, C-cation, and A-anion) for neutral, cation, and anion Au–Cu_n clusters, respectively.

4.3.1.3 NEUTRAL $AuCu_n$ ($n = 1$–8) CLUSTERS

The ground state configurations and low lying isomers of neutral bimetallic $AuCu_n$ clusters are shown in Figure 4.1. The lowest energy structure of neutral Cu_{n+1} ($n = 1$–8) clusters is also presented in Figure 4.1. The pure copper Cu_{n+1} clusters have planar structure till $n = 6$. It has been observed that replacement of single Cu atom by the Au atom in the ground state configuration of Ag clusters also exhibits the lowest energy structure of $AuCu_n$ cluster. Figure 4.1 reveals that the optimized ground state configurations of $AuCu_n$ clusters are having similar structure to the lowest energy structure of the Ag clusters. For cluster Au–Cu, only linear structure is possible. The structure with symmetry group $C_\infty v$ has bond length of 2.40 Å which is in agreement with the experimental result.[32] For Au–Cu_2 cluster, we have optimized three low lying isomers. The optimized ground state configuration has equilateral triangular structure with symmetry group C_s. This structure has been obtained after substitution of one Au with Cu atom in the Cu_3 cluster. The second-most lowest energy structure N-3-2 with symmetry $D_\infty h$ is 0.217 eV higher than the most stable structure. The structures N-3-3 is also having linear structure with symmetry group $C_\infty v$ but it is having higher energy than N-3-2 and N-3-1 clusters. We have optimized three low lying isomers of Au–Cu_3 cluster in the energy range of 0.925 eV. The Y-shaped structure of Au–Cu_3 cluster with C_1 symmetry is more favorable than rhombus structure. The isomer N-4-2 is 0.136 eV higher than most stable structure N-4-1. For $AuCu_4$, we have optimized five low lying isomers in the energy range of

1.480 eV. The lowest energy structure N-5-1 is obtained from isomer N-4 with addition of Au atom in the side of the cluster. The isomer N-5-2 is obtained from N-4-3 after addition of one Cu atom but it is having 0.201 eV high energy than most stable structure N-5-1. The isomer N-5-3 is formed after capping from Cu atom to N-4-3 isomer, the structure N-5-3 is 0.427 eV higher than N-5-1. Similarly, N-5-5 isomer is obtained from N-4 with capping from gold atom; however, the isomer N-5-5 is 1.480 eV higher than N-5-1. The lowest energy structure of $AuCu_5$ cluster is formed with addition of one Cu and Au atom in both sides of N-4 isomer. The isomers N-6-2 and N-6-3 are very much close to the lowest energy structure and only 0.008 and 0.010 eV higher than most stable structure N-6-1, respectively. The other isomers N-6-4 and N-6-5 are also of 0.272 and 0.380 eV higher than N-6-1 structure.

We have optimized five low lying isomers of neutral $Au-Cu_6$ cluster in the energy range of 0--0.740 eV. The ground state configuration of $Au-Cu_6$ cluster is of 3D structure. The isomers N-7-1 and N-7-2 are formed by replacement of one Cu atom with Au atom in N-7 cluster, both isomers N-7-1 and N-7-2 look like similar with the pure Cu_7 cluster. The other low lying isomers N-7-3, N-7-4 and N-7-5 are planar structure and of high energy with respect to the most stable structure N-7-1.

The low lying isomers and ground state configuration of $Au-Cu_7$ are optimized in the range of 0–1.039 eV. The lowest energy structure of $Au-Cu_7$ is obtained with addition of one Cu from one side and one Au atom from opposite side to N-6 structure. The optimized isomers N-8-1, N-8-2, and N-8-3 are very close to each other and are lying in the energy range of 0–0.005 eV. The isomer N-8-3 is formed from N-6-1 structure with addition of two Cu atoms from both sides. The fourth low lying isomer N-8-4 is obtained from two layer of N-4-2 structure, it is having 0.367 eV high in energy than N-8-1. The isomer N-8-5 is of star-like planar structure with symmetry group C_1 but having 1.039 eV higher in energy than N-8-1 structure. The most stable configuration (N-9-1) of $Au-Cu_8$ cluster is formed from two layers of N-4 structure capped with Au atom. The structure N-9-1 is similar with the pure cluster Cu_9. The second lowest energy structure is also obtained after substitution of one Cu atom with Au atom in N-9 structure but it is having 0.163 eV higher in energy than most stable structure. The other isomers N-9-4 and N-9-5 are planar structure and obtained from addition of N-4 and N-4-3 structures connected back to back, these isomers are 0.870 and 1.061 eV higher than the most stable structure.

4.3.1.4   CATIONIC $AuCu_n^+$ (n = 1–8) CLUSTERS

The lowest energy structure and low lying isomers of cationic $Au–Cu_n^+$ and most stable structure of cationic copper clusters are shown in Figure 4.2. It has been observed from Figure 4.2 that the most stable geometry of $AuCu^+$, $AuCu_2^+$, $AuCu_3^+$, $AuCu_4^+$, and $AuCu_5^+$ clusters are having planar structure and looks similar to the ground state configurations of pure copper clusters. For cluster $AuCu_5^+$, we have optimized five low lying isomers in the energy range of 0.217 eV. It is worth noticing that most stable structure (C-6-1) and second-most stable structure (C-6-2) have energy difference of only 0.027 eV. For $AuCu_n^+$ clusters containing 7, 8, and 9, the ground state configurations becomes 3D structures, which is similar with pure Ag clusters. Four isomers (C-7-1, C-7-2, C-7-3, and C-7-4) for $Au–Cu_6^+$ in the energy range of 0.081 eV were optimized. We have also observed four different 3D isomers (C-8-1, C-8-2, C-8-3, and C-8-4) in the energy range of 0.40 eV for $Au–Cu_7^+$ cluster. The planar structure (C-8-5) for $Au–Cu_7^+$ cluster is also optimized but having 0.734 eV higher in energy than most stable structure (C-8-1). Three isomers (C-9-1, C-9-2, and C-9-3) are observed as 3D structure in the energy range of 0.19 eV in the case of $Au–Cu_8^+$ cluster. Two planar isomers (C-9-4 and C-9-5) are also optimized but have 0.326 and 0.598 eV high in energy than the most stable structure (C-9-1).

4.3.1.5   ANIONIC $AuCu_n^-$ (n = 1–8) CLUSTERS

The ground state configurations and low lying isomers of $Au–Cu_n^-$ (n = 1–8) clusters and the most stable structure of pure anionic copper clusters are shown in Figure 4.3. The interesting fact observed from Figure 4.3 shows that the ground state configuration of bimetallic $Au–Cu_n^-$ clusters are having similar structure to the most stable states of pure anionic Ag clusters, which indicates that doping of one Au atom to anionic Ag clusters cannot change the geometry of pure Ag cluster. Figure 4.3 reveals that the ground state configurations of $Au–Cu_n^-$ (n = 1–5) clusters are having planar structure and $Au–Cu_n^-$ (n = 6–8) are having 3D structure.

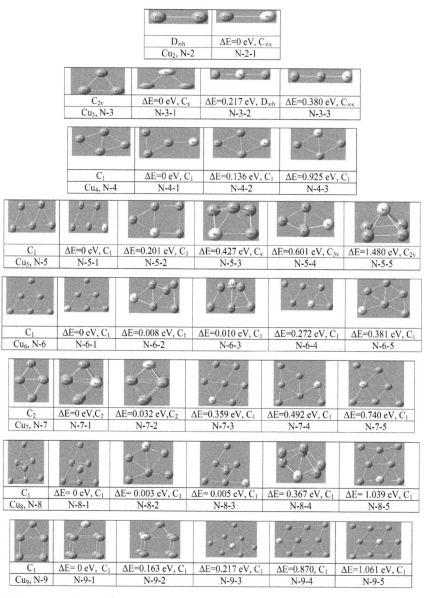

FIGURE 4.1   (See color insert.) The ground state configurations of neutral Cu_{n+1} and Au–Cu_n clusters, and low lying isomers of Au–Cu_n ($n = 1$–8) clusters. The yellow and red circles represent gold and copper atoms, respectively.

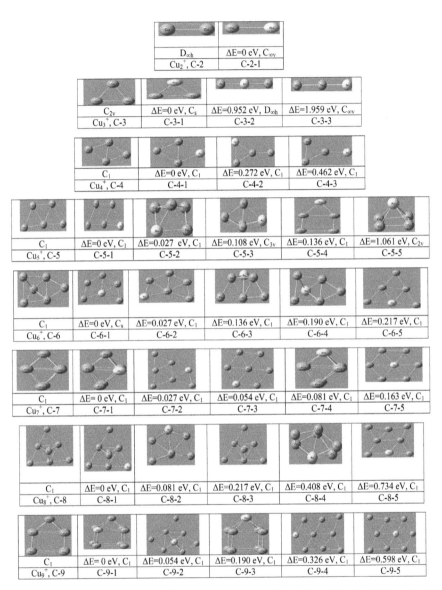

FIGURE 4.2 (See color insert.) The ground state configurations of cationic $[Cu_{n+1}]^+$ and $[Au–Cu_n]^+$ clusters, and low lying isomers of $[Au–Cu_n]^+$ ($n = 1$–8) clusters. The yellow and red circles represent gold and copper atoms, respectively.

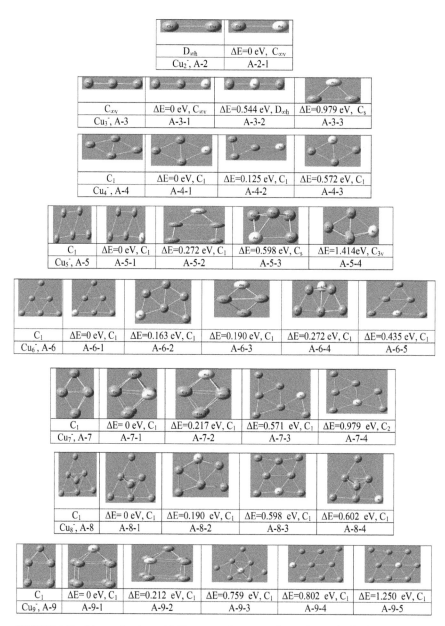

FIGURE 4.3 **(See color insert.)** The ground state configurations of anionic $[Cu_{n+1}]^-$ and $[Au\text{–}Cu_n]^-$ clusters, and low lying isomers of $[AuCu_n]^-$ ($n = 1\text{–}8$) clusters. The yellow and red circles represent gold and copper atoms, respectively.

4.4 HOMO–LUMO ENERGY GAPS AND DFT-BASED DESCRIPTORS

The HOMO–LUMO energy gap is a significant factor to compute the electronic properties of clusters. It signifies the energy required for an electron to jump from occupied orbital to unoccupied orbital. The HOMO–LUMO energy gaps have marked influence on chemical stability of the clusters. A small value of the energy gap indicates the maximum response to an external perturbation. For the ground state configurations of bimetallic neutral, cationic, and anionic $[AuCu_n]^\lambda$ ($\lambda = 0, \pm1$; $n = 1$–8) clusters, the computed electronegativity (χ), hardness (η), softness (S), electrophilicity index (ω), and dipole moment along with HOMO–LUMO gaps are presented in Tables 4.1, 4.2, and 4.3, respectively. The result from Tables 4.1, 4.2, and 4.33 reveals that $AuCu_5$, $AuCu_2^+$, and $Au–Cu_2^-$ clusters have large energy gaps in this molecular system, which are 3.374, 4.218, and 2.857 eV, respectively, whereas $AuCu_2$, $AuCu_5^+$, and $AuCu_7^-$ clusters exhibit low energy gaps. The HOMO–LUMO energy gaps of the lowest energy structure of $[AuCu_n]^\lambda$ ($\lambda = 0, \pm1$; $n = 1$–8) as a function of cluster size are displayed in Figure 4.4. From Figure 4.4, we have found that HOMO–LUMO energy gaps for neutral and charged clusters reveal odd–even oscillation behaviors, indicating that neutral clusters with an even number of total atoms have high value of HOMO–LUMO energy gap as compared with their neighbor with an odd number of atoms. Similarly, the charged clusters maintain high HOMO–LUMO gap for odd number of clusters and low energy gap for even number of clusters.

From Tables 4.1–4.3, we note that HOMO–LUMO energy gaps of clusters maintain direct relationship with hardness and inverse relationship with softness values. The computed HOMO–LUMO energy gap and hardness value runs hand in hand. This is an expected trend from experimental point of view, as the frontier orbital energy gap increases, their hardness value also increases. The similar kind of relationships is observed in case of computed electronegativity, electrophilicity index, and dipole moment along with their HOMO–LUMO energy gaps of clusters. The linear correlation between HOMO–LUMO energy gaps along with their computed softness of neutral, cationic, and anionic $[AuCu_n]\lambda$ clusters is lucidly plotted in Figure 4.5. The correlation coefficient of neutral, cationic, and anionic $[AuCu_n]\lambda$ clusters $R^2 = 0.979, 0.933$, and 0.944, respectively, validates our predicted model.

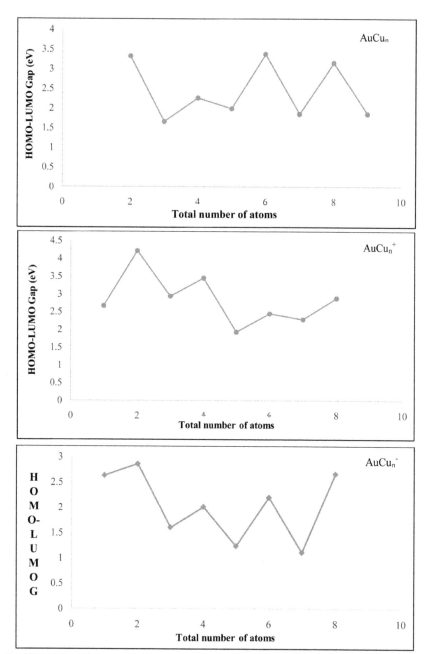

FIGURE 4.4   HOMO–LUMO energy gap as a function of cluster size of the lowest energy structure of $AuCu_n^\lambda$ ($\lambda = 0, \pm1$; $n = 1$–8) nanoalloy clusters.

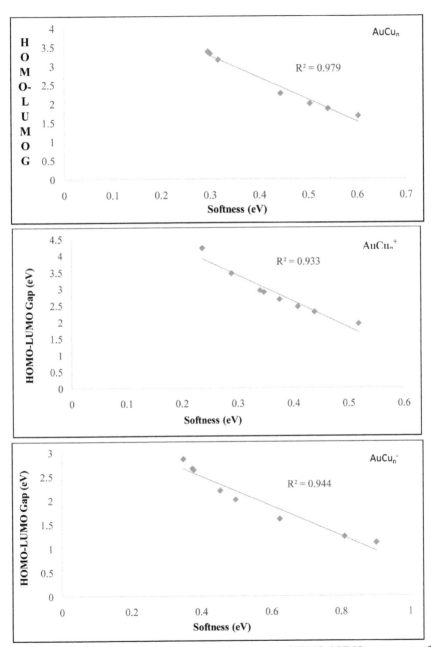

FIGURE 4.5 A linear correlation between softness versus HOMO–LUMO energy gap of $AuCu_n^\lambda$ ($\lambda = 0, \pm1$; $n = 1$–8) clusters.

TABLE 4.1 Computed DFT-based Descriptors of Neutral Au–Cu$_n$ (n = 1–8) Clusters.

Species	HOMO–LUMO gap (eV)	Electronegativity (eV)	Hardness (eV)	Softness (eV)	Electrophilicity index (eV)	Dipole moment (Debye)
Au–Cu	3.320	4.762	1.660	0.301	6.830	2.133
Au–Cu$_2$	1.660	3.741	0.830	0.602	8.443	2.142
Au–Cu$_3$	2.258	4.367	1.129	0.443	8.445	2.103
Au–Cu$_4$	1.986	4.231	0.993	0.503	9.013	2.519
Au–Cu$_5$	3.374	4.054	1.687	0.296	4.872	2.568
Au–Cu$_6$	1.850	3.510	0.925	0.540	6.659	1.763
Au–Cu$_7$	3.156	3.918	1.578	0.317	4.864	1.944
Au–Cu$_8$	1.850	3.320	0.925	0.540	5.956	2.173

TABLE 4.2 The Computed DFT-based Descriptors of Cationic $[Au–Cu_n]^+$ (n = 1–8) Clusters.

Species	HOMO–LUMO gap (eV)	Electronegativity (eV)	Hardness (eV)	Softness (eV)	Electrophilicity index (eV)	Dipole moment (Debye)
$Au–Cu^+$	2.667	11.673	1.333	0.375	27.417	3.979
$Au–Cu_2^+$	4.218	10.272	2.109	0.237	25.017	4.427
$Au–Cu_3^+$	2.939	9.061	1.469	0.340	27.938	5.477
$Au–Cu_4^+$	3.456	8.830	1.728	0.289	22.561	5.433
$Au–Cu_5^+$	1.932	8.313	0.966	0.518	22.735	1.321
$Au–Cu_6^+$	2.449	8.582	1.224	0.408	30.075	3.861
$Au–Cu_7^+$	2.286	7.728	1.143	0.438	26.127	4.441
$Au–Cu_8^+$	2.884	8.000	1.442	0.347	22.188	3.505

TABLE 4.3 The Computed DFT-based Descriptors of Anionic $[Au–Cu_n]^-$ ($n = 1$–8) Clusters.

Species	HOMO–LUMO Gap (eV)	Electronegativity (eV)	Hardness (eV)	Softness (eV)	Electrophilicity index (eV)	Dipole moment (Debye)
$Au–Cu^-$	2.631	-1.677	1.316	0.380	1.069	3.508
$Au–Cu_2^-$	2.857	-0.367	1.429	0.350	0.047	3.493
$Au–Cu_3^-$	1.603	-0.831	0.801	0.624	0.431	2.621
$Au–Cu_4^-$	2.008	-0.414	1.004	0.498	0.085	1.268
$Au–Cu_5^-$	1.235	-0.607	0.617	0.810	0.298	1.823
$Au–Cu_6^-$	2.199	-0.133	1.099	0.455	0.008	1.665
$Au–Cu_7^-$	1.110	-0.528	0.555	0.901	0.251	0.341
$Au–Cu_8^-$	2.661	0.269	1.331	0.376	0.027	1.617

4.5 CONCLUSION

The structural, electronic, and optical properties of $AuCu_n^\lambda$ ($\lambda = 0, \pm 1$; $n = 1$–8) nanoalloy clusters are systematically investigated invoking DFT-based descriptors. The conclusions are summarized as follows:

(i) The ground state configurations and low lying isomers of neutral and charged Au–Cu$_n$ ($n = 1$–8) clusters have been studied. It has been observed that $[Au\text{–}Cu_n]^\lambda$, ($\lambda = 0, \pm 1$; $n = 1$–8) clusters have planar structure till $n = 5$ and 3D structure for $n = 6, 7$, and 8. The DFT-based descriptors, namely, electronegativity, hardness, softness, electronegativity, and dipole moment along with HOMO–LUMO energy gaps are computed and reported. The result reveals that computed HOMO–LUMO energy gaps have direct relationship with hardness and inverse relationship with softness values.

(ii) The HOMO–LUMO energy gap of Cu–Au clusters have pronounced odd–even oscillation behavior as a function of cluster size, indicating that neutral clusters with even number of total atoms are more stable as compared to the clusters with odd number of total atoms. For charged clusters, the trend is reverse.

(iii) The high value of regression coefficient between softness and HOMO–LUMO energy gap supports our computational analysis.

KEYWORDS

- density functional theory
- bimetallic clusters
- Cu–Au
- HOMO–LUMO gap
- hardness
- odd–even oscillation

REFERENCES

1. Dyson, P. J.; Mingos, D. M. S. *Gold: Progress in Chemsitry, Bio-chemistry and Technology*; Schmidbaur, H., Ed.; Wiley: New York, 1999; p 511.
2. Teles, J. H.; Brode, S.; Chabanas, M. *Angew. Chem.* **1999,** *99*, 2589.
3. Valden, M.; Lai, X.; Goodman, D. W. *Science* **1998,** *281*, 1637.
4. Yoon, B.; Häkkinen, H.; Landman, U.; Wörz, A. S.; Antonietti, J. –M.; Abbet, S.; Judai, K.; Heiz, U. *Science* **2005,** *307*, 403.
5. Shaw III, C. F. *Chem. Rev.* (Washington, D. C.) **1999,** *99*, 2589.
6. Zorriasatein, S.; Joshi, K.; Kanhere, D. G. *J. Chem. Phys.* **2008,** *128*, 184314.
7. Darby, S.; Mortimer, T. V.; Johnston, R. L.; Roberts, C. *J. Chem. Phys.* **2001,** *116*, 1536.
8. Pyykkö, P.; Runeberg, N. *Angew. Chem. Int. Ed.* **2002,** *41*, 2174.
9. Neukermans, S.; Janssens, E.; Tanaka, H.; Silverans, R. E.; Lievens, P. *Phys. Rev. Lett.* **2003,** *90*, 033401.
10. Zhai, H. –J.; Li, J.; Wang, L. –S. *J. Chem. Phys.* **2004,** *121*, 8369.
11. Janssens, E.; Tanak, H.; Neukermans, S.; Silverans, R. E.; Lievens, P. *Phys. Rev. B* **2004,** *69*, 085402.
12. Rapallo, A.; Rossi, G.; Ferrando, R.; Fortunelli, A.; Curley, B. C.; Lioyd, L. D.; Tarbuck, G. M.; Johnston, R. L. *J. Chem. Phys.* **2005,** *122*, 194308.
13. Gao, Y.; Bulusu, S.; Zeng, X. –C. *Chem. Phys. Chem.* **2006,** *7*, 2275.
14. Wang, L. –M.; Bulusu, S.; Huang, W.; Pal, R.; Wang, L. –S.; Zeng, X. –C. *J. Am. Chem. Soc.* **2007,** *129*, 15136.
15. Rao, J. L.; Chaitanya, G. K.; Basavaraja, S.; Bhanuprakash, K.; Venkataramana, A. *Theochem* **2007,** *803*, 89.
16. Sun, Y.; Fournier, R. *Phys. Rev. A* **2007,** *75*, 063205.
17. Wang, L. –M.; Bulusu, S.; Zhai, H. –J.; Zeng, X. –C.; Wang, L. –S. *Angew. Chem.* **2007,** *46*, 2915.
18. Hafner, J.; Wolverton, C.; Ceder, G. *MRS Bull.* **2006,** *31*, 659.
19. Lee, H. M.; Ge, M.; Sahu, B. R.; Tarakeshwar, P.; Kim. *J. Phys. Chem.* **2003,** *107*, 9994.
20. Becke, A. D. *J. Chem. Phys.* **1993,** *98*, 5648.
21. Mielich, B.; Savin, A.; Stoll, H.; Preuss, H. *Chem. Phys. Lett.* **1989,** *157*, 200.
22. Wang, H. Q.; Kuang, X. Y.; Li, H. F. *Phys. Chem. Chem. Phys.* **2010,** *12*, 5156.
23. Jiang, Z. Y.; Lee, K. H.; Li, S. T.; Chu, S. Y. *Phys. Rev. B* **2006,** *73*, 235423.
24. Gaussian 03, Revision C.02, Frisch, M. J.; Trucks, G. W.; Schlegel, H. B.; Scuseria, G. E.; Robb, M. A.; Cheeseman, J. R.; Montgomery, Jr., J. A.; Vreven, T.; Kudin, K. N.; Burant, J. C.; illam, J. M.; Iyengar, S. S.; Tomasi, J.; Barone, V.; Mennucci, B.; Cossi, M.; Scalmani, G.; Rega, N.; Petersson, G. A.; akatsuji, H. N.; Hada, M.; Ehara, M.; Toyota, K.; Fukuda, R.; Hasegawa, J.; Ishida, M.; Nakajima, T.; Honda, Y.; Kitao, O.; Nakai, H.; Klene, M.; Li, X.; Knox, J.. E.; Hratchian, H. P.; Cross, J. B.; Bakken, V.; Adamo, C.; Jaramillo, J.; Gomperts, R.; Stratmann, R. E.; Yazyev, O.; Austin, A. J.; Cammi, R.; Pomelli, C.; Ochterski, J. W.; Ayala, P. Y.; Morokuma, K.; Voth, G. A.; Salvador, P.; Dannenberg, J. J.; Zakrzewski, V. G.; Dapprich, S.; Daniels, A. D.; Strain, M. C.; Farkas, O.; Malick, D. K.; Rabuck, A. D.; Raghavachari, K.; Foresman,

J. B.; Ortiz, J. V.; Cui, Q.; Baboul, A. G.; Clifford, S.; Cioslowski, J.; Stefanov, B. B.; Liu, G.; Liashenko, A.; Piskorz, P.; Komaromi, I.; Martin, R. L.; Fox, D. J.; Keith, T.; Al-Laham, M. A.; Peng, C. Y.; Nanayakkara, A.; Challacombe, M.; Gill, P. M. W.; Johnson, B.; Chen, W.; Wong, M. W.; Gonzalez, C.; Pople, J. A. Gaussian, Inc.: Wallingford CT, 2004.

25. Parr, R. G.; Yang, W. *Density Functional Theory of Atoms and Molecules*; Oxford University Press: Oxford, 1989.
26. Jug, K.; Zimmermann, B.; Calaminici, P.; Köster, A. M. *J. Chem. Phys.* **2002,** *116,* 4497.
27. Yang, M.; Jackson, K. A.; Koehler, C.; Frauenheim, T.; Jellinek, J. *J. Chem. Phys.* **2006,** *124,* 024308.
28. Massobrio, C.; Pasquarello, Z.; A.; Par, R. *Chem. Phys. Lett.* **1995,** *238,* 215.
29. Guvelioglu, G. H.; Ma, P.; He, X.; Forrey, R. C.; Cheng, H. *Phys. Rev. Lett.* **2005,** *94,* 026103.
30. Calaminici, P.; Köster, A. M.; Vela, A. *J. Chem. Phys.* **2000,** *113,* 2199.
31. Cao, Z.; Wang, Y.; Zhu, J.; Wu, W.; Zhang, Q. *J. Phys. Chem. B* **2002,** *106,* 9649.
32. Bishea, G. A.; Pinegar, J. C.; Morse, M. D. *J. Chem. Phys.* **1991,** *95,* 5630.

CHAPTER 5

CONFORMATIONAL STUDY OF BIMETALLIC TRIMERS Cu–Ag NANOALLOY CLUSTERS

PRABHAT RANJAN[1], AJAY KUMAR[1], and TANMOY CHAKRABORTY[2,*]

[1]*Department of Mechatronics Engineering, Manipal University Jaipur, Dehmi Kalan, Jaipur 303007, Rajasthan, India*

[2]*Department of Chemistry, Manipal University Jaipur, Dehmi Kalan, Jaipur 303007, Rajasthan, India*

Corresponding author. E-mail: tanmoychem@gmail.com; tanmoy.chakraborty@jaipur.manipal.edu

ABSTRACT

The study of bimetallic Cu–Ag nanoalloy clusters is of considerable interest due to unique electronic, optical, and magnetic properties and it has potential applications in medical sciences, fabrications, nanoscience, and catalysis. Density functional theory (DFT) is the most efficient technique of quantum mechanics to explore the electronic properties of materials. In this article, conformational analysis of three atoms of Cu–Ag clusters have been studied invoking DFT-based descriptors. We have computed DFT-based descriptors, namely, highest occupied molecular orbital (HOMO)–lowest unoccupied molecular orbital (LUMO) energy gap, electronegativity, hardness, softness, electrophilicity index, and dipole moment of trimers Cu–Ag clusters by changing the angle between the atoms. The study reveals that linear structure of bimetallic clusters is more stable than other conformations.

5.1   INTRODUCTION

It is a well-known fact that structure and property are very much closely related to each other.[1] There is no particular method to determine the symmetry structure of a molecule. Density functional theory (DFT)-based global descriptors like highest occupied molecular orbital (HOMO)–lowest unoccupied molecular orbital (LUMO) energy gap, hardness, softness, electronegativity, and electrophilicity index have been invoked to relate structure and property of molecules.[1–3] Even if there is not any distinctive relationship of structure and properties, the structure and property of molecules are interrelated. Nowadays, science and technology has been developed to a new level for understanding the biological and molecular activities of system.[4–7] Researchers are examining the descriptors in terms of mathematical values that pronounce the structure and physicochemical properties of molecular system.[8–12] But, for prediction of relationship between structure and property, it is very important to investigate the electronic structure and operational forces on the molecular system.[10,11] Although some researchers have tried to show the relationship between properties and reactivity of a molecule with reference to its equilibrium structure, a molecule rarely stays in its stationary equilibrium shape. The physical process of the dynamics of internal alternation begins the isomerization process mechanism and it produces countless conformations.[13,14] The conformational analysis have acute result on biological activities and response on the result of various stereochemical reactions.[1] Therefore, an insight of the relative energies and the mechanism of the advancement of conformational analysis will be able to predict information regarding the sensitivity, spatial arrangement of atoms, and their effects on the physicochemical properties and product dissemination in reaction.[15] Thus, study of conformations as to the source of barrier to internal alternation within a molecule is of high importance to theoretical, experimental, and biological scientists.

5.2   COMPUTATIONAL DETAILS

Since last couple of years, DFT has been dominant and effective computational technique for bimetallic and multi-metallic clusters. DFT methods are open to many new innovative fields in material science, physics, chemistry, surface science, nanotechnology, biology, and earth sciences.[16] Among all the DFT approximations, the hybrid functional Becke's three parameter Lee–Yang–Parr exchange correlation functional (B3LYP) has been proven very efficient and used successfully for bimetallic clusters.[17-20] In this section, conformational analysis of bimetallic nanoalloy clusters has been performed with the help of hybrid functional B3LYP exchange correlation functional. The basis set LanL2dz has high accuracy for metallic clusters which has been recently analyzed by researchers.[17,21,22] For optimization purpose, B3LYP exchange correlation with basis set LanL2dz has been adopted. All the modeling and structural optimization of compounds have been performed using Gaussian 03 software package[23] within DFT framework.

Invoking Koopmans' approximation,[24,25] we have calculated ionization energy (I) and electron affinity (A) of all the nanoalloys using the following ansatz:

$$I = -\varepsilon_{HOMO} \tag{5.1}$$

$$A = -\varepsilon_{LUMO} \tag{5.2}$$

Thereafter, using I and A, the conceptual DFT-based descriptors, namely, electronegativity (χ), global hardness (η), molecular softness (S), and electrophilicity index (ω) have been computed. The equations used for such calculations are as follows:

$$\chi = -\mu = \frac{I + A}{2} \tag{5.3}$$

Where, μ represents the chemical potential of the system.

$$\eta = \frac{I - A}{2} \tag{5.4}$$

$$S = \frac{1}{2\eta} \tag{5.5}$$

$$\omega = \frac{\mu^2}{2\eta} \tag{5.6}$$

5.3   RESULTS AND DISCUSSION

In this study, conformational analysis of three atoms of bimetallic clusters Cu–Ag, that is, Cu–Ag$_2$ and Cu$_2$–Ag have been performed. The orbital energies in form of HOMO–LUMO gap along with DFT-based descriptors have been investigated as a function of dihedral angles.

5.3.1   *Cu–Ag$_2$ NANOALLOY CLUSTERS*

(1) In Figure 5.1, Ag atom is fixed at the center and the other atoms, that is, Ag and Cu are located at corners and are not fixed.

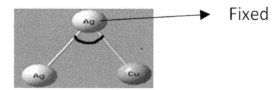

Fixed

FIGURE 5.1   Cluster Cu–Ag$_2$ with fixed Ag atom at the center.

The HOMO–LUMO energy gap along with computed DFT-based descriptors, namely, electronegativity, hardness, softness, electrophilicity index, and dipole moment have been analyzed and reported in Table 5.1. Table 5.1 and Figure 5.1 demonstrate that the preferred conformation of instant cluster has linear structure with Ag atom in the middle. The maximum HOMO–LUMO energy gap is obtained at the dihedral angle of 180°, while least gap is obtained at dihedral angle of 80°.

TABLE 5.1   Computed DFT-based Descriptors of Cluster Cu–Ag$_2$ with Fixed Ag Atom.

Angle (in degrees)	HOMO–LUMO Gap (eV)	Electro-negativity (eV)	Hardness (eV)	Softness (eV)	Electro-philicity Index (eV)	Dipole moment (Debye)	Symmetry
20	2.612	3.537	1.306	0.382	4.790	0.397	C$_s$
30	2.612	3.510	1.306	0.382	4.716	0.392	C$_s$
40	2.642	3.549	1.321	0.378	4.768	0.398	C$_s$
50	2.642	3.549	1.321	0.378	4.768	0.398	C$_s$

TABLE 5.1 *(Continued)*

Angle (in degrees)	HOMO–LUMO Gap (eV)	Electro-negativity (eV)	Hardness (eV)	Softness (eV)	Electro-philicity Index (eV)	Dipole moment (Debye)	Symmetry
60	2.642	3.549	1.321	0.378	4.768	0.398	C_s
70	2.612	3.537	1.306	0.382	4.790	0.393	C_s
80	1.442	3.469	0.721	0.693	8.345	0.628	C_s
90	2.721	3.564	1.360	0.367	4.669	0.502	C_s
100	2.721	3.564	1.360	0.367	4.669	0.502	C_s
110	2.721	3.564	1.360	0.367	4.669	0.502	C_s
120	2.721	3.564	1.360	0.367	4.669	0.502	C_s
130	2.721	3.564	1.360	0.367	4.669	0.502	C_s
140	2.748	3.578	1.374	0.363	4.658	0.501	C_s
150	2.721	3.564	1.360	0.367	4.669	0.501	C_s
160	2.721	3.564	1.360	0.367	4.669	0.502	C_s
170	2.721	3.564	1.360	0.367	4.669	0.502	C_s
180	3.020	3.632	1.510	0.331	4.368	0.140	$C_{\infty v}$

(2) In Figure 5.2, triangular structure of Cu–Ag$_2$ with fixed Ag atom at center is formed. The computed physicochemical properties of all the generated conformations are shown in Table 5.2 with the corresponding dihedral angles.

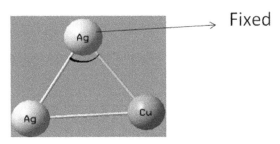

FIGURE 5.2 Triangular structure of Cu–Ag$_2$ with fixed Ag atom at the center.

Table 5.2 reveals that highest HOMO–LUMO energy gap is obtained at dihedral angle 20° and least HOMO–LUMO gap is obtained at dihedral angle 90°. We have not observed any change in the physicochemical properties of the instant cluster when dihedral angle varies from 100° to 170°, except there is a minor change at angle 160°.

TABLE 5.2 Computed DFT-based Descriptors of Triangular Cu–Ag$_2$ Cluster with Fixed Ag Atom at the Center.

Angle (in degrees)	HOMO– LUMO Gap (eV)	Electro- negativity (eV)	Hardness (eV)	Softness (eV)	Electro- philicity Index (eV)	Dipole moment (Debye)	Symmetry
20	2.748	3.550	1.374	0.363	4.588	0.397	C$_s$
30	2.612	3.537	1.306	0.382	4.790	0.398	C$_s$
40	2.666	3.537	1.333	0.375	4.692	0.399	C$_s$
50	2.642	3.549	1.321	0.378	4.768	0.398	C$_s$
60	2.642	3.549	1.321	0.378	4.768	0.398	C$_s$
70	2.582	3.525	1.291	0.387	4.812	0.388	C$_s$
80	2.639	3.550	1.319	0.378	4.777	0.398	C$_s$
90	1.768	3.496	0.884	0.565	6.912	0.374	C$_s$
100	2.696	3.576	1.348	0.370	4.744	0.502	C$_s$
110	2.696	3.576	1.348	0.370	4.744	0.502	C$_s$
120	2.696	3.576	1.348	0.370	4.744	0.502	C$_s$
130	2.696	3.576	1.348	0.370	4.744	0.502	C$_s$
140	2.696	3.576	1.348	0.370	4.744	0.502	C$_s$
150	2.696	3.576	1.348	0.370	4.744	0.502	C$_s$
160	2.693	3.578	1.346	0.371	4.752	0.504	C$_s$
170	2.696	3.576	1.346	0.370	4.744	0.503	C$_s$

(3) In Figure 5.3, Cu atom is fixed at the center and Ag atoms located at the corners are not fixed.

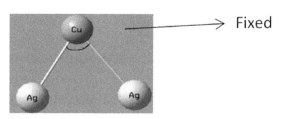

FIGURE 5.3 Cluster CuAg$_2$ with Cu atom fixed at the center.

TABLE 5.3 Computed DFT-based Descriptors of Cluster Cu–Ag$_2$ with Fixed Cu Atom at the Center.

Angle (in degrees)	HOMO–LUMO Gap (eV)	Electro-negativity (eV)	Hardness (eV)	Softness (eV)	Electro-philicity index (eV)	Dipole moment (Debye)	Symmetry
20	2.421	3.523	1.210	0.412	5.127	0.347	C$_s$
30	2.612	3.537	1.306	0.382	4.790	0.395	C$_s$
40	2.639	3.550	1.319	0.378	4.777	0.593	C$_s$
50	2.639	3.550	1.319	0.378	4.777	0.593	C$_s$
60	2.639	3.550	1.319	0.378	4.777	0.593	C$_s$
70	2.639	3.550	1.319	0.378	4.777	0.593	C$_s$
80	2.639	3.550	1.319	0.378	4.777	0.593	C$_s$
90	2.639	3.550	1.319	0.378	4.777	0.593	C$_s$
100	2.639	3.550	1.319	0.378	4.777	0.593	C$_s$
110	2.639	3.550	1.319	0.378	4.777	0.593	C$_s$
120	2.639	3.550	1.319	0.378	4.777	0.593	C$_s$
130	2.693	3.550	1.346	0.371	4.680	0.594	C$_s$
140	2.639	3.550	1.319	0.378	4.777	0.593	C$_s$
150	2.639	3.550	1.319	0.378	4.777	0.593	C$_s$
160	2.639	3.550	1.319	0.378	4.777	0.593	C$_s$
170	2.639	3.550	1.319	0.378	4.777	0.593	C$_s$
180	3.074	3.605	1.537	0.325	4.227	1.456	C$_{\infty v}$

Table 5.3 and Figure 5.3 demonstrate that the preferred conformation of instant cluster has linear structure with Cu atom in the middle. The maximum HOMO–LUMO energy gap is obtained at the dihedral angle of 180°, while least gap is obtained at dihedral angle of 20°. We have not observed any change in the physicochemical properties of the instant cluster when dihedral angle varies from 40° to 170°.

(4) In Figure 5.4, triangular structure of Cu–Ag$_2$ with Cu atom fixed at the center is formed. The computed physicochemical properties of all the generated conformations are presented in Table 5.4 with the corresponding dihedral angles.

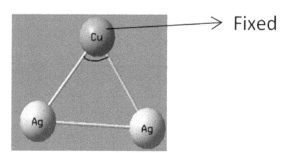

FIGURE 5.4   Triangular structure of CuAg$_2$ cluster with Cu atom fixed at the center.

TABLE 5.4   Computed DFT-based Descriptors of Triangular Structure of Cu–Ag$_2$ Cluster with Fixed Cu Atom.

Angle (in degrees)	HOMO–LUMO gap (eV)	Electro-negativity (eV)	Hardness (eV)	Softness (eV)	Electro-philicity index (eV)	Dipole moment (Debye)	Symmetry
20	0.761	3.455	0.380	1.312	15.673	0.059	C$_{2v}$
30	0.761	3.455	0.380	1.312	15.673	0.061	C$_{2v}$
40	0.761	3.455	0.380	1.312	15.673	0.060	C$_{2v}$
50	0.761	3.455	0.380	1.312	15.673	0.060	C$_{2v}$
60	2.639	3.550	1.319	0.378	4.777	0.113	C$_{2v}$
70	2.617	3.561	1.308	0.382	4.846	0.113	C$_{2v}$
80	2.617	3.561	1.308	0.382	4.846	0.113	C$_{2v}$
90	2.617	3.561	1.308	0.382	4.846	0.113	C$_{2v}$
100	2.617	3.561	1.308	0.382	4.846	0.113	C$_{2v}$
110	2.639	3.550	1.319	0.378	4.777	0.113	C$_{2v}$
120	2.639	3.550	1.319	0.378	4.777	0.113	C$_{2v}$
130	2.617	3.561	1.308	0.382	4.846	0.113	C$_{2v}$
140	2.617	3.561	1.308	0.382	4.846	0.114	C$_{2v}$
150	2.617	3.561	1.308	0.382	4.846	0.113	C$_{2v}$
160	2.617	3.561	1.308	0.382	4.846	0.113	C$_{2v}$
170	2.617	3.561	1.308	0.382	4.846	0.113	C$_{2v}$

Table 5.4 and Figure 5.4 demonstrate that the instant cluster has maximum HOMO–LUMO energy gap at dihedral angle of 100° and 110°, while least gap is obtained when cluster varies from dihedral angles 20–50°.

5.3.2 Cu$_2$–Ag NANOALLOY CLUSTERS

(1) In Figure 5.5, Ag atom is fixed at the center and Cu atoms located at corners are not fixed.

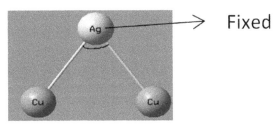

FIGURE 5.5   Cluster Cu$_2$–Ag with Ag atom fixed at the center.

TABLE 5.5   Computed DFT-based Descriptors of Cluster Cu$_2$–Ag with Ag Atom Fixed at the Center.

Angle (in degrees)	HOMO–LUMO Gap (eV)	Electro-negativity (eV)	Hardness (eV)	Softness (eV)	Electro-philicity index (eV)	Dipole moment (Debye)	Symmetry
30	1.142	3.374	0.571	0.875	9.961	1.236	C_{2v}
40	1.142	3.374	0.571	0.875	9.961	1.236	C_{2v}
50	1.142	3.374	0.571	0.875	9.961	1.235	C_{2v}
60	1.142	3.374	0.571	0.875	9.961	1.236	C_{2v}
70	1.088	3.428	0.544	0.918	10.799	0.684	C_{2v}
80	1.115	3.414	0.557	0.896	10.452	0.682	C_{2v}
90	1.142	3.374	0.571	0.875	9.961	1.235	C_{2v}
100	2.176	3.455	1.088	0.459	5.485	0.326	C_{2v}
110	2.693	3.496	1.346	0.371	4.538	0.520	C_{2v}
120	2.693	3.496	1.346	0.371	4.538	0.520	C_{2v}
130	2.693	3.496	1.346	0.371	4.538	0.520	C_{2v}
140	2.693	3.496	1.346	0.371	4.538	0.520	C_{2v}
150	2.693	3.496	1.346	0.371	4.538	0.520	C_{2v}
160	2.693	3.496	1.346	0.371	4.538	0.520	C_{2v}
170	2.693	3.496	1.346	0.371	4.538	0.520	C_{2v}
180	3.129	3.578	1.564	0.319	4.091	0	$D_{\infty h}$

The orbital energies in the form of HOMO–LUMO gap along with computed DFT-based descriptors, namely, electronegativity, hardness, softness, electrophilicity index, and dipole moment have been analyzed as a function of dihedral angles and reported in Table 5.5. Table 5.5 and Figure 5.5 demonstrate that the preferred conformation of instant cluster has linear structure with Ag atom in the middle. The maximum HOMO–LUMO energy gap is obtained at the dihedral angle of 180°, while least gap is obtained at dihedral angle of 70°. The instant cluster has constant HOMO–LUMO gap and DFT-based descriptors for dihedral angles varying between 30°, 60° and 110–170°.

(2) In Figure 5.6, triangular structure of Cu_2–Ag is formed with fixed Ag atom at center and two Cu atoms are placed at the corners of the triangle. The computed physicochemical properties of all the generated conformations are shown in Table 5.6 with the corresponding dihedral angles.

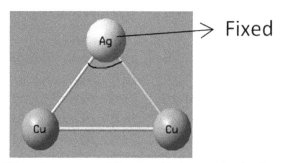

FIGURE 5.6   Triangular structure of Cu_2–Ag with Ag atom fixed at the center.

TABLE 5.6   Computed DFT-based Descriptors of Triangular Structure of Cu_2–Ag with Ag Atom Fixed at the Center.

Angles (in degrees)	HOMO– LUMO Gap (eV)	Electro- negativity (eV)	Hardness (eV)	Softness (eV)	Electro- philicity index (eV)	Dipole moment (Debye)	Symmetry
20	1.142	3.374	0.571	0.875	9.961	1.236	C_{2v}
30	1.170	3.387	0.585	0.854	9.808	1.236	C_{2v}
40	1.170	3.387	0.585	0.854	9.808	1.236	C_{2v}
50	1.170	3.387	0.585	0.854	9.808	1.236	C_{2v}

TABLE 5.6   *(Continued)*

Angles (in degrees)	HOMO– LUMO Gap (eV)	Electro- negativity (eV)	Hardness (eV)	Softness (eV)	Electro- philicity index (eV)	Dipole moment (Debye)	Symmetry
60	1.170	3.387	0.585	0.854	9.808	1.235	C_{2v}
70	1.115	3.414	0.557	0.896	10.452	0.683	C_{2v}
80	1.115	3.414	0.557	0.896	10.452	0.684	C_{2v}
90	1.115	3.414	0.557	0.896	10.452	0.682	C_{2v}
100	2.176	3.455	1.088	0.459	5.485	0.332	C_{2v}
110	2.693	3.496	1.346	0.371	4.538	0.520	C_{2v}
120	2.693	3.496	1.346	0.371	4.538	0.520	C_{2v}
130	2.693	3.496	1.346	0.371	4.538	0.520	C_{2v}
140	2.693	3.496	1.346	0.371	4.538	0.520	C_{2v}
150	2.693	3.496	1.346	0.371	4.538	0.520	C_{2v}
160	2.693	3.496	1.346	0.371	4.538	0.520	C_{2v}
170	2.693	3.496	1.346	0.371	4.538	0.520	C_{2v}

Table 5.6 reveals that least HOMO–LUMO energy gap is obtained at dihedral angles varying between 70° and 90° and maximum HOMO–LUMO gap is obtained at dihedral angle varying from 110° to 170°.

(3) In Figure 5.7, one Cu atom is fixed at the center and two atoms, that is, Ag and other Cu atoms placed at the corners are not fixed.

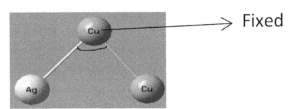

FIGURE 5.7   Cluster Cu_2–Ag with Cu atom fixed at the center.

TABLE 5.7 Computed DFT-based Descriptors of Cluster Cu_2–Ag with Cu Atom Fixed at the Center.

Angle (in degrees)	HOMO–LUMO Gap (eV)	Electro-negativity (eV)	Hardness (eV)	Softness (eV)	Electro-philicity index (eV)	Dipole moment (Debye)	Symmetry
20	1.496	3.387	0.748	0.668	7.668	0.593	C_s
30	1.496	3.387	0.748	0.668	7.668	0.598	C_s
40	1.550	3.387	0.775	0.644	7.399	0.541	C_s
50	1.550	3.387	0.775	0.644	7.399	0.560	C_s
60	1.550	3.387	0.775	0.644	7.399	0.560	C_s
70	1.550	3.387	0.775	0.644	7.399	0.560	C_s
80	1.550	3.387	0.775	0.644	7.399	0.561	C_s
90	1.904	3.428	0.952	0.525	6.171	0.326	C_s
100	1.550	3.387	0.775	0.644	7.399	0.561	C_s
110	2.421	3.469	1.210	0.412	4.970	0.433	C_s
120	2.612	3.482	1.306	0.382	4.643	0.475	C_s
130	2.584	3.469	1.292	0.386	4.656	0.470	C_s
140	2.557	3.482	1.278	0.390	4.742	0.468	C_s
150	2.557	3.482	1.278	0.390	4.742	0.470	C_s
160	2.557	3.482	1.278	0.390	4.742	0.470	C_s
170	2.557	3.482	1.278	0.390	4.742	0.470	C_s
180	3.178	3.553	1.589	0.314	3.973	0.143	$C_\infty v$

Table 5.7 and Figure 5.7 demonstrate that the preferred conformation of instant cluster has linear structure with Cu atom in the middle. The maximum HOMO–LUMO energy gap is obtained at the dihedral angle of 180°, while least gap, that is, 1.496 eV is obtained at dihedral angles 20° and 30°.

(4) In Figure 5.8, triangular structure of Cu_2–Ag is formed with Cu atom fixed at the center and other compounds, that is, Ag and Cu atoms are located at the corners of cluster. The computed physico-chemical properties of all the conformations are presented in Table 5.8 with the corresponding dihedral angles.

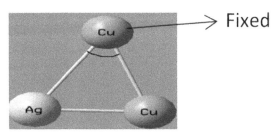

FIGURE 5.8   Triangular structure of cluster Cu_2Ag with Cu atom fixed at the center.

TABLE 5.8   Computed DFT-based Descriptors of Triangular Structure of Cluster Cu_2–Ag with Cu Atom Fixed at the Center.

Angles (in Degrees)	HOMO–LUMO Gap (eV)	Electro-negativity (eV)	Hardness (eV)	Softness (eV)	Electro-philicity index (eV)	Dipole moment (Debye)	Symmetry
30	1.496	3.387	0.748	0.668	7.668	0.591	C_s
40	1.550	3.387	0.775	0.644	7.399	0.541	C_s
50	1.550	3.387	0.775	0.644	7.399	0.560	C_s
60	1.550	3.387	0.775	0.644	7.399	0.560	C_s
70	1.550	3.387	0.775	0.644	7.399	0.560	C_s
80	1.578	3.401	0.789	0.633	7.330	0.534	C_s
90	1.931	3.414	0.965	0.517	6.036	0.321	C_s
100	2.013	3.428	1.006	0.496	5.837	0.319	C_s
110	2.448	3.455	1.224	0.408	4.876	0.438	C_s
120	2.584	3.469	1.292	0.386	4.656	0.471	C_s
130	2.612	3.482	1.306	0.386	4.643	0.474	C_s
140	2.584	3.469	1.292	0.386	4.656	0.471	C_s
150	2.584	3.469	1.292	0.386	4.656	0.471	C_s
160	2.584	3.469	1.292	0.386	4.656	0.471	C_s
170	2.584	3.469	1.292	0.386	4.656	0.471	C_s

Table 5.8 and Figure 5.8 demonstrate that the instant cluster has maximum HOMO–LUMO energy gap at dihedral angle 130°, while least gap is obtained at 30°. The instant cluster has constant energy gap between dihedral angles 40–70° and 140–170°.

5.4 CONCLUSION

In this chapter, we have studied the conformational effect of bimetallic nanoalloy clusters using DFT methodology. We have observed that linear structure of bimetallic clusters is more stable than other conformations. The reason for the stability of small linear structures may be understood from the molecular orbital concepts. For three atoms bimetallic clusters like $CuAg_2$, and Cu_2Ag, the 5-s orbitals play dominant role in bonding. The HOMO of the neutral cluster is anti-bonding and thus destabilized at the atomic level 7.50 eV. The destabilization is much greater for triangular structure than for the linear structure, leading to the stability of the latter. In this work, we have observed large HOMO–LUMO gap for linear structure, except $Ag–Au_2$ (where Ag atom is fixed at the center) and $Cu_2–Au$ (where one Au atom is fixed at the center) clusters.

KEYWORDS

- **density functional theory**
- **conformational analysis**
- **bimetallic**
- **Cu–Ag**
- **HOMO–LUMO energy gap**
- **hardness**

REFERENCES

1. Ghosh, D. C. *Int. J. Mol. Sci.* **2006,** *7,* 289.
2. Freitas, M. P. D.; Ramalho, T. D. C. *Ciênc. Agrotec.* **2013,** *37,* 485.
3. Castro, A. T.; Figuero-Villar, J. D. *Int. J. Quantum Chem.* **2002,** *89,* 135.
4. Gerbst, A. G.; Nikolaev, A. V.; Yashunsky, D. V.; Shashkov, A. S.; Dmitrenok, A. S.; Nifantiev, N. E. *Sci. Rep.* **2017,** *7.*
5. Alvarez-Ros, M. C.; Palafox, M. A. *Pharmaceuticals* **2014,** *7,* 695.
6. Krishnamurty, S.; Stefanov, M.; Mineva, T.; Begu, S.; Devoisselle, J. M.; Goursot, A.; Zhu, R.; Salahub, D. R. *J. Phys. Chem. B* **2008,** *112,* 13433.
7. Bao, G.; Kamm, R. D.; Thomas, W.; Hwang, W.; Fletcher, D. A.; Grodzinsky, A. J.; Zhu, C.; Morfad, M. R. K. *Mol. Cell Biomech.* **2010,** *3,* 91.
8. Agrawal, M.; Kumar, A.; Gupta, A. *RSC Adv.* **2017,** *7,* 41573.

9. Mary, Y. S.; Raju, K.; Panicker, C. Y.; Al-Saadi, A. A.; Thiemann, T.; Alsenoy, C. V. *Spectrochim. Acta, Part A* **2014,** *128*, 638.
10. Srivastava, A. K.; Dwivedi, A.; Kumar, A.; Gangwar, S. K.; Misra, N.; Sauer, S. P. A. *Cogent Chem.* **2016,** *2*, 1149927.
11. Sarmah, P.; Deka, R. C. *J. Comput. Aided Mol. Des.* **2009,** *23*, 343.
12. Pitzer, R. M. *Acc. Chem. Res.* **1983,** *16*, 207.
13. Schleyer, P. V. R.; Kaupp, M.; Hampel, F.; Bremer, M.; Mislow, K. *J. Am. Chem. Soc.* **1992,** *114*, 6791.
14. Ghosh, D. C.; Rahman, M. A. *Chem. Environ. Res.* **1996,** *5*, 73.
15. Freeman, F.; Tsegai, Z. M.; Kasner, M. L.; Hehre, W. *J. Chem. Educ.* **2000,** *77*, 661.
16. Hafner, J.; Wolverton, C.; Ceder, G. *MRS Bull.* **2006,** *31*, 659.
17. Wang, H. Q.; Kuang, X. Y.; Li, H. F. *Phys. Chem. Chem. Phys.* **2010,** *12*, 5156.
18. Becke, A. D. *J. Chem. Phys.* **1993,** *98*, 5648.
19. Lee, C.; Yang, W.; Parr, R. G. *Phys. Rev. B: Condens. Matter* **1988,** *37*, 785.
20. Mielich, B.; Savin, A.; Stoll, H.; reuss, H. *Chem. Phys. Lett.* **1989,** *157*, 200.
21. Jiang, Z. Y.; Lee, K. H.; Li, S. T.; Chu, S. Y. *Phys. Rev. B* **2006,** *73*, 235423.
22. Zhao, Y. R.; Kuang, X. Y.; Zheng, B. B.; Li, Y. F.; Wang, S. J. *J. Phys. Chem. A* **2011,** *115*, 569.
23. Gaussian 03, Revision C.02, Frisch, M. J.; Trucks, G. W.; Schlegel, H. B.; Scuseria, G. E.; Robb, M. A.; Cheeseman, J. R.; Montgomery, Jr., J. A.; Vreven, T.; Kudin, K. N.; Burant, J. C.; illam, J. M.; Iyengar, S. S.; Tomasi, J.; Barone, V.; Mennucci, B.; Cossi, M.; Scalmani, G.; Rega, N.; Petersson, G. A.; akatsuji, H. N.; Hada, M.; Ehara, M.; Toyota, K.; Fukuda, R.; Hasegawa, J.; Ishida, M.; Nakajima, T.; Honda, Y.; Kitao, O.; Nakai, H.; Klene, M.; Li, X.; Knox, J.. E.; Hratchian, H. P.; Cross, J. B.; Bakken, V.; Adamo, C.; Jaramillo, J.; Gomperts, R.; Stratmann, R. E.; Yazyev, O.; Austin, A. J.; Cammi, R.; Pomelli, C.; Ochterski, J. W.; Ayala, P. Y.; Morokuma, K.; Voth, G. A.; Salvador, P.; Dannenberg, J. J.; Zakrzewski, V. G.; Dapprich, S.; Daniels, A. D.; Strain, M. C.; Farkas, O.; Malick, D. K.; Rabuck, A. D.; Raghavachari, K.; Foresman, J. B.; Ortiz, J. V.; Cui, Q.; Baboul, A. G.; Clifford, S.; Cioslowski, J.; Stefanov, B. B.; Liu, G.; Liashenko, A.; Piskorz, P.; Komaromi, I.; Martin, R. L.; Fox, D. J.; Keith, T.; Al-Laham, M. A.; Peng, C. Y.; Nanayakkara, A.; Challacombe, M.; Gill, P. M. W.; Johnson, B.; Chen, W.; Wong, M. W.; Gonzalez, C.; Pople, J. A. Gaussian, Inc.: Wallingford CT, 2004.
24. Koopmans, T. *Physica* **1934,** *1*, 104.
25. Parr, R. G.; Yang, W. *Density Functional Theory of Atoms and Molecules;* Oxford University Press: Oxford, 1989.

CHAPTER 6

RESEARCH PROGRESS FOR THREE-DIMENSIONAL RECONSTRUCTION OF NANOFIBROUS MEMBRANES FROM TWO-DIMENSIONAL SCANNING ELECTRON MICROSCOPE IMAGES

BENTOLHODA HADAVI MOGHADAM[1], SHOHREH KASAEI[2], and A. K. HAGHI[1,*]

[1]Textile Engineering Department, University of Guilan, Rasht, Iran

[2]Department of Computer Engineering,
Sharif University of Technology, Tehran, Iran

*Corresponding author. E-mail: AKHaghi@yahoo.com

ABSTRACT

An efficient three-dimensional (3D) image analysis method is proposed for 3D reconstruction of nanofibrous membranes using scanning electron microscope (SEM) imaging. The 3D reconstruction of nanofibrous membranes is obtained from the 2D information of SEM images. It is based on 3D reconstruction from two views of single 2D SEM images for characterization of nanofibrous membranes. The proposed method exhibits very realistic 3D surfaces for 3D visualization of nanofibrous membranes. The performance of the proposed method has been evaluated on different regions of nanofibrous membranes. The resulting points are reconstructed and the error is evaluated and discussed. As the results show, a high accuracy is obtained in the case of low magnification and small angle between two viewpoints of SEM images.

6.1   INTRODUCTION

Nanofibrous membranes are an important class of nanomaterial that have gained increasing attention due to the large surface area per mass ratio, small pore sizes, flexibility, fine fiber diameter, high porosity, and their production and application in the development of filter media. Nanofibrous membranes typically contain two kinds of structure, two- or three-dimensional (3D) nanofibrous network assemblies while the three-dimensional structure provides good handling characteristics (Fig. 6.1).[1-4] SEM analysis of electrospun nanofibrous membrane by incorporating different image analysis methods is one of the renowned methods for fabric, nonwoven, and for researchers in the 2D characterization of woven membranes, but it returned only partial information about the 3D structure of the membrane because it is limited to relatively small fields of view.[5]

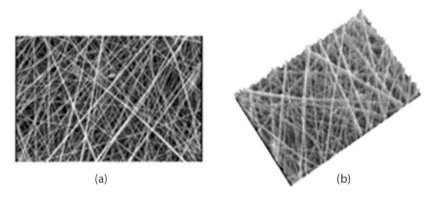

(a)   (b)

FIGURE 6.1   Image of electrospun nanofibrous membrane: (a) 2D, (b) 3D.

There are nondestructive 3D image analysis techniques that have been used for 3D reconstruction of an electrospun nanofibrous membrane, such as laser scanning confocal microscope (LSCM), micro-computed tomography (micro-CT), X-ray diffraction microscopy, focused ion beam-scanning electron microscope (FIB-SEM) and 3D electron back-scatter diffraction (EBSD).[5-10]

Over the years, there have been few studies directed toward the 3D reconstruction of nanofibrous membranes. Kazemi et al.[11] proposed a model for the nanostructured fibrous network using adaptive image analysis. In this study, the images are captured from different regions of a membrane and the number of pixels within a distinct grayscale level are calculated by selecting a specific grayscale interval. Each local sub-image is determined using the threshold values based on neighboring pixels within a specific radius. Sambaer et al.[12] proposed the realistic 3D structure model as mutually connected tubes having different diameters of producing polyurethane (PU)-based nanofiber electrospun mat. The 3D structure is obtained using rotation in the depth of particular average diameter circle along every corresponding centerline pixel. Jaganathan et al.[13] obtained the 3D geometries based on fiber-level information via digital volumetric imaging (DVI) technique. Also, the 3D geometries are converted into stereo lithography (STL) files for meshing such geometries using a computer-aided design (CAD) technique. Zobel et al.[14] assumed that 3D reconstruction of fiber-webs happens when fibers have square cross-sections and bend over each other according to a simple set of rules. Faessel et al.[15] developed a 3D model of random fiber networks using the Visualization Toolkit (VTK) libraries. In order to obtain a finite element mesh of a unit volume of the material, the 3D model of the network is discretized by shell elements. Also, morphological data of real networks are extracted from X-ray micro-tomographic observations. Soltani et al.[16] proposed the 3D model for reconstruction of fibrous networks using sets with random arrangement. Also, the direct 3D model of the needled nonwoven fabrics generated by X-ray micro-CT is used in conjunction with a computational fluid dynamics (CFD) model for the simulation of transverse permeability. Hosseini et al.[17,18] generated 3D fibrous geometries resembling the microstructure of a fibrous medium, by the C++ computer program to produce fibrous structures of different fiber diameters, porosities, thicknesses, and orientations. Ji et al.[19] used LSCM for 3D reconstruction of the nanofibrous scaffold. Reingruber et al.[20] extracted the 3D information about the membrane structure from serial sectioning and imaging of the membrane embedded in all three directions. Ostadi et al.[21] generated a binary 3D model of the complex structures using the software techniques and threshold-tuning the greyscale X-ray images. The actual 3D structure of the fuel cell is captured through X-ray and FIB-SEM nano-tomography.

The abovementioned methods are very expensive, time-consuming, and not readily accessible instruments. Also, recreating detailed geometry becomes very difficult which cannot be fully automated, which itself poses a substantial source of errors. However, among these methods, 3D reconstruction from statistical information is the most attractive option for characterization of nanofibrous membranes. 3D reconstruction of a number of perspective images is a challenging task in computer vision due to the loss of depth in the process of the photographing image. The 3D reconstruction of 2D images, from the information obtained from a 2D image analysis, is a relatively new research area.[10]

The key task of 3D reconstruction is to recover high-quality and detailed 3D models from two or more views of the image, which may be taken from widely separated viewpoints. Hence, due to the sensitivity of the issue, in order to overcome these problems, we have provided a novel, precise, and economical technique based on the 3D model from two views of the single SEM image for geometrical characteristics of the nanofibrous membrane. This method is a cost-effective way for 3D reconstruction of the nanofibrous membrane because only one camera is used.

6.2   PROPOSED METHOD

Recently the problem of interactive 3D reconstruction of a number of perspective images has been one of the fundamental problems of computer vision, while reconstruction from two views from single images is the simplest one.[22–28] The method's algorithm for a 2D image pair is based on the following stages:

(a) Detection of feature points in two images;
(b) Finding matched points in two views of images;
(c) Triangulating the 3D points into a 3D mesh;
(d) Computing 3D points from 2D matched points;
(e) Mapping a 2D image as a texture on the surface.

Fundamental problem in computer vision: Matching can provide valuable information about the similarity between the images for 3D reconstruction from multiple images.[29–31]

The second challenge is the determination of the 3D location of each image point. If we know the calibration parameters of the camera, but not the point positions, we could find the points through triangulation; conversely, if we don't know the calibration parameters of the camera, we cannot write the projection relations using normalized image, but we have to include the calibration matrixes.[29-33]

We, therefore, often refer to reconstruction from uncalibrated cameras as projective reconstruction. Obviously, if we want to compute projective reconstructions, we just have to find at least one possible solution. In its most general form, the problem of computing a 3D reconstruction from two views of the camera can be formulated as finding the projection-matrixes, M

$$
\begin{pmatrix} x \\ y \\ 1 \end{pmatrix} = \begin{pmatrix} m_{11} & m_{12} & m_{13} & m_{14} \\ m_{21} & m_{22} & m_{23} & m_{24} \\ m_{31} & m_{32} & m_{33} & m_{34} \end{pmatrix} \begin{pmatrix} X \\ Y \\ Z \\ 1 \end{pmatrix} \tag{6.1}
$$

Equation 6.1 describes the pixel coordinates in the 3D image while we know the image coordinates in the two views and the twelve coefficients m_i of the projection matrix. According to eq 6.2, we can obtain two new equations with three additional unknowns in 3D

$$
x = \frac{m_{11}X + m_{12}Y + m_{13}Z + m_{14}}{m_{31}X + m_{32}Y + m_{33}Z + m_{34}}
$$

$$
\tag{6.2}
$$

$$
y = \frac{m_{21}X + m_{22}Y + m_{23}Z + m_{24}}{m_{31}X + m_{32}Y + m_{33}Z + m_{34}}
$$

We can always let the projection point of the first view of the camera be at the origin with the image plane at the unit distance along the Z-axis. The projection point of the second view of the camera is at X_0, Y_0, and Z_0, and it has rotation. We can always include the scale factor λ in the unknown calibration matrix (K). According to eq 6.3, we have the projection matrixes for two different views of an image,

$$
\begin{pmatrix} x^a \\ y^a \\ 1 \end{pmatrix} = \lambda K \begin{pmatrix} 1 & 0 & 0 & 0 \\ 0 & 1 & 0 & 0 \\ 0 & 0 & 1 & 0 \end{pmatrix} \begin{pmatrix} X \\ Y \\ Z \\ 1 \end{pmatrix}
$$

$$
\tag{6.3}
$$

$$
\begin{pmatrix} x^b \\ y^b \\ 1 \end{pmatrix} = \lambda K R \begin{pmatrix} 1 & 0 & 0 & -X_0 \\ 0 & 1 & 0 & -Y_0 \\ 0 & 0 & 1 & -Z_0 \end{pmatrix} \begin{pmatrix} X \\ Y \\ Z \\ 1 \end{pmatrix}
$$

Finding projective and affine invariants is an important problem in 3D reconstruction from uncalibrated cameras. For this reason, regardless of the camera parameters, we can assume that we have a camera with known internal calibration, and the invariants of this projection are thus independent of the camera parameters. The calibration matrix (K) contains the position of image center (width/2, height/2) in the image, also, as in eq 6.4, known as the principal point (u_0, v_0), and the focal length, f, of the camera.

$$K = K_a = K_b = \begin{bmatrix} f & 0 & u_0 \\ 0 & f & v_0 \\ 0 & 0 & 1 \end{bmatrix} \tag{6.4}$$

It is possible to compute external parameters of the camera system, rotation of the second camera relative to the first, using projected image data (as in eq 6.5).

$$R_x = \begin{pmatrix} 1 & 0 & 0 \\ 0 & \cos(\theta) & \sin(\theta) \\ 0 & -\sin(\theta) & \cos(\theta) \end{pmatrix}$$

$$R_y = \begin{pmatrix} \cos(\theta) & 0 & \sin(\theta) \\ 0 & 1 & 0 \\ -\sin(\theta) & 0 & \cos(\theta) \end{pmatrix}$$

$$R_z = \begin{pmatrix} \cos(\theta) & \sin(\theta) & 0 \\ -\sin(\theta) & \cos(\theta) & 0 \\ 0 & 0 & 1 \end{pmatrix} \tag{6.5}$$

6.3 IMPLEMENTATION

This study assumed that images are obtained by perspective projection and the camera is always uncalibrated, with its internal parameters unknown. The ideas in this study can be seen as reversing the rules for 3D modeling using 2D images with no information on the cameras of scanning electron microscope (SEM) being used. Generally, for 3D reconstruction of a single 2D image, at least two views of the image are needed. In this work, we have prepared two different views in five different regions of the nanofibrous membrane in different magnification of camera for 3D reconstruction of single 2D images by rotating the object in three positions while the camera is fixed (Fig. 6.2).

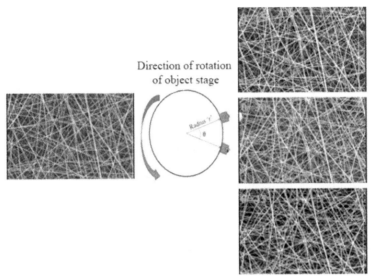

FIGURE 6.2   Three different views of nanofibrous membrane.

The first step in our method is the generation of corresponding points. For this purpose, pairs of keypoints are matched across three different views of the nanofibrous membrane. Thus, we must provide the usual way to find the best match in the other image (Fig. 6.3).

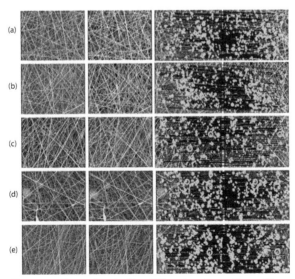

FIGURE 6.3   Corresponding points in five different regions of nanofibrous membrane.

The scale-invariant feature transform (SIFT) keypoint extractor is used here to find the corresponding points. An important advantage of this method is that it generates large number of features that densely cover the image over the full range of scales. One needs at least seven corresponding points for the reconstruction, or eight, if the problem is to be solvable linear.

6.4 EXPERIMENTAL RESULTS

According to Table 6.1, the low magnification with longer working distance leads to a better detection of feature points and improving the depth of focus. It is important for improving the depth of focus at low magnification because insufficient focusing of regions far away from the image center may lead the reconstruction procedure to fail at the edges. Improving the depth of focus is particularly important for the microscopes which do not compensate for the beam rotation induced by a change of focus since the function of dynamic focusing cannot be used in such microscopes. In addition, the low magnification with a long working distance has some problem and it may lead to decrease in the field distortion parallel to the tilt axis.[34]

Generally, an increasing angle between of two views decreases the number of matched points. Because, the larger the angle, the bigger the perspective distortions. So, it is getting more difficult for the matching algorithm to identify points correctly and the probability of the wrong match increases. These differences are then a source of error in depth estimation. Therefore, the angle of rotation should be kept as small as possible in order to overcome the problem of the quick disappearance of the objects in successive views. For this reason, least angle is the appropriate angle for generation of two views of the image.

TABLE 6.1 Number of Matched Keypoints in Five Different Regions of Nanofibrous Membrane.

No.	Angle	Number of matched points		
		Magnification		
		× 2500	× 5000	× 10,000
	15	357	302	199
1	30	68	119	79
	45	54	86	87

TABLE 6.1 *(Continued)*

No.	Angle	Number of matched points		
		Magnification		
		× 2500	× 5000	× 10,000
	15	342	306	268
2	30	100	126	149
	45	72	106	110
	15	248	213	197
3	30	63	116	119
	45	58	94	113
	15	335	301	245
4	30	88	101	127
	45	64	78	131
	15	440	390	234
5	30	117	137	95
	45	58	103	93

For obtaining a 3D point cloud, we need to compute the 3D position associated with each match. To this end, the Delaunay triangulation is an excellent way that is computed using Computational Geometry Algorithms Library (CGAL). It provides our data Points as grouped into sets of points that belong to the same global surface and evaluates the projection of each triangle of the Delaunay triangulation in the chosen views and computes the mean of the color variance of the pixels in this triangle and obtain the 3D position of the feature points to modify a generic model using a geometrical deformation (Fig. 6.4).[35]

The texture mapping of the 3D model is a method for adding surface texture to a computer-generated model in computer graphics. The textures of the 3D objects are generated using an automatic projection method. The texture coordinates are equivalent to the XYZ coordinates in the 3D spaces. Basically, the texture mapping specifies the connection between the points in a 3D object with the position of the image on the object. The key task of 3D reconstruction is to recover high-quality and detailed 3D models from two or more views of the image, which may be taken from widely separated viewpoints.

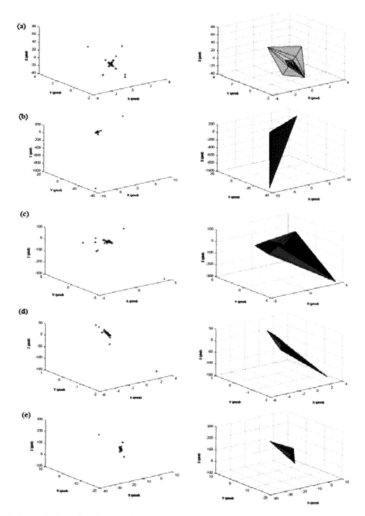

FIGURE 6.4 Point cloud and triangle surface in five different regions of nanofibrous membrane.

To the best of our knowledge, the implemented methodology is the first to include photo-consistency testing into a Digital Surf Mountains Map software (Mountains Map, Digital Surf, Besancon, France) volumetric reconstruction method. The following sub-sections describe the steps of the novel algorithm, pointing out the main differences relative to the standard voxel-based volumetric methods. Once the 3D dataset is acquired from each optical instrument, it is then processed using the Digital Surf Mountain Map (v5) software. This software is a common software solution widely used in research and industries for analysis of 3D surface data acquired from various instruments. The data collected by SEM were reconstructed into a 3D image using MountainsMap® Imaging Topography software (Fig. 6.5).

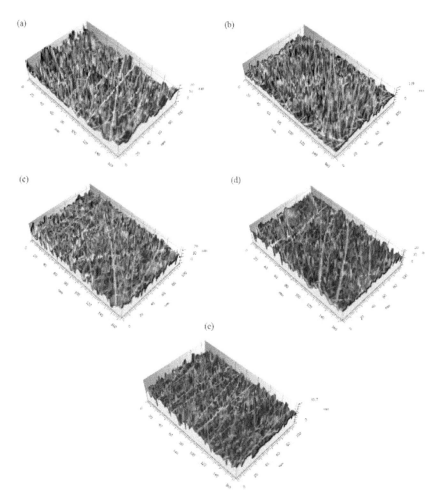

FIGURE 6.5 3D reconstruction of nanofibrous membrane in five different regions of nanofibrous membrane.

The first part of this study revealed that the number of points constituting a 3D surface model formed by 3D reconstruction from two views of single 2D SEM image. The verification step is very important in evaluation of 3D reconstruction. This step focused on point cloud accuracy. The point cloud is the first output during the 3D reconstruction.

In order to verify a reconstructed point and calculate the accuracy achieved in the different region of nanofibrous membrane, we used the root mean square error (RMSE; as in eq 6.8). Root mean square error is the standard error measure used in the Geo-correction. RMSE is often used to assess the difference between the observed value (x_t) and the estimated value (x_{nt}) for each user-defined point; this parameter finds the mean squared value and then finds the square root of that means

$$\text{RMSE} \quad x = \sqrt{\frac{\sum_{t=1}^{N}(x_{nt} - x_t)^2}{N}} \tag{6.6}$$

$$\text{RMSE} \quad y = \sqrt{\frac{\sum_{t=1}^{N}(y_{nt} - y_t)^2}{N}} \tag{6.7}$$

$$\text{RMSE} \quad xy = \sqrt{\sum_{t=1}^{N}\left((\text{RMSE} \ x)^2 + (\text{RMSE} \ y)^2\right)} \tag{6.8}$$

TABLE 6.2   Root Mean Square Error (RMSE) in Five Different Regions of Nanofibrous Membrane.

Check points no.	RMSE X (m)		RMSE Y (m)	
	X1	X2	Y1	Y2
1	1.22	1.23	0.19	0.06
2	1.26	3.6	0.23	0.32
3	1.31	0.83	1.54	2.25
4	1.69	3.62	0.19	0.87
5	0.016	1.4	0.23	0.024
Mean RMSE	1.57		0.59	
RMSE XY	1.67			

According to Table 6.2, this model allows us to assume that the 3D reconstruction result has a high accuracy in different regions of nanofibrous membrane.

6.5 CONCLUSION

In this research, a new method to reconstruct the 3D structure of nanofibrous membranes based on reconstruction of 3D point clouds was proposed. The method simultaneously represents very realistic nanostructures and is fast and simple to implement. The method was applied to a sample of nanofibrous membrane in five different regions of mat, with three different angles between two viewpoints on SEM image. A set of 3D points is computed from two SEM images of a surface, and then a 3D surface is reconstructed from these points using the Delaunay triangulation. A very realistic 3D surface is obtained by applying texture mapping using the Digital Surf Mountain Map (v5) software. The resulting points were reconstructed and the errors were evaluated and discussed. The advantage of this system is that the 2D images do not need to be calibrated in order to obtain a reconstruction and a few points must be picked in the image to define the image positions. It, thus, appears to be an efficient way to build 3D models from SEM images of any type of nanofibrous membrane for which no information about the camera is known. The accuracy of the reconstruction obtained depends on the number of viewpoints of images used, the positions of each viewpoint considered, the angle between of two viewpoints of SEM image, and the complexity of the object's shape. Therefore, high accuracy was obtained in the low magnification and small angle between two viewpoints of SEM image.

KEYWORDS

- **nanofibrous membrane**
- **scanning electron microscope**
- **three-dimensional image analysis**
- **three-dimensional reconstruction**
- **scale-invariant feature transform keypoint extractor**

REFERENCES

1. Barhate, R. S.; Ramakrishna, S. *J. Membrane. Sci.* **2007**, *296*, 1.
2. Gopal, R.; Kaur, S.; Ma, Z.; Chan, C.; Ramakrishna, S.; Matsuura, T. Electrospun Nanofibrous Filtration Membrane. *J. Membrane. Sci.* **2006**, *281*, 581.
3. Zander, N. E. *Polymers* **2013**, *5*, 19.
4. Bhardwaj, N.; Kundu, S. C. *Biotechnol. Adv.* **2010**, *28*, 325.
5. Liang, Z. R.; Fernandes, C. P.; Magnani, F. S.; Philippi, P. C. *J. Pet. Sci. Eng.* **1998**, *21*, 273.
6. Bagherzadeh, R.; Latifi, M.; Najar, S. S.; Kong, L. *J. Biomed. Mater. Res. A* **2013**, *101*, 765.
7. Miao, J.; Ishikawa, T.; Johnson, B.; Anderson, E. H.; Lai, B.; Hodgson, K. O. *Phys. Rev. Lett.* **2002**, *89*, 088303.
8. Rollett, A. D.; Lee, S. B.; Campman, R.; Rohrer, G. S. *Annu. Rev. Mater. Res.* **2007**, *37*, 627.
9. Groeber, M. A.; Haley, B. K.; Uchic, M. D.; Dimiduk, D. M.; Ghosh, S. *Mater. Charact.* **2006**, *57*, 259.
10. Pierantonio, F.; Masiero, A.; Bezzo, F.; Beghi, A.; Barolo, M. *IFAC Proc.* **2011**, *44*, 12066.
11. Pilehrood, M. K.; Heikkil, P. *WJNSE* **2013**, *3*, 6.
12. Sambaer, W.; Zatloukal, M.; Kimmer, D. *Chem. Eng. Sci.* **2011**, *66*, 613.
13. Jaganathan, S.; Tafreshi, H. V.; Pourdeyhimi, B. *Chem. Eng. Sci.* **2008**, *63*, 244.
14. Zobel, S.; Maze, B.; Tafreshi, H. V.; Wang, Q.; Pourdeyhimi, B. *Chem. Eng. Sci.* **2007**, *62*, 6285.
15. Faessel, M.; Delisée, C.; Bos, F.; Castéra, P. *Compos. Sci. Technol.* **2005**, *65*, 1931.
16. Soltani, P.; Johari, M. S.; Zarrebini, M. *Powder Technol.* **2014**, *254*, 44.
17. Hosseini, S. A.; Tafreshi, H. V. *Chem. Eng. Sci.* **2010**, *65*, 2249.
18. Hosseini, S. A.; Tafreshi, H. V. *Powder Technol.* **2010**, *201*, 153.
19. Ji, Y.; Ghosh, K.; Shu, X. Z.; Li, B.; Sokolov, J. C.; Prestwich, G. D.; Rafailovich, M. H. *Biomaterials.* **2006**, *27*, 3782.
20. Reingruber, H.; Zankel, A.; Mayrhofer, C.; Poelt, P. *J. Membrane. Sci.* **2011**, *372*, 66.
21. Ostadi, H.; Rama, P.; Liu, Y.; Chen, R.; Zhang, X. X.; Jiang, K. *J. Membrane. Sci.* **2010**, *351*, 69.
22. Kushal, A. M.; Bansal, V.; Banerjee, S. *In Proceedings Indian Conference in Computer Vision, Graphics and Image Processing*, **2002**, p. 1–6.
23. Sturm, P. F.; Maybank, S. J. The 10th British Machine Vision Conference (BMVC '99), **1999**, *265*.
24. Criminisi, A.; Reid, I.; Zisserman, A. *Int. J. Comput. Vis.* **2000**, *40*, 123.
25. Debevec, P. E. Ph.D. Thesis, University Of California At Berkeley, 1996.
26. Grossmann, E.; Ortin, D.; Santos-Victor, J. The 5th Asian Conference on Computer Vision, **2002**, *23.*
27. Liebowitz, D.; Criminisi, A.; Zisserman, A. *Comput. Graph. Forum* **1999**, *18*, 39.
28. Wilczkowiak, M.; Boyer, E.; Sturm, P. 8th International Conference on Computer Vision (ICCV '01), **2001**, *1*, 142.
29. Carlsson, S. Lecture Notes, *Taught at KTH Stockholm Spring* **2007**.

30. Lowe, D. G. *Int. J. Comput. Vis.* **2004,** *60*, 91.
31. Brown, M.; Lowe, D. G. *BMVC.* **2002,** *4*, 1–10.
32. Facciolo, G.; Limare, N.; Meinhardt-Llopis, E. *IPOL* **2014,** *16*, 344.
33. Mery, D. Springer International Publishing: Switzerland, 2015, p 35–51.
34. Pouchou, J. L.; Boivin, D.; Beauchêne, P.; Le Besnerais, G.; Vignon, F. *Microchim. Acta.* **2002,** *139*, 135.
35. Choi, B. K.; Shin, H. Y.; Yoon, Y. I.; Lee, J. W. *CAD* **1988,** *20*, 239.

CHAPTER 7

NANOTECHNOLOGY RESEARCH, INDUSTRIAL ECOLOGY, AND THE VISIONARY FUTURE

SUKANCHAN PALIT*

43, Judges Bagan, Post Office-Haridevpur, Kolkata 700082, West Bengal, India

E-mail: sukanchan68@gmail.com, sukanchan92@gmail.com

ABSTRACT

Human scientific pursuit and the progress of human civilization are today in the path of newer vision and new scientific rejuvenation. The world of science and technology are today faced with immense difficulties and hurdles. Global climate change, global warming, and frequent environmental disasters are challenging the vast scientific firmament of immense might and vision. In this chapter, the author deeply discusses and uncovers the need of nanotechnology in its application in ecology and environment. Environmental protection globally is the need of the hour. Loss of biodiversity and frequent environmental catastrophes are veritably urging the civil society, scientists, and engineers to gear forward toward newer techniques and newer environmental engineering tools to protect the environment. The author, with deep scientific conscience, focuses on the necessity of traditional and nontraditional environmental engineering tools in the furtherance of the cause of industrial ecology. The vast vision and the challenges of science, the world of technological marvels, and the futuristic vision of scientific validation will go a long way in the true realization of environmental engineering and nanotechnology. The crisis of mankind today is the inadequate supply of pure drinking water in developing and developed countries around the world. This chapter will veritably open a new window in the domain of both industrial ecology and applications of nanotechnology. Nano-science and nanotechnology are the wonders of science and engineering today. Application of nano-technology in the field of preservation of ecology and environment thus assumes immense importance. The author rigorously points out toward the scientific intricacies and the scientific difficulties in the furtherance of science of industrial ecology and nanotechnology.

7.1 INTRODUCTION

Human civilization and scientific progress are moving toward a newer era of vast scientific regeneration. Protection of ecological biodiversity and the need of environmental protection are the veritable needs of civiliza-tion's progress today. Modern science and technological advances are in the state of immense scientific disaster and vast barriers. Nuclear science, space technology, and nanotechnology are the wonders of science and engineering today. Importance of nanotechnology to environmental protec-tion, green engineering, industrial ecology, and sustainability engineering is slowly growing, thus surpassing one scientific boundary over another. Technological masterpieces, such as environmental engineering, chemical engineering, and petroleum engineering, are in a state of immense disaster today. The needs of scientific advancements are immense, path-breaking, and, at the same time, scientifically thought provoking. Scientific and engi-neering achievements, as regards environmental protection, needs to be envisioned and reframed with the passage of visionary timeframe. Today, the branches of industrial ecology and sustainability need to be inter-linked. Environmental and energy sustainability are the immediate needs of the hour as regards progress of human civilization. The term "industrial ecology" implies a strong and definite relationship to the field of ecology. A deep and basic understanding of ecology is useful in understanding and promoting industrial ecology, which veritably garners different ecological concepts. This is the scientific discipline that is vastly concerned with the relationships between organisms and their past, present, and future envi-ronments. These intertwined relationships include physiological responses of individuals, structure and dynamics of populations, interactions among species, organization of biological communities, and processing of energy and matter in ecosystems.

7.2   THE VISION AND THE AIM OF THIS STUDY

Science, technology, and engineering are today in the path of immense scientific regeneration. The status of environment is extremely dismal today. Mankind and scientific progress today stands in the midst of deep scientific introspection and immense scientific hurdles. The need for industrial ecology is today immense and groundbreaking. In industrial ecology, one focus (or object) of study is the interrelationship among firms, as well as among their products and processes, at the local, regional, national, and global system platform. Systems engineering and industrial ecology are the two opposite sides of the visionary coin. Social ecology is an inherent pillar of successful realization of sustainability. Social ecology deeply integrates and abounds the study of human and natural ecosystems through understanding the interrelationships of culture and nature. The contribution of science and engineering in industrial ecology are immense and needs to be envisioned and re-envisaged with the passage of scientific history and time. In this chapter, the author deeply and rigorously points out toward the scientific success in the application of industrial ecology to the furtherance of science and technology globally. Sustainable development, whether it is energy, environmental, social, or economic, needs to be intertwined with industrial ecology and nanotechnology today. The difficulties and barriers of industrial ecology are immense as it is a latent and immature domain of scientific rigor and progress. The vision and the aim of this study go beyond scientific imagination and targets the doctrines of industrial ecology and nanotechnology in the furtherance of science and technology globally.

7.3   THE VAST SCIENTIFIC DOCTRINE OF INDUSTRIAL ECOLOGY

The word "ecology" is derived from the word "oikos," meaning "household," combined with the root "logy," meaning, the "study of." Thus, ecology is, literally, the study of households including the plants, animals, microbes, and people that live together as interdependent beings on Spaceship Earth. Ecology can be broadly defined as the study of the interactions between the abiotic and biotic components of a system. Industrial ecology may be defined as the interactions between industrial and ecological systems; consequently, it addresses the environmental concerns on both the abiotic and biotic components of the ecosphere. Today, globally environmental protection science and drinking water treatment stands in the crucial juncture of deep comprehension and vast vision. The science of ecology and industrial ecology needs to be rebuild and re-envisioned as respects its application in developing and developed countries around the globe. Water purification, drinking water treatment, and industrial wastewater treatment are today the parameters toward the economic growth and scientific advancement of a nation. Industrial ecology should be connected to sustainable development and proliferation of science and engineering globally. The vast domain of "industrial ecology" has immense overlapping with the field of sustainability and environmental protection. Environmental and energy sustainability are the ultimate needs of the human civilization today. The protection of water, air, and energy are the scientific imperatives of the human scientific progress today. Industrial ecology also implies and encompasses the vast domain of sustainability science and management. In this chapter, the author deeply delves into the murky depths of environmental protection, environmental engineering science, and the vast domain of water pollution control with the sole vision of the furtherance of science and technology.

7.4 INDUSTRIAL ECOLOGY: A BRIEF DEFINITION

There is no single definition of ecology today. However, most definitions cover the following areas:

- A deep systems view of the interactions between industrial, ecological, and environmental systems;
- The study of material and energy flows;
- An interdisciplinary approach;
- A vast orientation toward a futuristic vision;
- A change from linear processes to cyclic processes;
- An effort to reduce the environmental impacts to ecological systems;
- An integration of industrial systems and ecology;
- A vision toward sustainable natural systems;
- The emancipation of industrial and natural systems.

7.4.1   GOAL, ROLES, AND OBJECTIVES

Human scientific vision in the field of industrial ecology today is in the midst of deep scientific introspection and ardor. The goals, roles, and objectives in the field of industrial ecology applications are immensely visionary and scientifically inspiring. Industrial ecology, industrial and systems engineering, and water pollution control are today linked to each other with immense might, scientific determination, and grit. An industrial ecologist's goals are to target and adapt a thorough scientific understanding of the natural system in order to enhance industrialization in both developed and developing nations around the world. A practical and visionary goal of industrial ecology is to illuminate the environmental impact per person and per dollar of economic activity and gear toward development and improvement of human activity. Technological hiatus and scientific impasse in global water and air research and development initiatives are immense and thought provoking. This chapter opens a new chapter of innovation and a newer window of scientific instinct in the field of green engineering, green chemistry, and industrial ecology in decades to come.

7.5   THE VISION OF NANOTECHNOLOGY APPLICATIONS TO HUMAN SOCIETY

Nanotechnology and nano-engineering are the wonders of human society today. It is linked with every branch of science and engineering. Membrane separation processes and novel separation processes are changing the entire scientific firmament of chemical process engineering and environmental engineering. Human society is veritably plundered with the ever-growing concerns of provision of clean drinking water, industrial wastewater treatment, and the vast world of water-purification impasse. Technology and engineering has literally no answers to the heavy metal groundwater contamination. Here comes the vast importance of membranes, nanofiltration, and reverse osmosis. Industrial ecology and green engineering are changing the face of human civilization and human scientific progress today. A deep scientific introspection in the field of desalination and water treatment is the utmost need of the hour. Thus, human society will usher in a newer era in the field of scientific forays in water research and development.

7.6   GREEN ENGINEERING, GREEN CHEMISTRY, AND THE VISIONARY FUTURE

Green engineering and green chemistry are today in the midst of rejuvenation and revamping. Energy and environmental sustainability are the utmost needs of scientific research pursuit today. Sustainable development, whether it is energy, environmental, social, and economic, are changing the face of human scientific progress today. The future of global sustainable development is bright as well as scientifically thought provoking. Scientific articulation, scientific finesse, and deep scientific discerning are the needs of research endeavor in industrial ecology and green engineering today. The author deeply cautions the ever-growing water purification issues and the global water-shortage problems. Green chemistry and green technology will surely open up new doors of scientific ardor and vision in the field of global scientific research and development initiatives.

7.7   APPLICATION OF NANOTECHNOLOGY IN THE VAST DOMAIN OF INDUSTRIAL ECOLOGY

Application of nanotechnology in the vast emancipation of industrial ecology is the necessity of human scientific progress today. The field of industrial ecology and systems engineering are veritably transforming the face of scientific endeavor. The challenges and the vision of industrial ecology, chemical process synthesis, and systems engineering are replete with immense scientific grit, profundity, and ingenuity. In this chapter, the author reiterates the difficulties and impasse in the field of nanotechnology applications in human society. This well-researched treatise is today opening up newer thoughts and newer scientific enigma in the field of industrial ecology applications. Mankind and human scientific progress will, thus, ensure a newer beginning and a newer emancipation in the field of energy and environmental sustainability applications.

7.8 RECENT SCIENTIFIC ADVANCES IN THE FIELD OF NANOTECHNOLOGY, NANOFILTRATION, AND INDUSTRIAL ECOLOGY

Technology, engineering, and science of nanotechnology and nanofiltration are today witnessing immense scientific upheaval and rejuvenation. Difficulties and vision of novel separation processes, such as membrane science and the vast scientific ingenuity of water purification will all lead a long and visionary way in the true emancipation of environmental sustainability today. Nanofiltration is the area of research pursuit, which is in the process of newer scientific discernment and vision. Arsenic and heavy metal groundwater contamination today stands in the crucial juxtaposition of deep scientific introspection and vast scientific barriers. The author, in this chapter, pointedly focuses on the scientific needs, scientific determination, and the scientific girth in water purification techniques and the vast scientific rigor behind it.

Shannon et al.[1] discussed, with lucid and cogent insight, science, technology, and engineering of water purification in the coming decades and the coming visionary centuries ahead. One of the deeply pervasive issues afflicting citizens around the world is inadequate access to clean water and improper sanitation. Arsenic and heavy metal contamination of drinking water and groundwater around the world are deeply challenging the global scientific firmament.[1] In order to vehemently address these problems, a tremendous amount of research needs to be envisioned to identify robust newer methods of purifying water at lower cost and lower energy while at the same time minimizing the use of hazardous chemicals.[1] Technology has literally no answers to the global monstrous issues of water shortage and water contamination. In this chapter, the authors pinpoint some of the science and technology being envisioned and envisaged to improve the disinfection and decontamination of water.[1] The vast intricate problems worldwide associated with the lack of clean and fresh water are well-known: 1.2 billion people lack proper access to safe drinking water, 2.6 billion people have little or no sanitation, and millions of people die annually—3900 children a day—from diseases transmitted through unsafe drinking water or human excreta.[1] Intestinal parasitic infections and diarrheal diseases caused by waterborne bacteria and enteric viruses have become a leading cause of malnutrition owing to poor digestion of food eaten by people.[1] The authors in this chapter deeply delineates the domain of disinfection, decontamination, reuse, and reclamation and the vast intricate domain of desalination.[1] The status of global environment today is extremely dismal and thought provoking. Scientific research endeavor in the field of water purification is witnessing immense scientific upheavals and scientific travails. In this well-researched treatise, the authors deeply highlighted some of the science and next-generation technologies pursued: to disinfect water; to remove current and emerging pathogens without intensive use of chemicals or production of toxic by-products; to sense, transform and remove low-concentration contaminants; and to reuse wastewater and desalinate water from sea and inland saline aquifers.[1] An important goal for provision of safe water is to disinfect water from traditional and emerging pathogens, without creating more problems due to the disinfection itself.[1] According to the chapter, a recent increase in research and development initiatives in water pollution control and water purification offers immense hope in mitigating the impact of contaminated water around the world. Humankind's immense scientific girth and prowess, the needs of the human society, and the futuristic vision of science and engineering will all lead a long and visionary way in the true realization of environmental protection science and environmental sustainability. Sustainable development as regards energy and environment are at deep stake and at a state of immense disaster. In South Asia, particularly India and Bangladesh, water shortage and contamination of drinking water with heavy metal and arsenic are leading human civilization to an immense catastrophe.[1] The authors with deep scientific conscience focuses on the immediate necessities of global water research and development initiatives and the participation of the civil society, scientists, engineers, and the common man.[9–12] Traditional methods of water disinfection, decontamination, and desalination can veritably address many of the pervasive problems with quality and proper supply of clean drinking water. Even in highly industrialized nations around the world, the costs and the time needed to develop state of the art water infrastructure make it extremely arduous to address these monstrous water purification problems.[1] Further to these problems, intensive chemical treatment techniques (such as those involving ammonia, chlorine compounds, hydrochloric acid, sodium hydroxide, ozone, permanganate, alum and ferric salts, coagulation and filtration aids, anti-scalants, corrosion control chemicals, and ion exchange resins and reagents) and residuals resulting from treatment can proliferate the water pollution control and environmental protection challenges.[1]

Waterborne pathogens have a devastating effect on public health engineering particularly in Sub-Saharan Africa and South East Asia.[1] Human scientific endeavor in water pollution control, thus, stands in the midst of deep scientific comprehension and finesse. Here, comes the importance of integrated water resource management in many poor, developing, and industrialized nations around the globe. This chapter targets the success of technology and engineering in addressing global water shortage issues today.[1] Civilization and scientific prowess and splendor stands in the midst of deep introspection and scientific subtleties. The authors in this chapter deeply addresses the true scientific facts and the scientific ingenuities in the field of water purification and environmental protection.[1]

Ayres et al.[2] with immense scientific far-sightedness, targets the goals and definitions of industrial ecology. This well-researched treatise deeply targets context and history of industrial ecology, the vast methodology, economics and industrial ecology, industrial ecology at the national or the regional level, industrial ecology at the materials level, and the applications and policy implications.[2] The name "industrial ecology" veritably conveys some of the contents of the field. Industrial ecology is industrial in that it focuses on product design and the intricate manufacturing processes.[2] It largely views firms as agents for environmental protection improvements because they possess the technological and scientific finesse that is highly critical to the successful execution of environmentally informed design of products and processes. Technology, engineering, and science of ecology and environmental engineering are highly developed today and are in the path of newer vision and regeneration. In such a crucial juncture of human history and time, the need of the science of industrial ecology assumes immense importance. The authors, in this book, rigorously points toward the history, the concepts, and the futuristic vision of industrial ecology in present-day human civilization. Also, this book elucidates on the concept of industrial metabolism.[2] The science of industrial metabolism needs to be envisioned and deeply envisaged as modern science and human civilization moves forward. The other salient features of this book are the process analysis approach to industrial ecology.[2] Materials flow analysis and economic modeling are the other pillars of this book. Global biogeochemical cycles and industrial ecology approach as in developed and developing countries are enumerated in deep details in this treatise.[2]

Lifset et al.[3] with immense scientific foresight, elucidated on the goals and definitions of industrial ecology. Setting the mission and the scientific boundaries of the vast unknown domain of industrial ecology is an extremely difficult task. However, in a name as industrial ecology, which is highly provocative, the deliberations of goals and definitions are critical. Here comes the necessity of a deeply visionary chapter as this. In this introductory chapter, the authors identify key topics, characteristic tools, and approaches. The definition of "industrial ecology" is extremely difficult to explain. Industrial ecology is ecological in two ways.[3] Industrial ecology looks to nonhuman "natural" ecosystems as models of industrial activity and the advancements of industrial human civilization. Many biological ecosystems are especially effective at recycling resources and thus are highly regarded as exemplars for efficient cycling of materials and energy in industry.[3] Industrial ecology integrates the concepts of energy and material recycling in a conglomerate of different industries. This idea and the visionary concept is redefined and re-envisioned in this chapter. The most conspicuous example of industrial reuse and recycling is an increasingly famous industrial district in Kalundborg, Denmark.[3] The district contains a cluster of industrial facilities including an oil refinery, a power plant, a pharmaceutical fermentation plant, and a wallboard factory. The network of exchanges has been dubbed as a robust "industrial symbiosis" and is similar as in biological symbiosis.[3] Secondly, an industrial ecology places human technological activity—industrial technology in the widest sense, examining the sources of resources used in society and the sinks that may act to absorb and detoxify wastes.[3] Waste regeneration and energy recycling are the other visionary pivots of the science of "industrial ecology."[2] The authors rigorously points toward the innovations and the scientific instincts in the field of research endeavor in the field of industrial ecology.[3]

Shon et al.[4] deeply targeted nanofiltration as a visionary area of nanotechnology in water purification. The application of membrane technology in water and wastewater treatment is highly enhanced due to stringent water-quality standards.[4] Nanofiltration (NF) is widely used as a membrane separation process for water treatment and wastewater treatment process particularly in desalination application.[4] NF has vastly replaced reverse osmosis in a wide variety of applications due to low energy applications and a cheaper scientific research pursuit. Membrane filtration is a pressure driven process in which membrane acts as a selective barrier to restrict the passage of pollutants such as organics, nutrients, turbidity, microorganisms, inorganic metal ions, and other oxygen depleting compounds.[4] Technological and scientific prowess, the needs of water purification technologies, and the futuristic vision of membrane science will surely unravel the vast scientific truth and the scientific profundity behind engineering and science today.[4] Applications of membrane science are huge colossus with a definite vision of its own. In this chapter, the author deeply comprehends the importance of nanofiltration in water treatment and industrial wastewater treatment. The membrane process is classified into four broad categories depending on their pore sizes: microfiltration, ultrafiltration, nanofiltration, and reverse osmosis membranes. Technological and scientific validation of membrane separation processes are yet to be uncovered at its fullest as science and engineering surges forward.[4]

Loganathan et al.[5] discussed with immense scientific conscience and insight defluoridation of drinking water using adsorption processes. Excessive intake of fluoride, mainly through drinking water, is an extremely important health hazard affecting people in the developing and developed countries around the world.[5] Technological and scientific vision in the field of drinking water treatment needs to be restructured and re-organized with the passage of scientific history and time.[9–12] There are several methods for the defluoridization of drinking water, of which adsorption processes are generally considered to be extremely promising and attractive because of their effectiveness, convenience, ease of operation, simplicity of design, and for highly economic reasons.[5] The authors in this chapter presents a comprehensive and a critical literature review on various adsorbents used for defluoridation, their relative effectiveness, mechanisms and thermodynamics of adsorption, and suggestions and deliberations are made on the choice of adsorbents for various conditions and circumstances.[5] Effects of pH, temperature, kinetics, and co-existing anions on fluoride adsorption are also extensively reviewed.[5] Future research trends needs to be targeted to explore highly efficient, low-cost adsorbents that can be easily generated for reuse and recycling without significant loss of adsorptive capacity. The authors in this chapter deeply discussed with immense foresight adsorption mechanisms, factors influencing adsorption, different types of adsorbents, and adsorption thermodynamics.[5] The authors in deep details elucidated on the success of defluoridation techniques in the provision of pure drinking water to the common mass.[5] The vast vision of drinking water treatment, the futuristic vision of treatment techniques and the scientific ingenuity of environmental protection will surely lead a long way in uncovering the scientific truth behind drinking water treatment and industrial wastewater treatment.[5]

Barrow[6] with immense scientific conscience and insight discussed in details the present and the future of environmental management and development. Environmental management is a science which is rapidly developing and is being increasingly applied in developing and developed nations around the world. This book investigates environmental management, with the focus on developing nations and the development process worldwide.[6] The process of development takes place in the environment, using resources, producing wastes, and causing visible impacts on the human society. The book explores theory and approaches in the implementation of environmental management in developing countries around the globe, resource management issues by sector, and environmental tools and policies.[6] Environmental accounting, green economics and business, and the future of environmental management in developing countries around the world are deliberated in details in this book.[6] Environmental management and development are both difficult to define and today it needs to be envisioned with each step of human history and time.[6] Environmental management can be a goal or vision, an attempt to steer a process , the application of a set of tools, a more philosophical exercise to establish a newer outlook and vision.[6] Development can be a goal or vision and also an application of a set of tools. Both environmental management and development are domains of science and engineering which demand a multidisciplinary view and make it more possible for different disciplines, religions, classes, ethnic groups, political outlooks, and genders. The world today has a rich North and a poor South with pockets of poverty in the former and some rich people in the latter. Thus, the imminent need of environmental management and development.[6] Technological finesse, the world of scientific validation, and the necessity of environmental management will surely pave the way of development in rich and poor countries around the world. The author in this book deeply reiterates the importance of the science of environmental protection in the furtherance of science and technology globally.[6]

The Royal Society Report[7] emphasizes on opportunities and uncertainties in the field of nanoscience and nanotechnologies. The authors in this report deeply discussed hopes and concerns about nanoscience and nanotechnologies, nanomanufacturing, and the industrial applications of nanotechnologies, possible adverse health, environmental and safety impacts, social and ethical issues, and regulatory issues.[7] The remit of this study is vast and versatile. The report carries out a forward outlook to see how the technologies might be used in future. Also, the authors of this report identifies what health and safety, environmental, and ethical and societal implications or uncertainties may arise from the use of innovative technologies, both current and future.[7] Technological and scientific validation of nanotechnology are the utmost needs of the human society today and needs to be restructured. Among the current and potential use of nanotechnologies, the authors of the report targets nanomaterials, metrology, electronics, optoelectronics, and Information and Communication Technologies (ICT).[7] This report also highlights health and environmental impacts of nanotechnology on human society and human scientific progress. The challenges and the vision of nanotechnology are vast and versatile today. This report is a watershed text in the field of nano-vision and nanotechnology. Scientific verve, technological endurance, and deep scientific ardor are the pillars of this well-researched treatise.[7]

7.9   ENVIRONMENTAL PROTECTION, DRINKING WATER TREATMENT, AND INDUSTRIAL WASTEWATER TREATMENT

The status of environment is extremely dismal today. Global concerns of drinking water treatment and industrial wastewater treatment are of highest order today. Mankind's immense scientific progress and academic rigor are at a deep risk with the growing concerns of arsenic and heavy metal drinking water contamination. Technology and engineering, in similar manner, are at immense risk. Human scientific endeavor's immense prowess, scientific one-upmanship, and the vast scientific ardor will all lead a long and visionary way in the true realization of environmental protection and environmental sustainability. Sustainable development in the fields of energy, environment, social, and economic, needs to be revamped and re-organized with the surge of science and engineering. In the similar vein, drinking water shortage and global water emancipation needs to be restructured as civilization moves forward. Industrial ecology and industrial metabolism are the utmost needs of human scientific progress today. In this entire treatise, the author deeply pronounces the scientific success, the scientific vision, and the deep scientific ingenuity in industrial ecology applications in human society.

7.10   INDUSTRIAL ECOLOGY AND SUSTAINABILITY

Sustainability, as envisioned by Dr. Gro Harlem Brundtland, former Prime Minister of Norway needs to be restructured and re-envisaged as science and engineering surges ahead. Industrial ecology, in the similar vein, is in the process of newer scientific regeneration. Global water shortage issues and global sustainability are the two opposite sides of the visionary coin. In this chapter, the author relentlessly puts forward to the scientific domain the immediate scientific needs of industrial ecology and sustainability in the true emancipation and true perseverance of human civilization. Arsenic and heavy metal groundwater and drinking water contamination today stands in the crucial juncture of scientific might, scientific girth, and vast scientific determination. This chapter opens up newer thoughts and newer vision in the field of industrial ecology and sustainable development in decades to come.

7.11   ENVIRONMENTAL SUSTAINABILITY AND NANOTECHNOLOGY

Environmental protection science and environmental sustainability today are in the midst of deep scientific vision and vast scientific profundity. Nanotechnology and environmental engineering are today linked with each other by an unsevered umbilical cord. Human scientific progeny, the deep scientific divination, and the futuristic vision of engineering and science will all be the pallbearers toward a newer era in the field of nanotechnology today.[9–12] Sustainable development as regards energy and environment are the utmost needs of human scientific rigor today. The academic and scientific rigor in the field of sustainable development needs to be re-envisioned and re-addressed as human civilization moves forward toward a newer era of scientific might and vision. Nano-science and nanotechnology are today linked with every branch of science and engineering. Today is the world of nanotechnology, space technology, and nuclear engineering. Technology and engineering science are so advanced today that mankind needs to envision itself about energy and environmental sustainability. Today, both nanotechnology and sustainable development are linked by an unsevered umbilical cord. Here comes also the importance of industrial ecology and industrial metabolism.

Process systems engineering and systems engineering are the utmost needs for the progress of industrial ecology. The author in this chapter deeply pronounces the success of industrial ecology and nanotechnology in solving major problems in environmental engineering science. The world of science and engineering are in the midst of devil and the deep sea. Frequent environmental disasters and industrial disasters globally are urging the scientific domain to delve deep into the questions of chemical process safety, environmental protection, and industrial ecology. This chapter opens up newer thoughts and newer futuristic trends in the field of environmental protection, systems engineering, ecological engineering, process systems engineering, and the vast new world of industrial ecology. Ecological engineering and environmental sustainability are two opposite sides of the coin as science and engineering surges forward toward newer thoughts and newer knowledge dimensions. The remit of this study is to open newer windows of scientific innovation and scientific instinct in the field of industrial ecology, industrial metabolism, and the vast world of nano-science and nano-engineering.

7.12   PRESENT RESEARCH TRENDS IN THE FIELD OF INDUSTRIAL ECOLOGY AND NANOTECHNOLOGY

Nanotechnology and industrial ecology are the two promising branches of scientific endeavor today. Industrial ecology, systems engineering, and industrial engineering are highly interlinked branches of research pursuit today. The interfaces of industrial ecology with mining and metallurgical engineering are immensely bright and visionary. In this chapter, the author deeply pronounces the scientific understanding, the scientific truth, and the vast needs of industrial ecology toward the proliferation of engineering science today. Nanotechnology is diversifying today to other branches of science and engineering. Nanofiltration is a burgeoning area of membrane science, drinking water treatment, and industrial wastewater treatment. Thus, nanotechnology is integrated with other branches of engineering and science. Ayres et al.[2] discussed with minute details the future of the domain of industrial ecology. Some of the areas the authors[2] touched upon are industrial ecology and cleaner production, material flow analysis, substance flow analysis methodology, process analysis approach to industrial ecology, environmental accounting, economic modeling, transmaterialization, industrial ecology in the national and regional level, and industrial ecology at the materials level.[2] These are the areas which needs to be reorganized and restructured as scientific progress of ecology moves into a newer knowledge dimension.[2]

Fischer-Kowalski[8] deeply discussed with immense scientific far-sightedness the history of industrial metabolism. The scholarly forays of "industrial metabolism" can be traced back for more than 150 years, across various scientific interfaces , and even beyond the scope and vision of industrial societies.[8] The application of the term of "metabolism" to human society and human scientific progress cuts across the great hiatus between humanities and social sciences. In the 1860s, when this divide was relatively less rigid, the concept of metabolism from biology transformed the vast domain of social sciences.[8] Later with the progress of human civilization, social science concept became more restricted. Today, technology and engineering are in the path of scientific regeneration. Human research and development forays, in the similar manner, needs to be envisioned and re-envisaged with the progress of scientific vision and time.[8] The awakening of environmental awareness and the first sound views of economic growth during the late 1960s urged a revival of interest in society's metabolism under a newer scientific perspective.[8]

7.13   FUTURE RESEARCH TRENDS IN INDUSTRIAL ECOLOGY AND FUTURE FLOW OF THOUGHTS

Research trends in industrial ecology needs to be addressed more toward the visionary world of energy and environmental sustainability. Technology and engineering have today few answers toward the growing scientific barriers of environmental protection such as heavy metal and arsenic groundwater contamination.[9–12] Human civilization, thus, is in a deep risk as provision of clean drinking water faces immense challenges. Future research trends in industrial ecology should be directed toward a holistic scientific understanding and deep scientific profundity in the field of sustainable development. Future flow of scientific thoughts should gear toward a greater emancipation of environmental protection science and environmental sustainability.[9–12] The needs of human society are immense as well as vulnerable today. Here comes the challenge and vision of nano-science and nanotechnology. Water purification and membrane science are two opposite sides of the coin today. Nanofiltration is a deeply advanced area of membrane separation process. The author with immense far-sightedness delves deep into the science of nanotechnology and its applications in industrial ecology and environmental engineering. Nano-vision and nano-engineering are the coinwords of research pursuit globally today. This chapter opens up newer issues and newer challenges in the field of environmental engineering, chemical engineering, and water process engineering. Industrial ecology will then surely open up newer innovations and newer engineering alternatives to the ever-growing concerns of environmental sustainability.[9–12]

7.14   FUTURE RECOMMENDATIONS OF THIS STUDY

The science of industrial ecology needs to be envisioned and reframed as global research and development initiatives move forward. Industrial ecology is a latent and undeveloped area of science and engineering today. Nanotechnology is a vast and highly developed area of science and engineering today. Today is the age of nuclear science, space technology, and nanotechnology. The author deeply elucidates the necessity of both industrial ecology and nanotechnology in tackling issues concerning environmental engineering, chemical process engineering, and other diverse areas of engineering and technology.[9–12] Principles of industrial ecology also encompasses chemical engineering and industrial and systems engineering. The world of challenges and the vision of industrial ecology also involves industrial metabolism. Future recommendations of this study should be directed toward larger vision of systems engineering, production engineering, and industrial engineering.[9–12] Sustainability should be integrated with industrial ecology as there is a greater need of scientific emancipation of environmental engineering and chemical process engineering. Water engineering and water purification techniques are the other side of the visionary coin of industrial ecology today. Future of science and engineering of nanotechnology should be toward greater emancipation of water science, water purification, and green engineering. Human civilization today stands in the crucial juxtaposition of deep scientific comprehension and unending vision. Environmental protection and environmental sustainability should be the pillars of scientific research pursuit today. Sustainability, whether it is energy, environmental, social, or economic, needs to be more readdressed and re-envisaged as science and engineering moves toward a newer era. The author deeply stresses on these areas of scientific research pursuit vehemently.[9–12]

7.15   CONCLUSION AND SCIENTIFIC PERSPECTIVES

Scientific endeavor and modern science are in the midst of deep scientific introspection and vast comprehension. Scientific perspectives in the field of industrial ecology and nanotechnology are in the path of newer regeneration and vast articulation. Nanotechnology and nanofiltration are huge pillars of science today. Water shortage has urged the scientific domain to innovate newer techniques and newer visionary domains of scientific discernment and profundity. Water purification, drinking water treatment, and industrial wastewater treatment are the utmost needs of human civilization today. Thus also comes the importance of the science of industrial ecology. The future of industrial ecology needs to be integrated with energy and environmental sustainability. The author in this chapter deeply elucidates the scientific success, the vast scientific ingenuity, and the scientific intricacies in the application of industrial ecology to modern science and present-day human civilization. The world today stands in the midst of deep pessimism and introspection as environmental disasters devastates the scientific and technological firmament. In this chapter, the author brings forward to the reader the imminent needs of industrial ecology, energy, and environmental sustainability and the vast domain of water process engineering in the furtherance of science and technology. Scientific perspectives in the field of chemical process engineering and petroleum engineering science are equally in the path of devastation with the evergrowing challenges of chemical process safety and fossil fuel depletion. This treatise opens up newer vision and newer scientific optimism in the field of engineering science and industrial ecology in decades to come. The challenges to science, technology, and engineering today are vast and versatile. The author deeply comprehends the necessity and the usefulness of the domain of industrial ecology in the furtherance of the science of sustainability. Today, technology and engineering have a few answers to the frequent environmental disasters which includes global water shortage and contamination of groundwater. The arsenic groundwater contamination in South Asia particularly India and Bangladesh is an environmental disaster of monstrous proportions. The author pointedly focuses on the need of environmental engineering, water process engineering, and water treatment techniques in mitigating this global water shortage issue. This chapter will surely open up newer thoughts and newer deliberations in the field of ecology, industrial ecology, water pollution control, and environmental protection in years to come.

KEYWORDS

- **industrial**
- **ecology**
- **nanotechnology**
- **sustainability**
- **water**
- **vision**

REFERENCES

1. Shannon, M. A.; Bohn, P. W.; Elimelech, M.; Georgiadis, J. G.; Marinas, B. J.; Mayes, A. M. Science and Technology for Water Purification in the Coming Decades. *Nature* **2008,** *452,* 301–310.
2. Ayres, R. U.; Ayres, L. W. *A Handbook of Industrial Ecology;* Edward Elgar Publishing Limited: United Kingdom, 2002.
3. Lifset, R.; Graedel, T. E. Industrial Ecology: Goals and Definitions (Chapter 1). In *A Handbook of Industrial Ecology;* Ayres, R. U., Ayres, L. W., Eds.; Edward Elgar Publishing Limited: United Kingdom, 2002; pp 3–15.
4. Shon, H. K.; Phuntsho, S.; Chaudhary, D. S.; Vigneswaran, S.; Cho, J. Nanofiltration for Water and Wastewater Treatment: A Mini Review, *Drink. Water Engineer. Sci.* **2013,** *6,* 47–53.
5. Loganathan, P.; Vigneswaran, S.; Kandasamy, J.; Naidu, R. Defluoridation of Drinking Water Using Adsorption Processes. *J. Hazard. Mater.* **2013,** *248–249,* 1–19.
6. Barrow, C. J. *Environmental Management and Development;* Routledge, Taylor and Francis: New York, 2005.
7. *The Royal Society Report, 2004; Nanoscience and Nanotechnologies: Opportunities and Uncertainties;* The Royal Society and The Royal Academy of Engineering, UK, 2004.
8. Fischer-Kowalski, M. Exploring the History of Industrial Metabolism (Chapter 2). In *A Handbook of Industrial Ecology;* Ayres, R. U., Ayres, L. W., Eds.; Edward Elgar Publishing Limited: United Kingdom, 2002; pp 16–26.

9. Palit, S. Application of Nanotechnology, Nanofiltration and Drinking and Wastewater Treatment: A Vision for the Future (Chapter 17). In *Water Purification;* Grumezescu, A. M., Ed., Academic Press (Elsevier): USA, 2017.

10. Palit, S. Nanofiltration and Ultrafiltration: The Next Generation Environmental Engineering Tool and A Vision for the Future. *Inter. J. Chem. Tech. Res.* **2016,** *9* (5), 848–856.

11. Palit, S. Filtration: Frontiers of the Engineering and Science of Nanofiltration: A Far-Reaching Review. In *CRC Concise Encyclopedia of Nanotechnology;* Ortiz-Mendez, U., Kharissova, O. V., Kharisov, B. I., Eds., Taylor and Francis, 2016; pp 205–214.

12. Palit, S. Advanced Oxidation Processes, Nanofiltration, and Application of Bubble Column Reactor. In *Nanomaterials for Environmental Protection;* Boris, I., Kharisov, O. V., Kharissova, R., Dias, H. V., Eds., Wiley: USA, 2015; pp 207–215.

CHAPTER 8

NANOTECHNOLOGY RESEARCH, GREEN ENGINEERING, AND SUSTAINABILITY: A VISION FOR THE FUTURE

SUKANCHAN PALIT*

43, Judges Bagan, Post Office-Haridevpur, Kolkata 700082, India

*E-mail: sukanchan68@gmail.com, sukanchan92@gmail.com

ABSTRACT

The world of science and engineering today is witnessing immense challenges and drastic changes. Green engineering, green technology, and environmental protection are interrelated to each other in the areas of scientific vision and scientific might. Sustainability, in a similar manner, is linked with green engineering and green technology by an unsevered umbilical cord. In this chapter, the author focuses on the scientific needs, the futuristic vision, and the immense scientific research pursuit in the field of green engineering, green chemistry, sustainable development, and environmental sustainability. Sustainable development and environmental management are today linked with each other by an umbilical cord. This chapter presents profoundly the scientific success in the field of nanotechnology and green engineering with the sole purpose of furtherance of science and technology. Mankind's immense scientific prowess, the world of scientific validation, and the futuristic vision of environmental engineering science all lead to a long and visionary way in the true emancipation of environmental sustainability and green engineering. Nanotechnology applications in environmental protection are the cornerstone of the research endeavor in this chapter. Global climate change, frequent environmental disasters, and the depletion of fossil fuel resources are today challenging the vast scientific fabric of might and vision. The author deeply envisions the need of nanotechnology in environmental protection and the success of green engineering in the advancement of science and engineering globally.

8.1   INTRODUCTION

The world of science and technology are witnessing immense challenges globally. Environmental protection and the domain of environmental engineering science in a similar manner are in the process of newer scientific rejuvenation. Global warming, frequent environmental catastrophes, and the depletion of fossil fuel resources are veritably challenging the domains of chemical process engineering, petroleum engineering, and environmental engineering science. The progress of human civilization and the immense scientific progress are in a state of immense scientific disaster today. Water purification, drinking water treatment, and industrial wastewater treatment are the utmost needs of human scientific research pursuit today. Social, economic, environmental, and energy sustainability are changing the face of human civilization today. In this chapter, the author deeply comprehends the necessity of green engineering, green technology, and green chemistry in the scientific advancement of human civilization. This chapter opens up newer avenues of green engineering, newer arenas of scientific imagination, and scientific profundity in green chemistry and green nanotechnology. The state of environment is dismal and immensely retrogressive today. Thus there is a need for a comprehensive treatise in the application of green engineering and green nanotechnology to human society. This chapter delves deep into the murky depths of human endeavor in nanoscience and nanotechnology and veritably opens up newer challenges and newer opportunities in the vast scientific endeavor in the field of green nanotechnology.

8.2   THE AIM AND OBJECTIVE OF THIS STUDY

Human scientific advancements and the vast world of scientific and technological validation are groundbreaking and need to be re-envisioned and re-envisaged with the passage of scientific history and time. The aim and objective of this study are to target the utmost needs of green engineering, green nanotechnology, and green chemistry in environmental protection and diverse areas of science and engineering. The other arenas of research endeavor are the world of environmental and energy sustainability and its applications in human society. Technological ardor, scientific vision, and vast scientific profundity are the necessities of innovation and invention today. This treatise veritably opens up newer avenues and newer aisles in the field of energy and environmental sustainability in years to come. The state of environmental engineering science and environmental protection are immensely dismal today. In this chapter, the author pointedly focuses on the scientific success, the scientific subtleties and the scientific profundity in nanotechnology applications in environmental engineering and the vast world of environmental sustainability. The sole objective of this well-researched treatise is to envision and re-envisage the domain of environmental sustainability and the contribution of environmental engineering science to human society and human scientific endeavor. The challenges and the vision of sustainable development are immense and groundbreaking today. Now, environmental management and sustainable development are also linked by an unsevered umbilical cord. The civil society, educationists, and scientists should gear forward toward a newer vision of sustainable development in energy and environment. Here comes the vast importance of renewable energy and the vast domain of energy engineering. The author, in this chapter, deeply focuses on both environmental management and application of nanotechnology in the furtherance of science and engineering globally.

8.3   THE SCOPE OF THIS STUDY

The over-arching goal of this study is to target the world of sustainable development and green engineering in the furtherance of science and technology globally. Today, global fossil fuel resources would soon be depleted and environmental disasters will destroy the very scientific fabric of mankind. Thus the branches of chemical process engineering, environmental engineering, and petroleum engineering are highly challenged as science and engineering surges forward. Technology adroitness, scientific far-sightedness and the world of scientific validation will surely lead a long and visionary way in the true emancipation of science and technology in the global scenario. Today is the age of nuclear science and space technology. Thus the scope of study goes beyond scientific imagination and encompasses diverse areas of science and engineering such as environmental engineering and petroleum engineering science. The vast scope of this study is ground-breaking and surpassing vast and versatile scientific boundaries. Recent advances in environmental protection science and environmental engineering are deliberated in minute details in this chapter apart from the application of green engineering and nanotechnology to human society. Human civilization today stands in the midst of scientific vision and scientific fortitude. The world of science and technology in a similar manner are in the midst of deep scientific introspection and foresight. This study will surely lead a long and visionary way in the true realization of environmental and energy sustainability apart from green engineering and nano-engineering. Energy engineering and environmental engineering are the scientific truth of today's research pursuit. Renewable energy technology such as biomass energy, solar energy, and wind energy are changing the face of human civilization. This chapter presents a detailed overview of the needs of sustainable development and environmental management in the scientific progress of mankind. Provision of basic human needs such as water, food, energy, and shelter is in a state of immense devastation. Thus water purification, drinking water treatment, and industrial wastewater treatment are the scientific needs of human society today. Thus the scope of this study goes beyond environmental protection tools and also envelopes traditional and non-traditional environmental engineering techniques such as advanced oxidation processes and membrane science. The world today stands perplexed with the ever-growing concerns for heavy metal and arsenic groundwater and drinking water contamination. Science and engineering thus need to be re-envisaged and scientifically revamped as regards provision of pure drinking water to the human masses. The author in this chapter rigorously points toward the success, the vision, and the immense scientific determination in the research pursuit in the field of green engineering and nanotechnology.

8.4   THE SCIENTIFIC DOCTRINE OF SUSTAINABILITY

Mankind and human scientific endeavor today stand in the midst of deep scientific introspection and regeneration. Loss of ecological biodiversity, frequent environmental disasters, global warming, and depletion of fossil fuel resources are challenging the scientific firmament of vision, might, and determination. Sustainability is the process of change, in which the exploitation of resources, the direction of investments, the wide orientation of technological development and institutional change are all in perfect harmony and increase both current and future potential to meet human needs and aspirations. The scientific doctrine of sustainability is path-breaking today and needs to be envisioned and re-envisaged with the passage of scientific history and time. The visionary definition of "sustainability" as propounded by Dr. Gro Harlem Brundtland, former Prime Minister of Norway needs to be applied with all scientific vision and scientific introspection. Brundtland Report of the World Commission on Environment and Development (1992) with immense scientific conscience and foresight introduced the concept of sustainable development. Sustainability can also be defined as a socio-ecological procedure characterized by the scientific endeavor of a common ideal.

8.5 GREEN ENGINEERING, GREEN CHEMISTRY, AND GREEN NANOTECHNOLOGY

Science and technology of green engineering and green science are today advancing at a rapid pace. The world of green engineering is today linked to environmental and energy sustainability by an unsevered umbilical cord. Nanotechnology is today a wonder of science globally and is replete with immense vision and academic rigor. Systems engineering, green engineering and industrial ecology are linked with each other today with immense scientific might and vision. In a similar vein, chemical process engineering and environmental engineering science are linked to green engineering. Also, material science and product engineering are the hallmarks of scientific endeavor in green materials and smart materials. Human technological advancements, the scientific vision, and redemption will surely lead a long and visionary way in the revelation of scientific truth in green material science today. In such a discussion, the utmost need is of scientific understanding in the field of green nanotechnology which is a relatively new branch of scientific pursuit. The world stands perplexed and mesmerized with the ever-growing concerns for health effects of nanomaterials and nanotechnology, thus the immediate need for green nanotechnology and green nanomaterials. Process technology, chemical process engineering, and process design and synthesis needs to be revamped and re-envisaged as material science, polymer science and composites science surges forward toward a newer knowledge dimension. The procedure of manufacturing of nanomaterials needs to be readdressed as regards green process, green engineering, and green nanotechnology. Environmental protection and environmental regulations need to be adhered to as regards the manufacturing procedures for nanomaterials or green nanomaterials. Water engineering and membrane science are today linked by an unsevered umbilical cord. Here comes the utmost need green nanomembranes and green membrane separation processes. Water science and water technology need to be metamorphosed into green processes and green process synthesis. In this chapter, the author deeply elucidates the scientific success, the scientific redemption and the vast scientific ingenuity in the materials synthesis processes of green nanomaterials and green nanotechnology. Technological progeny, the deep engineering prowess and the futuristic vision of green nanomaterials processes will veritably lead a long and visionary way in the true realization and true scientific truth of green nanotechnology.

8.6   RECENT SCIENTIFIC ENDEAVOR IN THE FIELD OF SUSTAINABILITY

Sustainable development as regards energy, environmental, social, and economic is changing the face of human civilization today. Technology has a few answers to the growing concerns of global warming and climate change. Industrial ecology, systems engineering, and human factors engineering are the fountainhead of global water research and development today. The failures of sustainable development need to be re-organized and revamped with every step of scientific history.

OECD Report[1] discussed and deliberated with great scientific conscience nanotechnology for green innovation and green scientific vision. This chapter brings forward to the readers' discussions and projects undertaken by the OECD Working Party of Nanotechnology (WPN) relevant to the development and use of nanotechnology for green innovation, vision, and deep scientific girth.[1] The purpose of this chapter is to provide a vast background information for future work and future endeavor by the WPN on the application of nanotechnology to green engineering emancipation.[1] The report deeply discusses strategies for green innovation through nanotechnology and the impact of green nanotechnology to human progress and advancement of scientific rigor. The utmost need of addressing global challenges, in domains such as energy, environment, and health has become immensely important. The global need for environment and energy is expected to rise by more than 30% between 2010 and 2035.[1] Scientific vision in the field of sustainable development needs to be refurbished as civilization moves forward. More than 800 million people are currently without the reach of pure drinking water. The vast technological advancements, the futuristic vision of energy, and environmental sustainability and the utmost needs of human progress such as water, energy, shelter, food, and education will lead a long and visionary way in the true unraveling of engineering and science today. This report opens up newer vision and newer future thoughts in the field of green nanotechnology innovations in years to come. Green innovation is innovation which veritably reduces environmental impacts: by increasing energy efficiency, by reducing greenhouse emissions, and by minimizing non-renewable raw materials for example.[1] The roads leading to sustainable development are today bright and ground-breaking.[1] The vision, the scientific truth, and the future challenges will surely delve deep into the world of energy and environmental sustainability.[1] Since it began its work in 2007, the OECD WPN has increasingly developed a number of sustainable projects addressing a number of issues of science, technology, and innovation relating to the development of nanotechnology.[1] Green nanotechnology can have multiple roles and vast impacts across the whole value chain of a product being used as a tool to support technology and product development for example:

(1) Nanotechnology can play a pivotal and fundamental role to enhance a key functionality of a product (e.g., nanotechnology-enabled batteries).[1]

(2) Nanotechnology can enhance the size-dependent phenomena of nanotechnology (e.g., electric cars using nanotechnology-enabled batteries).[1]

(3) Nanotechnology can improve and embolden sustainable and green processes.[1]

Global water shortage and global scientific impasse in the field of water research and development initiatives are in a state of immense conundrum and vast scientific difficulties. When reviewing global strategies for science, technology, and innovations, the forays for green innovation are apparent. Human vision in innovation and scientific instinct needs to be revamped if green innovation in nanotechnology needs to be enhanced and envisioned.[1] This report delves deep into the policy environment for green nanotechnology. Green nanotechnology operates in a complex landscape of regulations, fiscal, and legislative policies and allied measures for green growth, green rejuvenation, and green unfurling in science technology and innovation.[1]

Kuhlman et al.[2] deeply comprehended with great scientific conscience and great vision the concept and vision of sustainability. Sustainability as a concept has its origin in the Brundtland Report.[2] This visionary document was a balance between a better life on one hand and the limitations imposed by nature on the other hand. Technology, engineering, and science are today in the path of newer scientific overhauling globally today.[2] In the course of scientific history and time, sustainability metamorphosed into three dimensions such as social, economic, and environmental. Since 1987, the concept of sustainability shifted its meaning drastically.[2] This chapter vehemently argues the real contradiction which exists between long-term sustainability and short-term welfare. The concept of sustainability originated from the domain of forestry.[2] The answers toward better way of human life are originated from the basic definition of sustainability:

"Development that meets the needs of the present without compromising the ability of future generations to meet their own needs."

The Brundtland report targeted and envisioned two concerns: development and environment. Happiness, well-being, and welfare are the other cornerstones of this well-developed research endeavor. The authors in this report elucidated with deep insight and conscience the immense necessity of sustainable development whether it is social, economic, energy, and environmental in the furtherance of science and engineering globally.[2]

United Nations Development Programme Report[3] with vast scientific farsightedness delineated sustainability and inequality in human development. The success of human civilization and human scientific progress lies in the hands of energy and environmental sustainability today. This chapter vastly utilizes the theoretical and empirical links between inequality in human development on one hand and sustainability on the other hand.[3]

Science and engineering today stands in the midst of deep introspection and scientific fortitude. This report widely pronounces on the scientific success, the scientific ingenuity and the vision of engineering science in the furtherance and scientific proliferation. Inequality in various dimensions of human development and human progress is analyzed with respect to both weak and strong sustainable development, where weak sustainability gives importance to substitutability among different forms of development while strong sustainability rejects substitutability.[3] This chapter deeply elucidates the links between inequality in human development on the one hand and sustainability on the other hand.[3] The goal of human development is the ability to do things freely and be the person they want to be with greater emancipation of sustainability.[3] Inequality in human development is the inequality in such capability and the freedom to sustain.[3] Human progress and human scientific advancement's immense prowess will go parallel if equality and sustainability can be achieved. Inequality, poverty, and social sustainability are the domains of immense introspection and vision today.[3] In this chapter, a greater visionary deliberation is deeply done with respect to both sustainability and human development. This chapter deeply comprehends the immediate distinction of strong and weak sustainability in the furtherance of human civilization.[3]

Markulev et al. (2013)[4] deeply elucidated with cogent insight the economic approach of sustainability. Mankind and human scientific research forays are changing the face of sustainable development whether it is social, economic, energy, or environmental.[4] At its most general level, sustainability refers to the ability to continue an activity or a process indefinitely. Sustainability has different definitions in different scientific and knowledge dimensions.[4] Sustainability means different things to different people and has deeply entered the vernacular process in varied ways. Success of human civilization and vast success of human academic today depends on both energy and environmental sustainability. The authors discussed with vast scientific foresight sustainability from an economic perspective, economic interpretations of sustainability, the definition and measuring of well being, intergenerational equity, the difference between natural capital and other forms of capital, application of sustainability concepts, understanding of ecological systems, and the futuristic vision of sustainability applications to human society.[4] Policy descriptions in industrial waste management are the other pillars of this chapter. The main goal of this chapter is to target how sustainability can be applied in practice in human society. The author discussed with immense scientific girth and vision the need of environmental sustainability in the furtherance of human scientific progress.[4]

Scientific regeneration and vast scientific understanding are the needs of environmental protection and environmental sustainability today. In this entire chapter, the author deeply discussed the necessity of green engineering and sustainable development toward the advancement of human civilization today. The author discussed some of the important areas of sustainability applications in human society today with the sole vision of furtherance of science and technology.

8.7 RECENT SCIENTIFIC RESEARCH PURSUIT IN THE FIELD OF GREEN ENGINEERING

Green engineering and green chemistry are the hallmarks of scientific research pursuit globally today. Every branch of engineering and science are integrated with green engineering and green chemistry. Green nanotechnology is another area of scientific vision today. Green engineering today is highly advanced area and research pursuit is linked to environmental science and environmental sustainability.

Verma et al.[5] discussed with lucid and cogent insight green nanotechnology. The vision, the challenges and the goals of green nanotechnology are immense and groundbreaking today. Nanotechnology encompasses a unique phenomenon that enables diverse applications in different arenas. Nanotechnology promises a deep and sustainable future by its growth in green chemistry to develop a deep scientific understanding in green nanotechnology. Green nanotechnology infers the application of green chemistry and green engineering principles in the field of nanotechnology to transform molecules in the nanorange.[5] The scientific challenges in the field of green nanotechnology, the futuristic vision of green engineering and the needs of human civilization are all the visionary torchbearers toward a newer era in the field of nano-science and nanotechnology.[5] The authors in this chapter delineated in minute details nanoparticle synthesis by green route, challenges in green synthesis, and new approaches in green nanotechnology. The green nanoparticles synthesized via green chemical principle provides important applications to prevent industrial waste, synthesized less hazardous chemicals, renewable feedstock, and reduce derivatives.[5] Science and engineering of green nanotechnology are today in the path of regeneration and vision.[5] This chapter reviewed the success of engineering science and technology in the scientific revelation of green synthesis and green engineering.[5]

Lu et al.[6] deeply discussed with vast scientific far-sightedness the sustainable future of green nanomaterials. With vast applications that surpass scientific frontiers, from electronics to medicine, to advanced manufacturing, to cosmetics—nanotechnology has the immense scientific potential to change lifestyles, jobs, and urban and rural economies. Most of the materials and processes used not only depend on non-renewable resources but also vastly create hazardous chemical wastes. In this chapter, the authors discussed with vast scientific conscience the integration of green chemistry techniques with nano-vision. This chapter vastly opens up new avenues in the world of green synthesis and application of natural products chemistry.[6] Green nanotechnologies are now vastly finding their way from the laboratory to industrial applications. The author in this well-researched chapter reviews the recent advances, challenges, and the targets in true emancipation of green nanomaterials.[6]

Demirdoven et al.[7] described and discussed with immense vision green nanomaterials with examples of applications. Nanotechnology implies material applications in nanoscale. Technology validation, scientific forays in nanomaterials, and the futuristic vision of nanotechnology will surely lead a visionary way in the true unraveling of science and engineering.[7] During the last few decades, a vast range of smart materials revolutionized the vast landscape of science and technology. Nanotechnology, in most cases, serves as a visionary heading for all material investigations at the nanoscale.[7] Human civilization's vast scientific understanding, the futuristic vision of green nanomaterials and the vast scientific needs of human society are all the pallbearers toward a newer era in the field of green nanotechnology today. The potential contribution to sustainability makes nanotechnology one of the key technologies in green building arena.[7] This chapter classifies the application of the area of green buildings as (1) safety and security, (2) indoor quality, (3) material surface advancement, (4) energy generation and storage, and (5) environmental impact and control.[7] This chapter is a watershed text in the field of green buildings and environmental design.[7] Environmental sustainability will surely open new frontiers of scientific innovation and newer scientific instinct in the areas of science and engineering in decades to come.[7]

8.8 ENVIRONMENTAL PROTECTION AND THE VISION FOR THE FUTURE

Environmental protection and environmental engineering science are today in the avenues of vision and scientific redemption. The vast challenges and the instinctive vision of green nanotechnology are in the process of newer regeneration. The vision for the future in the field of industrial pollution control, the drinking water treatment and industrial wastewater treatment are immense and path-breaking today globally. The needs and the vision of environmental science are in a similar manner immense, visionary and scientifically inspiring. Science and technology of environmental protection are today globally huge pillars of might and vision. The world of engineering science stands crucially in the midst of deep scientific and engineering introspection. Globally scientific and research order are faced with unbelievable scientific challenges and vast difficulties. Environmental engineering, water engineering, and novel separation processes are veritably challenging the scientific firmament globally. Arsenic and heavy metal groundwater contamination are in the midst of vision and vast scientific girth and determination. Mitigation of groundwater and drinking water contamination are of immense importance as environmental sustainability and global water research and development surges forward. The subtleties of science, the vast technological impasse in environmental protection and the failures in environmental engineering will all lead a long and visionary way toward true emancipation of sustainable development and scientific progress today.

8.9 RECENT SCIENTIFIC ADVANCES IN ENVIRONMENTAL ENGINEERING

Environmental sustainability and environmental protection are the fountainheads of scientific research pursuit today. Arsenic and heavy metal groundwater and drinking water are banes to human civilization and human scientific progress. South Asia particularly India and Bangladesh are in the threshold of world's largest environmental engineering crisis that is global water scarcity. In this chapter, the author with vast and lucid scientific conscience deeply relates the challenges and the vision in scientific articulation and scientific introspection in water purification.

Hashim et al.[8] reviewed remediation technologies for heavy metal contaminated groundwater. The contamination of groundwater is a matter of immense concern to the public health globally. In this chapter, 35 approaches for groundwater treatment have been deeply reviewed and classified under three large categories that is, chemical, biochemical/biological/biosorption and physico-chemical treatment techniques.[8] The scientific forays in chemical engineering and water process engineering today stands amidst introspection and comprehension.[8] Heavy metal groundwater remediation is a must for human society and human scientific progress today.[8] The authors in this chapter with vast scientific fortitude elucidate the necessity of water purification in achieving the goals of environmental sustainability. Selection of a suitable technology for contamination remediation at a particular site is one of the visionary and ever-growing challenges due to extreme soil chemistry and aquifer characteristics. In the past years, iron-based technologies, microbial remediation, biological sulfate reduction, and various adsorbents played a vital and crucial role in groundwater remediation.[8] But today, science and engineering have progressed further.

Mandal et al.[15] discussed with immense lucidity arsenic around the world in a review chapter. Scientific vision and deep scientific fortitude are at extreme risk as civilization faces this monstrous environmental disaster.[15] Technology has few answers to the marauding issue of arsenic groundwater contamination. This review deals with environmental origin, occurrence, episodes, and impact on human health due to arsenic. Water science and water technology today thus stands in the midst of vast introspection and scientific articulation. Groundwater contamination by arsenic is an ever-growing threat to Mankind.[15] It can also enter food chain causing widespread distribution throughout the plant and animal kingdom. The authors discussed occurrence, anthropogenic sources, manmade sources, metabolisms, and toxicity of arsenic. A deep evaluation of arsenic toxicity on human health is deeply deliberated in this chapter.[15]

Technology and engineering should be directed toward green chemistry, green nano-materials, and green engineering. Water technology and environmental protection should be merged with green engineering. Thus the status of human scientific progress will usher in a newer era of immense might and scientific vision.

8.9.1 *SCIENTIFIC AND TECHNOLOGICAL VALIDATION AND THE VISION OF WATER PURIFICATION*

Civilization and water science today stands amidst deep scientific comprehension and scientific marvels. Poor and developing nations around are in a state of deep disaster. The environmental catastrophe of arsenic drinking water contamination has no ends and is ever-growing. Validation of water science and technology assumes immense importance in the crucial juncture of this disaster. The vision of water purification, drinking water treatment, and industrial wastewater treatment is groundbreaking and scientifically inspiring. Global research and development initiatives should be targeted toward the civil society, the scientific and the engineering domain. In this chapter, the author targets with immense vision and deep scientific ingenuity the needs of human society which are energy and environmental sustainability. Technological failures and scientific difficulties in water technology are immense. Here comes the importance of water technology and water pollution control. Global warming is due to excessive industrial pollution. The author thus pointedly focuses on green engineering and science of sustainability in the furtherance of human scientific process.

8.10 WATER PURIFICATION AND SUSTAINABILITY

Water purification and environmental sustainability are the two opposite sides of the visionary coin today. Industrial wastewater engineering and drinking water decontamination are the vision and the challenges of civilization today.[9-12] In South Asia, particularly India and Bangladesh are in the throes of the world's largest environmental disaster—the marauding arsenic groundwater and drinking water contamination. Technological challenges and scientific vision in the field of water purification and groundwater remediation are immense, versatile, and groundbreaking. Human technological challenges in the field of groundwater arsenic remediation are at deep stake and are in the process of newer vision and rejuvenation.[13,14] Civilization's scientific prowess and the deep scientific vision in water process engineering needs to be revamped and rejuvenated as environmental protection science and industrial pollution control ushers in a new era of might and determination. The true vision of Dr. Gro Harlem Brundtland, former Prime Minister of Norway on the science of sustainability are today in a newer avenue of scientific girth and determination.[13,14] In the coming decades and the coming years, the world will face immense environmental issues such as climate change, global warming, and water scarcity. In this chapter, the author pointedly focuses on the needs of water purification science and its integration with nanotechnology and sustainability. Sustainable development as regards energy, environmental, social or economic needs to be envisioned and readjudicated with every step of human history and human vision. Human vision and human factors engineering are veritably linked with each other as regards water purification and drinking water treatment. Industrial ecology and systems engineering are the other areas of scientific vision globally. In this well-researched treatise, the author brings strongly to the scientific forefront the scientific redemption, the scientific impasse and the needs of scientific validation in application of water purification technologies to human society.[9–14]

8.11   HEAVY METAL GROUNDWATER REMEDIATION AND THE ROAD AHEAD

Developing and developed countries around the world are today faced with a monstrous and unrelenting environmental engineering disaster that is groundwater heavy metal and arsenic contamination. In India and other South Asian countries, water shortage is a bane to scientific progress and scientific achievements. The vision, the challenges, and the targets of research pursuit need to be readdressed and re-envisaged as environmental engineering, chemical process engineering, and diverse areas of science and technology surges forward. Arsenic groundwater remediation is an important necessity toward global emancipation in environmental sustainability today. Technological impasse and subsequent regeneration in environmental engineering science and water process engineering are the enigmatic issues facing scientific domain today. In this chapter, the author rigorously points toward the scientific difficulties, the challenges and the vast scientific profundity in environmental protection science and the vast world of water purification technologies. The world of green engineering and green nanotechnology will surely usher in a newer era in the field arsenic drinking water mitigation in developing nations in years and decades to come.

8.12   GREEN ENGINEERING AND GREEN NANOTECHNOLOGY ADVANCES

The science of nanotechnology today is in the process of newer scientific rejuvenation. Green chemistry and green nanotechnology are of utmost need today as protection of environment assumes immense importance. Mankind's immense scientific grit and determination, the futuristic vision of nanotechnology, and the immediate needs of human society are all the forerunners toward a newer visionary era in the field of science and engineering. Advances in green nanotechnology, green engineering, and green chemistry are veritably changing the face of human civilization today. Nanotechnology advances are today in the path of newer scientific regeneration. Green chemistry, green nanotechnology, and environmental protection are the utmost needs of this decade. The world of science and engineering are totally devastated as regards environmental engineering and chemical process safety. Here comes the importance of vast scientific forays in the field of environmental engineering science, process technology, and safety engineering. Also comes the immense importance of green engineering and green nanotechnology advances. Human scientific endeavor today stands in the midst of immense environmental engineering crisis and the face of frequent environmental catastrophes such as global climate change. The author in this chapter deeply reiterates the human scientific finesse and the scientific fortitude in the field of application of green nanotechnology and green engineering in the future scientific vision and scientific struggle of human society. Green nanotechnology advances are the needs of human civilization globally. The vision and the challenges of science and engineering are today surpassing vast and versatile scientific boundaries. This enormous challenge of green nanotechnology applications to human society is enumerated in deep details in this well-researched treatise. The needs of human society are green engineering and green nanotechnology today. The author deeply targets the intense necessity and the visionary drive in the application of green nanotechnology to human scientific progress and the human academic rigor.

8.13   FUTURE RESEARCH TRENDS AND FUTURE FLOW OF SCIENTIFIC THOUGHTS

Technology and engineering science today are in some sense retrogressive as human civilization surges forward. Provision of basic needs such as water, electricity, food, and shelter are in a state of immense catastrophe and deep scientific introspection. Today heavy metal and arsenic groundwater and drinking water contamination in developing and developed nations around the world is the world's largest environmental disaster. The developing nations of India and Bangladesh are in the threshold of a disaster of monstrous proportions with the ever-growing concerns for arsenic and heavy metal groundwater contamination of drinking water. Groundwater remediation technologies should be the prime objectives of environmental engineering science today. This chapter opens up newer vision and newer objectives in the field of green engineering, nanotechnology and sustainability science in decades to come. Application of nanotechnology in human society and mankind should be the hallmark of science today. Civil society, engineers, and scientists need to target environmental and energy sustainability in the true realization and true emancipation of science and engineering. Future research trends should be targeted toward more research and development initiatives in the field of environmental engineering and environmental protection science.

8.14   CONCLUSION AND VAST SCIENTIFIC PERSPECTIVES

The status of the science of environmental protection is immensely dismal and deeply thought-provoking. Global water crisis, climate change, and the crisis in petroleum resources are changing the face of human scientific research pursuit. The scientific perspectives of green engineering, green nanotechnology, and green chemistry are path-breaking and are surpassing vast and versatile scientific frontiers. Provision of pure drinking water is a scientific challenge to human civilization today. The crisis of science and engineering of water technology is unimaginable and need to be vehemently addressed with the passage of history and time. Nanotechnology is a wonder of science today. Application of nanotechnology in diverse areas of science and engineering are equally wonders of scientific advancements. Future research trends and future scientific thoughts should be targeted toward more scientific emancipation of water science and technology. Green engineering and green technology should be the cornerstones of research pursuit. Green chemistry is another area of scientific vision. Perspectives of science and engineering need to be vehemently addressed as regards application of nanotechnology. Membrane science and nanofiltration should be integrated with integrated water resource management globally. Technological fervor, scientific foresight, and vast area of scientific validation are all the forerunners toward a newer era in the field of green nanotechnology applications to Mankind. In this research work, the author deeply comprehends the scientific ingenuity, the scientific success and the deep vision of sustainability applications and the sustainable development of human civilization. Energy crisis and water shortage are the two major disasters of human scientific progress today. These issues are deeply elucidated in this chapter. Scientific comprehension and introspection will thus lead a long and effective way in the true realization of energy and environmental sustainability.

KEYWORDS

- green
- nanotechnology
- sustainability
- vision
- global
- water
- purification

REFERENCES

1. OECD, *Nanotechnology for Green Innovation, OECD Science, Technology and Industry Policy Papers*, No.5; OECD Publishing: France, 2013.
2. Kuhlman, T.; Farrington, J. What is Sustainability? *Sustainability* **2010,** *2,* 3436–3448.
3. Neumayer, Eric. UNDP Human Development Research Paper, 2011/04, Sustainability and Inequality in Human Development, Nov 2011.

4. Markulev, A.; Long, A.; *On Sustainability: An Economic Approach, Productivity Commission Staff Research Note*; Australian Government Productivity Commission: Commonwealth of Australia, 2013.

5. Verma, A.; Sharma, M.; Tyagi, S. Green Nanotechnology, Research and Reviews. *J. Pharm. Pharm. Sci.* **2017,** *5* (4), 60–66.

6. Lu, Y.; Ozcan, S. Green Nanomaterials: On Track for a Sustainable Future. *Nano Today*, 2015, *10*, 417–420.

7. Demirdoven, J. B.; Karacar, P. Green-nanomaterials with Examples of Applications, Green Age Symposium; 15–17 April 2017, Mimar Sinan Fine Arts University Faculty of Architecture: Istanbul, Turkey, 2015.

8. Hashim, A.; Mukhopadhyay, S.; Sahu, J. N.; Sengupta, B. Remediation Technologies for Heavy Metal Contaminated Groundwater. *J. Env. Manag.* **2011,** *92*, 2355–2388.

9. Shannon, M. A.; Bohn, P. W.; Elimelech, M.; Georgiadis, J. A.; Marinas, B. J. *Science and Technology for Water Purification in the Coming Decades*; Nature Publishing Group, London, United Kingdom, 2008; pp 301–310.

10. Palit, S.; Nanofiltration and Ultrafiltration—The Next Generation Environmental Engineering Tool and a Vision for the Future. *Int. J. Chem. Tech. Res.* **2016,** *9* (5), 848–856.

11. Palit, S. Filtration: Frontiers of the Engineering and Science of Nanofiltration: A Far-reaching Review. In *CRC Concise Encyclopedia of Nanotechnology*; Ubaldo, Ortiz-Mendez; Kharissova, O. V.; Kharisov, B. I., Eds.; Taylor and Francis, 2016; pp 205–214.

12. Palit, S. Advanced Oxidation Processes, Nanofiltration, and Application of Bubble Column Reactor, In *Nanomaterials for Environmental Protection*; Boris, I. Kharisov; Oxana, V. Kharissova; Rasika Dias, H. V., Eds,; Wiley: USA, 2015; pp 207–215.

13. Palit, S. Advanced Environmental Engineering Separation Processes, Environmental Analysis and Application of Nanotechnology: A Far-reaching Review (Chapter-14), In *Advanced Environmental Analysis: Application of Nanomaterials*; Chaudhery, Mustansar Hussain; Boris, Kharisov, Eds.; Royal Society of Chemistry Detection Science, 2017.

14. The Royal Society and The Royal Academy of Engineering Report, Nanoscience and Nanotechnologies, July 2004.

15. Mandal, B. K.; Suzuki, K. T. Arsenic Round the World: A Review. *Talanta* **2002,** *58*, 201–235.

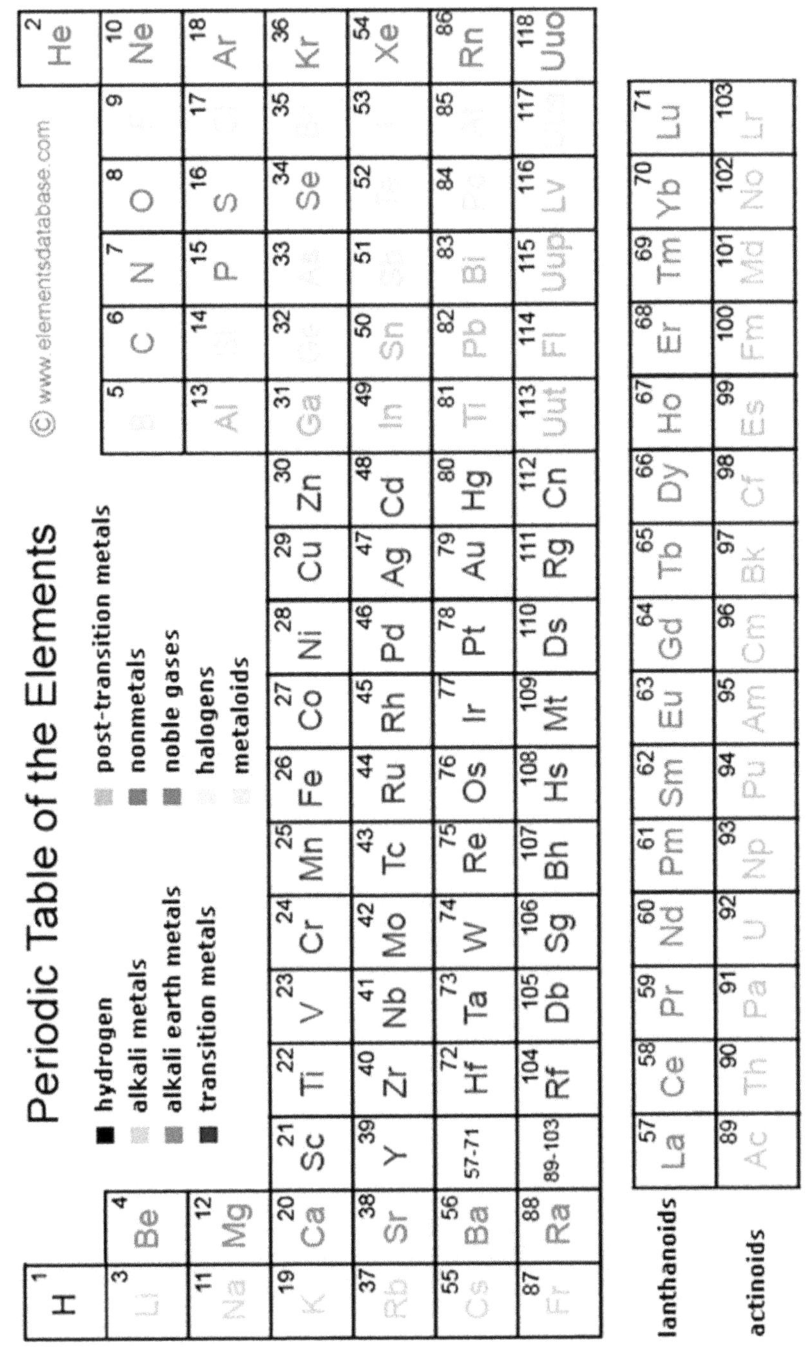
Periodic Table of the Elements
© www.elementsdatabase.com
hydrogen
alkali metals
alkali earth metals
transition metals
post-transition metals
nonmetals
noble gases
halogens
metaloids
lanthanoids
actinoids

FIGURE 2.1 In emergence of some heavy PTEs, most were formed merging NSs: Pt/Au factory in sky.

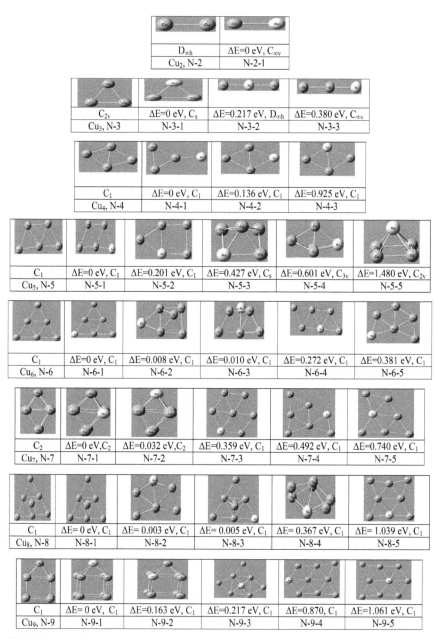

FIGURE 4.1   The ground state configurations of neutral Cu_{n+1} and $Au–Cu_n$ clusters, and low lying isomers of $Au–Cu_n$ ($n = 1$–8) clusters. The yellow and red circles represent gold and copper atoms, respectively.

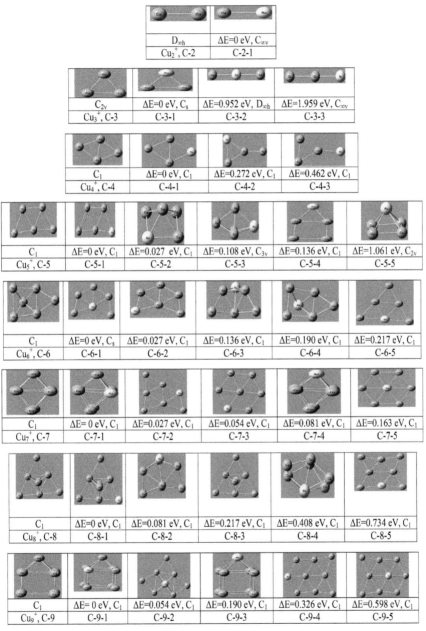

FIGURE 4.2   The ground state configurations of cationic $[Cu_{n+1}]^+$ and $[Au–Cu_n]^+$ clusters, and low lying isomers of $[Au–Cu_n]^+$ ($n = 1–8$) clusters. The yellow and red circles represent gold and copper atoms, respectively.

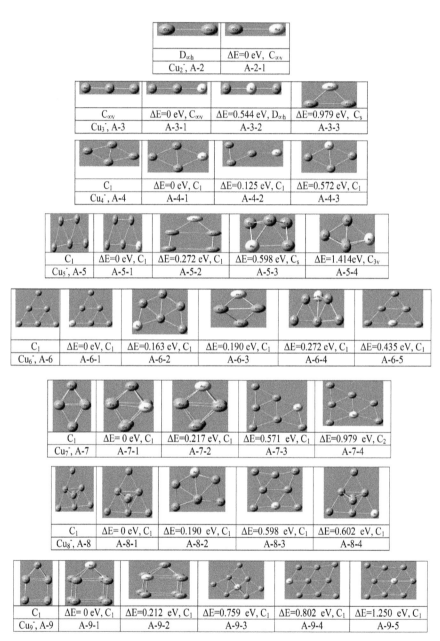

FIGURE 4.3 The ground state configurations of anionic $[Cu_{n+1}]^-$ and $[Au-Cu_n]^-$ clusters, and low lying isomers of $[AuCu_n]^-$ ($n = 1-8$) clusters. The yellow and red circles represent gold and copper atoms, respectively.

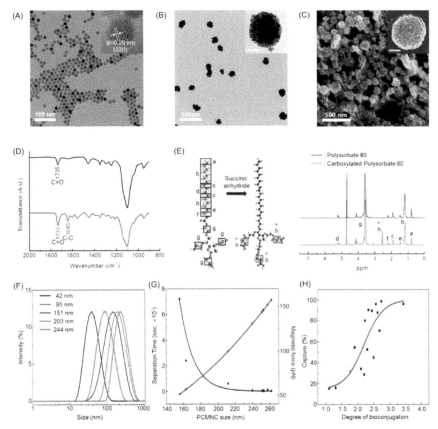

FIGURE 10.11 TEM images of (A) monodispersed MNPs with oleic acid (inset shows a HRTEM image) and (B) magnetic nanoclusters with polysorbate 80. (C) SEM images of magnetic nanoclusters. Insets show TEM and SEM images at high magnification. Inset scale bar is 100 nm. (D) FT-IR spectra of oleic acid-coated MNPs (black line) and polysorbate 80-coated MNCs (red line). (E) 1H NMR spectrum of carboxylated polysorbate 80. (F) Size distributions of MNCs analyzed by DLS. (G) Relationship between magnetic separation time (black line) and magnetic force under specific field gradients (blue line). (H) Capture efficacy corresponds to the ELISA values for degree of bioconjugation.
Source: Reproduced with permission from Ref. [63].

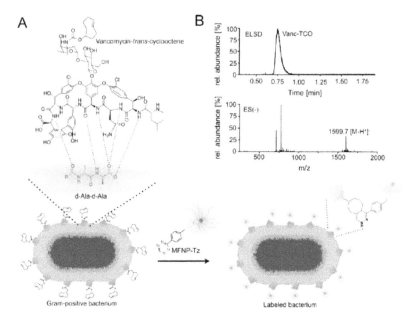

FIGURE 10.12 Chemistry underlying bioorthogonal magnetofluorescent nanoparticle (MFNP) labeling. (A) Vancomycin-transcyclooctene (vanc-TCO) targets Gram-positive bacteria by binding onto their membrane subunits. Following incubation with MFNP-Tz, bacteria are labeled and can be detected via fluorescent or magnetic sensors. (B) HPLC (top) and ESI-MS (bottom) traces of vanc-TCO verifying its identity and purity.
Source: Reproduced with permission from Ref. [77].

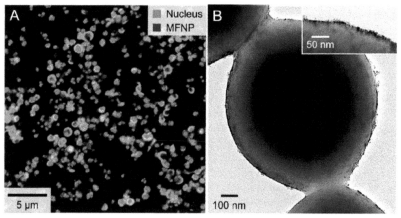

FIGURE 10.13 Bacterial labeling. (A) Confocal microscopy and (B) transmission electron microscopy of *S. aureus* labeled with vanc-TCO and MFNP-Tz. Inset in top right of (B) shows labeling at a higher magnification.
Source: Reproduced with permission from Ref. [77].

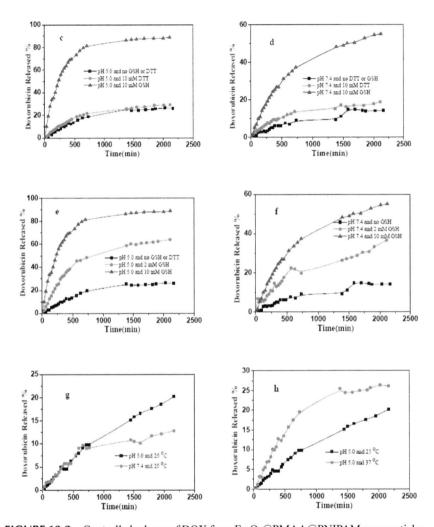

FIGURE 13.2 Controlled release of DOX from Fe$_3$O$_4$@PMAA@PNIPAM nanoparticles under different reductive agent DTT/GSH concentrations and pH values at 37°C: (a) 10 mM DTT and (b) 10 mM GSH; different pH values and reductive agent DTT/GSH concentrations at 37°C: (c) pH 5.0 and (d) pH 7.4; different pH values and GSH concentrations at 37°C: (e) pH 5.0 and (f) pH 7.4; different pH and temperatures (g and h). *Source*: Reprinted with permission from Ref. [24], copyright (2015) American Chemical Society.

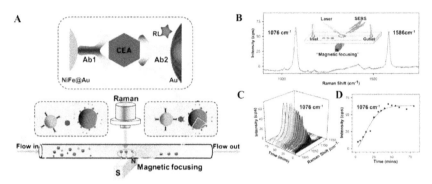

FIGURE 13.3 (a) Illustration of SERS detection of cancer biomarker CEA using functional nanoprobes consisting of Au-coated NiFe magnetic nanoparticle (NiFe@Au), Ab1, capture antibody; Ab2, detection antibody; and RL, Raman label B; (b) illustration of magnetic focusing of NiFe@Au NPs in amicrofluidic channel (~0.5 mm); (c) SERS spectra in 1000–1150 cm^{-1} for an aqueous solution of MBA-labeled NiFe@Au NPs in the channel as a function of time when a magnet was applied to the focal point; (d) kinetic plot of peak intensity at 1076 cm^{-1} (dots). The fitting curve was based on Avrami theoretical model with a rate constant of 0.0075 min$^{-1.6}$. *Source*: Reprinted with permission from Ref. [117], copyright (2015) American Chemical Society.

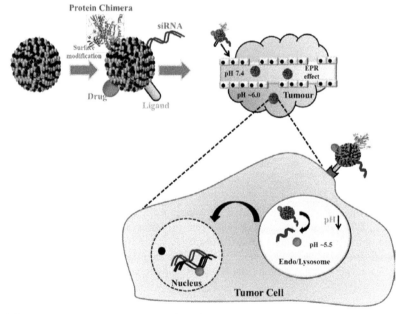

SCHEME 13.4 Core–shell nanoparticle for siRNA and drug delivery based on pH; targeted core–shell nanoparticles deliver siRNA and drug into the tumor site via EPR effect followed by pH-sensitive behavior of delivery.

CHAPTER 9

FORWARD OSMOSIS, NANOFILTRATION, AND CARBON NANOTUBES APPLICATIONS IN WATER TREATMENT: WINDOWS OF INNOVATION AND VISION FOR THE FUTURE

SUKANCHAN PALIT*

43, Judges Bagan, Post Office-Haridevpur, Kolkata 700082, India

*E-mail: sukanchan68@gmail.com, sukanchan92@gmail.com

ABSTRACT

The world of environmental engineering science and water process engineering is today full of avenues for newer scientific regeneration. Membrane science and novel separation processes need to be re-envisioned and deeply re-envisaged with every step of scientific history and time. Nanotechnology has links with diverse branches of science and engineering today. In this chapter, the author deeply focuses on the application of forward osmosis, nanofiltration, and other avenues of nanotechnology to the vast domain of water purification. Water purification encompasses drinking water treatment as well as industrial wastewater treatment. Environmental sustainability and its varied applications are elucidated in great details in this chapter. Human civilization's immense scientific prowess and Mankind's girth and determination will all lead to a long and effective way in the true emancipation and true realization of sustainable development, which includes energy, environmental, social, and economic development. Water treatment and industrial pollution control today stands in the midst of immense vision and deep scientific introspection. The challenges and the vision of water treatment and nanotechnology are discussed in minute details in this chapter. The author also focuses on forward osmosis, which is a new avenue of scientific endeavor. In a nutshell, the author elucidates on the vast application of environmental sustainability to the future of human civilization and human scientific progress. There are other areas of scientific endeavor, such as carbon nanotubes applications in water treatment, which the author has discussed in details. The author stresses upon the application of nanotechnology in water purification in this well-researched treatise.

9.1   INTRODUCTION

Human civilization and human scientific research pursuit in the field of nanotechnology are gearing forward toward newer vision and newer scientific regeneration. Technology and engineering science are high challenges as regards environmental protection, environmental sustainability, and water treatment. This chapter opens up new avenues of research pursuit and scientific genre in the field of membrane science, forward osmosis, and nanofiltration. Novel separation processes and water process engineering are today interlinked. Scientific regeneration and scientific rejuvenation are the cornerstones of research and development initiatives globally. In the similar vision, membrane science and drinking water treatment and industrial wastewater treatment are integrated to each other with an umbilical cord. Technology, engineering, and science today have practically no answers to the ever-growing environmental crisis of arsenic and heavy metal groundwater contamination. Human civilization today stands in the midst of deep scientific divination and vision today. Carbon nanotubes application in water treatment and purification is the absolute need of the hour as science and engineering surge forward toward a newer visionary era. Scientific challenges in the field of nanotechnology applications in water purification are immense and path-breaking. The author in this chapter deeply reiterates the success of carbon nanotubes applications in the true emancipation of science and engineering. This well-researched treatise opens up newer chapters in scientific innovations in the field of forward osmosis, nanofiltration, and nanotechnology applications in water purification, drinking water treatment, and industrial wastewater treatment. The futuristic vision of membrane science, the immense scientific prowess of human civilization, and the visionary innovations in civilization's basic needs will surely open up newer areas of innovation and instinct in the field of environmental protection and environmental sustainability in years to come. The author, in this chapter, deeply comprehends the innovations and recent advances in the field of environmental engineering science and membrane separation tools and opens up new windows of scientific revelation and vast scientific vision.

9.2   THE AIM AND OBJECTIVE OF THIS STUDY

The world of environmental protection today stands in the midst of scientific profundity, divination, and vast vision. The aim and objective of this study are to investigate the recent advances in the field of forward osmosis, nanofiltration, and carbon nanotubes applications in water treatment. This chapter goes beyond that vision and opens up newer areas in the fields of environmental sustainability and environmental protection. Technological vision, scientific validation, and deep scientific ingenuity are the pillars of environmental engineering science today. This treatise focuses on the scientific success, the vast scientific needs, and the scientific forbearance in the field of nanotechnology and nanotubes applications in water purification. The sole aim and objective of this treatise is to focus on scientific issues and deep engineering intricacies in applications of forward osmosis and nanofiltration to water purification. Today millions of people around the world are faced with water shortage and arsenic and heavy metal groundwater contamination. This chapter widely opens the windows of scientific innovation and scientific truth in the field of carbon nanotubes application in water pollution control. Today provision of clean drinking water is an absolute need with the progress of human civilization. The author deeply focuses on the scientific success of forward osmosis and nanofiltration in environmental protection. A deep glance on carbon nanotubes applications in water purification is done with minute details. Today, the technological ardor in the field of nanotechnology is immense and path-breaking. This treatise successfully comprehends the needs of the human society such as water and sustainability. The definition of "sustainability" as explained by Dr. Gro Harlem Brundtland, former Prime Minister of Norway needs to be re-envisioned and restructured as science and technology surge forward. The vision and targets of successful sustainable development as regards energy and environment are vastly discussed in this chapter.

9.3   THE SCIENTIFIC DOCTRINE OF FORWARD OSMOSIS AND NANOFILTRATION

Forward osmosis and nanofiltration are today in the avenues of scientific vision and vast scientific profundity. Water purification science and sustainable development are two opposite sides of the visionary coin. Today globally the needs of human scientific regeneration are immense and path-breaking. Application of forward osmosis and nanofiltration in water treatment are of pivotal importance. In the similar manner, carbon nanotubes applications in water purification and industrial wastewater treatment are facing new orders of the day. Desalination of seawater in developing and developed nations around the world is challenging the vast scientific firmament. Human civilization's immense scientific prowess in the domain of 'sustainability' thus needs to be redefined if the global water shortage issues are to be deeply confronted. Technology and engineering of membrane separation processes thus need to be reframed and overhauled as civilization faces immense challenges and vision today. Provision of clean drinking water is itself an area of vast neglect globally along with global warming. Thus this chapter opens up newer vision and newer futuristic thoughts in years to come.

9.3.1   FORWARD OSMOSIS AND THE NEWER CHALLENGES

Forward osmosis is an osmotic process, which like reverse osmosis, uses a semi-permeable membrane to result in the separation of water from dissolved solutes. The driving force for this separation is an osmotic pressure gradient, such that a high "draw" solution of high-concentration, is used to induce a net flow of water through the membrane into the draw solution thus effectively separating the feed water from its solutes. Technological validation, deep scientific vision, and the futuristic vision of engineering science will all lead a long and visionary way in the true realization of novel separation processes such as forward osmosis and nanofiltration. In contrast to forward osmosis, the reverse osmosis process uses hydraulic pressure as the driving force for separation, which serves to counteract the osmotic pressure gradient. In forward osmosis process, there may be solute diffusion in both directions depending on the concentration of the draw solution and the feed water. Application areas are the processing of energy drinks, desalination, landfill leachate treatment, and feed water pretreatment for thermal desalination. The vision and the challenges of science and technology of membrane separation processes are vast and versatile. This chapter opens up new doors of scientific ingenuity in the field of forward osmosis and other membrane science areas with deep details.

9.3.2   NANOFILTRATION, THE SCIENTIFIC DOCTRINE AND THE VISION AHEAD

Nanofiltration is a relatively recent membrane filtration process used most often with low total dissolved solids water such as surface water and fresh groundwater with the sole purpose of softening and removal of disinfection by-product precursors such as natural organic matter and synthetic organic substances. Nanofiltration is widely used in food processing applications such as dairy, for simultaneous concentration and partial demineralization. Scientific doctrine and scientific vision in the field of nanofiltration and membrane science applications are immense, ground-breaking, and thought-provoking. The world of challenges in novel separation processes, the scientific divination and the futuristic vision of environmental engineering will all lead a long and effective way in the true realization of environmental protection and water purification. Here comes the importance of membrane science and nanofiltration. Technology, engineering, and science of membrane technology are today surpassing vast and versatile scientific frontiers. This environmental engineering tool will surely open newer vision and newer scientific research thoughts in years to come. Nanofiltration is a membrane-filtration based method that uses nanometer-sized through-pores that pass through the membrane. Nanofiltration membranes have pore sizes from 1 to 10 nm, smaller than used in microfiltration and ultrafiltration. Historically, nanofiltration and other membrane technology used for molecular separation were applied entirely on aqueous solutions. Today, nanofiltration and reverse osmosis are in the path of newer scientific regeneration. The original uses of nano-filtration were water treatment and in particular water softening. Now, nanofiltration can be used in diverse areas of science and engineering. In recent years, the use of nanofiltration has been veritably extended to other industries such as milk and juice production. Nanofiltration membranes can be applied in the areas of pharmaceuticals, fine chemicals and flavor, and fragrant industries.

Science and technology in the present century are moving fast surpassing one visionary boundary over another. Technology today has no answers to the intricacies of water science and environmental protection. Scientific impasse and vast scientific provenance in the field of water treat-ment will all lead a visionary way in the true emancipation of scientific truth in environmental engineering and chemical process engineering.

9.4   THE VISION OF APPLICATION OF CARBON NANOTUBES IN WATER TREATMENT

Carbon nanotubes are allotropes of carbon with cylindrical nanostructures. These cylindrical carbon molecules have unusual properties, which are extremely valuable for nanotechnology, electronics, optics, and other domains of material science and technology. Owing to their extraordi-nary thermal conductivity, mechanical and electrical properties, carbon nanotubes find immense applications as additives to various structural materials.

Carbon nanotubes application in water treatment is the newer avenues of scientific girth and scientific determination. Engineered nanomaterials and biomaterials are linked with research endeavor in the field of water treatment. The world today stands tall in the midst of environmental crisis, groundwater heavy metal contamination, and water shortage. Global warming stands as a major hurdle to scientific vision. The forays of science and engineering should be directed toward provision of basic needs such as water, energy, food, shelter, and education. Here comes the vast importance of energy and environmental sustainability. Scientists and engineers around the world are immensely concerned about the efficacy and the importance of carbon nanotubes applications in water treatment. Scientific subtleties and scientific redefinition will enhance the research endeavor in the field of environmental protection and water purification.

In this chapter, the author deeply targets the scientific needs and the scientific ingenuity in the field of nanotechnology as well as environmental engineering. The definite and purposeful vision of the carbon nanotubes application in water treatment is vast and versatile. Human scientific and academic rigor in the field of environmental engineering is of highest order today as civilization moves forward. Today in the 21st century, the status of provision of basic human needs is at a dismal state. Sustainability application in human society in the similar manner is highly challenged. Thus the question of nanotechnology and nanoscience emancipation comes to the future. This well-researched treatise will surely open a newer dawn of scientific thoughts and scientific redeeming in decades to come.

9.5   GLOBAL WATER CRISIS, MEMBRANE SCIENCE, AND THE VISION OF WATER PURIFICATION

Science and technology are today huge colossus with a vast and definite vision of its own. The vast domain of membrane science today stands in the midst of vision and deep scientific comprehension. Global water scarcity is ever-growing and needs to be redefined with the progress of global engineering and technology. Environmental and energy sustainability in the similar vein is in the avenues of vision and deep scientific revelation. Globally membrane science is seen as a major environmental engineering tool with immense scientific potential. The contribution and participation of civil society, the futuristic vision of water purification, and the needs of scientific innovation will surely lead a long and effective way in the true realization of environmental engineering today. Human scientific progress in environmental protection is highly stressed and challenged. This chapter veritably opens up newer thoughts and newer vision in the vast and intricate domain of environmental protection and membrane science in years to come. Heavy metal groundwater contamination is changing the face of human scientific progress today. Human Mankind's vast scientific ingenuity and profundity are the imminent needs today. The author with immense scientific lucidity depicts the needs of human society such as water purification, sustainability, and nanotechnology. The immense and visionary marvel of science that is nanotechnology is veritably linked with diverse areas. This chapter reiterates and stresses the scientific vision behind nanotechnology as well as nanofiltration. Science and engineering will thus usher in a newer era of might and determination in years to come.

9.6   RECENT SCIENTIFIC ADVANCES IN FORWARD OSMOSIS

Forward osmosis and membrane science are today in the path of newer vision and scientific regeneration. Technology and engineering of membrane science are highly challenged and needs to be re-envisioned and reframed as science and engineering surges forward. Water purification, drinking water treatment, and industrial wastewater treatment are today in the midst of vision, scientific fortitude, and deep scientific revelation. In this section, the author deeply comprehends the recent scientific research in the field of forward osmosis which is a frontier area in the field of membrane science.

Akther et al.[1] discussed with deep scientific foresight recent advancements in forward osmosis desalination. Technological and scientific ardor, the futuristic vision of environmental protection, and the immense challenges of science are the torchbearers toward a newer era in the field of desalination. Desalination science is veritably revolutionizing the scientific firmament of environmental protection.[1] Forward osmosis is one of the evolving membrane technologies in desalination science with recent scientific thrust as a low energy process. The most important parts of forward osmosis are the membrane and draw solution since both of them play a pivotal part in its performance. This review targets the recent progress in forward osmosis, aiming at the prospects and vast challenges. The critical part of the review is a thorough discussion of hybrid forward osmosis systems, different forward osmosis membranes and draw solutes available coupled with their effects on forward osmosis performance.[1] The world of challenges and the scientific difficulties in membrane separation processes are immense, thought-provoking, and visionary. The other pivots of this well-researched treatise are the area of sustainable desalination. Human factors engineering and environmental engineering science are today linked veritably by an umbilical cord.[1] Desalination and environmental sustainability in the similar manner are also linked to each other. The world population has continued to grow exponentially at a drastic rate.[1] Human civilization is thus at a stake and research endeavor in science and engineering needs to be re-envisioned and re-organized with the passage of scientific history and time. The authors in this chapter discussed with deep scientific ingenuity advantages and potential uses of forward osmosis, major problems associated with forward osmosis, membrane fouling, and the vast world of sustainability in seawater desalination.[1] Forward osmosis desalination processes have a medley of prospective benefits if they are deployed on a massive scale. Technological divination, scientific vision, and vast scientific profundity are the pillars of forward osmosis applications in water science and desalination science. This is a watershed text and will open up new windows of innovation and scientific instinct.[1]

Alsvik et al.[2] discussed with immense lucidity and foresight pressure retarded osmosis and forward osmosis membranes. In the past four decades, membrane development and its scientific emancipation have occurred based on the demand in pressure driven processes. However, in the last decade, the interest in osmotically driven processes such as forward osmosis and pressure retarded osmosis has increased and moved toward newer scientific regeneration.[2] Recently, several promising membrane preparation methods for forward osmosis/pressure retarded osmosis applications have emerged. Preparation and fabrication of thin film composite membranes with customized polysulfone support, electrospun support, TFC membranes on hydrophilic support, and hollow fiber membranes have been deeply investigated for forward osmosis/pressure retarded osmosis applications.[2] Human civilization, human scientific progress, and the vast world of human factor engineering are the necessities of innovation and scientific instinct today.[2] This chapter elucidates on the scientific success, the scientific provenance, and the scientific redeeming in the field of pressure retarded osmosis and forward osmosis.

McCutcheon et al.[3] deeply depicted with profound literary impact forward osmosis and its vast applications. For decades, aqueous separations have veritably relied on hydraulic pressure to force water across membranes that retain suspended and dissolved solids. This pressure is generated by pumps that can veritably require a large amount of electricity and energy. This can have large drawbacks, including increased carbon footprint and high cost as energy prices continue to escalate. Here comes the importance of engineered osmosis or salinity-driven osmosis.[3] Technological vision, deep scientific motivation, and the scientific need of the human society will be the future torchbearers toward a newer visionary era in the field of membrane science and environmental protection. This chapter widens the scientific views and the scientific subtleties in the field of osmosis. The authors in this chapter discussed with immense scientific far-sightedness forward osmosis, direct osmosis concentration, osmotic dilution, pressure-retarded osmosis, and the vast world of membrane transport and concentration boundary layer. Research endeavor in the field of forward osmosis as human civilization progress is in the process of scientific rejuvenation and deep scientific revelation. This chapter will surely open up newer scientific issues and deep scientific intricacies in membrane separation techniques in years to come.[3]

Kochanov[4] discussed and described with deep scientific conscience and immense lucidity forward osmosis for seawater desalination and along with membrane development and process engineering. Chemical process engineering, environmental engineering, and environmental protection are linked with each other in today's world of science and engineering.[4] Scientific might, deep scientific grit, and determination are the pillars of this doctoral thesis.[4] Today forward osmosis is attracting immense scientific interest and this interest has enabled science and engineering to move forward at a drastic pace. Forward osmosis has been suggested as viable separation process, with numerous potential applications—from osmotic drug delivery and concentration of liquid foods to water purification and re-use in space.[4] The futuristic vision and marvels of science and engineering, the scientific success, and the deep scientific ingenuity are the veritable pillars of this thesis. The thesis focused on the development of membranes for water desalination by forward osmosis and estimation of the requirements for seawater desalination by a two-stage membrane based forward osmosis and nanofiltration process.[4] Science and engineering of membrane separation processes and forward osmosis are today in the process of newer scientific regeneration in spite of the immense difficulties and scientific travesty. In this thesis, the author deeply pronounces the intense and immediate necessity of a scientific deliberation in the field of membrane fouling, membrane development, and process technology.[4] The author deeply touched upon integrally skinned asymmetric membranes for forward osmosis, thin-film composite membranes on polymeric and defined ceramic supports, membrane fouling in osmosis and other membrane separation processes, a detailed investigation in the field of seawater desalination and the challenges and the vision of membrane technology applications in diverse areas of scientific endeavor.[4]

Human civilization and human scientific research pursuit are in the midst of deep scientific revelation and vast scientific grit and determination.

Nanotechnology applications in water treatment and carbon nanotubes applications in vast areas of scientific research pursuit are the pivots of research endeavor today. Scientists, engineers, governments, and civil society are today in the midst of deep scientific introspection and deep scientific overhauling. This chapter will surely open up newer thoughts and newer directions in the field of non-conventional separation techniques such as membrane science.

9.7   RECENT SCIENTIFIC FORAYS IN NANOFILTRATION

Technology and engineering science of nanotechnology today are in the crucial juncture of deep scientific comprehension and scientific vision. Global water shortage has urged the scientific domain to delve deep into the scientific intricacies and scientific difficulties of membrane separation processes. Nanofiltration and other branches of nanotechnology are the only solutions to the ever-growing global water crisis. In this section, the author deeply portrays the recent scientific advancements in the field of nanofiltration, nanotechnology, and other areas of membrane science. Today, nanofiltration and membrane science and global water research and development initiatives are two opposite sides of the visionary coin.

Hilal et al.[5] deeply discussed with immense scientific vision and lucidity the current state of play for cost-effective water treatment by membranes. This article presents a wider perspective on the recent development and application of membranes for the treatment of water. The authors discussed in details how membranes contribute immensely to the global challenge of sustainable supply of clean and pure drinking water. The primary theme is on desalination and how innovative science and technology has been effectively applied. The techniques such as advanced membrane materials, biomimetic membranes, hybrid systems, forward osmosis, and membrane distillation are discussed and described in details.[5] The authors in this chapter deeply discuss the challenges facing membranes, advanced membrane technologies, and the vast scientific emancipation in the field of membrane fouling. The energy efficient sustainable treatment of water is one of the key problems of the 21st century and with current global population increase, this problem is going to aggravate more.[5] According to the United Nations, there are over 748 million people that do not have access to clean drinking water and water demand from industry is expected to increase by 400% between 2000 and 2050 in the global scenario.[5] Desalination is drastically increasing the number of water resources that are available for the production of pure drinking water and membrane desalination and thermal processes are the most common techniques. Thermal processes require a large amount of energy to achieve evaporation and condensation of water.[5] In the recent years, the energy needed for desalination has decreased a lot. This is due to the development of more efficient pumps, energy recovery devices, process configuration, and membranes. In comparison to thermal processes, membrane desalination technologies have the immense advantage of comparatively low energy requirements and they are based on processes such as reverse osmosis, forward osmosis, electrodialysis, and nanofiltration that desalinate water by rejecting salt at a membrane barrier.[5] Technology and engineering of nanotechnology and membrane science today need to be revamped and re-organized as global water shortage destroys the vast scientific firmament. The authors with deep scientific revelation discussed the overall state of membrane applications in water treatment, drinking water treatment, and industrial wastewater treatment.

Chorawala et al.[6] discussed and described with cogent insight and immense lucidity applications of nanotechnology in wastewater treatment. Wastewater treatment is an ever-growing scientific issue today and is today surpassing vast and versatile scientific frontiers. Nanoparticles have a great potential to be used in water and wastewater treatment.[6] Some of the genuine and unique characteristics of it having high surface area are that they have the potential to remove toxic metal ions, disease containing microbes, inorganic, and organic solutes from water. Technology needs to be diversified and reorganized as regards environmental protection and water pollution control. This review includes recent developments in nanotechnology for water treatment and industrial wastewater treatment. This chapter covers nanomaterials that enable the applications, advantages, and limitations as compared to the existing processes.[6] In terms of wastewater treatment, nanotechnology science is applicable in detection and removal of different pollutants. Heavy metal pollutants pose a serious threat to environment because it is toxic to living organisms, including humans and not biodegradable.[6] Technology and engineering science are highly stressed and challenged as human civilization and human scientific endeavor moves forward.[6] Various methods such as photocatalysis, nanofiltration, adsorption, and electrochemical oxidation involve the use of titanium dioxide, zinc oxide, ceramic membranes, nanowire membranes, polymer membranes, carbon nanotubes, and other engineered nanomaterials are effectively used to solve water treatment issues. The authors deeply discussed nano-sorbents, nanocatalysts, nanostructured catalytic membranes, catalytic wet air oxidation, nanofibers, and membrane filtration technology.[6] The other areas of scientific deliberation are the effects of nanotechnology in human society.

Kunduru et al.[7] discussed with immense scientific farsightedness applications of nanotechnology methods in wastewater treatment. Technological validation, vast scientific motivation, and the futuristic vision of environmental protection will all lead a long and visionary way in the true realization of environmental engineering and process technology.[7] Water is the most powerful asset of human civilization and human scientific progress. This water problem will increase with time and thus concerted scientific efforts are needed with immense scientific re-envisioning. The authors discussed in details major limitations associated with conventional water purification methods and give an incisive overview of different nanomaterials in water and wastewater treatment. The authors with immense scientific vision describe metal-based nanoadsorbents, carbon-based nanoadsorbents, and polymeric nanoadsorbents.[7] The other areas of scientific research pursuit are membranes and membrane processes, photocatalysis, antimicrobial nanomaterials in disinfection, and microbial control.[7] The areas of nano-antimicrobial polymers are discussed in details in this chapter.[7]

Technological and scientific validations are the cornerstones of scientific vision and scientific determination today. The contribution of nanomaterials and engineered nanomaterials in water treatment are today ushering in a newer era in the domain of environmental protection. The challenges and the vision need to be re-envisaged and revamped as science and engineering moves forward. The author in this chapter deeply discusses and pointedly focuses on the success, the intricacies, and the challenges of nanotechnology applications in water treatment.

9.8   SCIENTIFIC ADVANCES IN CARBON NANOTUBES APPLICATIONS IN WATER PURIFICATION

Scientific research and development initiatives globally are today in the midst of deep scientific vision and scientific profundity. Nanotechnology applications in water pollution control in the similar vein need to be over-hauled with the passage of scientific history and the visionary timeframe. Drinking water treatment and industrial wastewater treatment are the visionary challenges of human scientific endeavor in modern civilization. There are immense needs of nanomaterials, engineered nanomaterials and carbon nanotubes for the future of human civilization. In this section, the author pointedly focuses on some of the recent advances in the application areas of carbon nanotubes.

Pandey et al.[8] deeply discussed with immense scientific comprehension types, methods, and applications of carbon nanotubes. Carbon nanotubes are nanostructures derived from rolled graphene planes and have various interesting chemical and physical properties.[8] Carbon nanotubes can be veritably conjugated with various biological macromolecules including drugs, proteins, and nucleic acid to increase the bio-functionalities. Carbon nanotubes exist as single and multi-walled structures. They present impor-tant properties such as high aspect ratios, ultra-lightweight, strength, high thermal conductivity, and electronic properties ranging from metallic to semi-conducting. Today, technological advancements in carbon nanotubes and nanotechnology are in the path of newer vision and newer scientific introspection. This chapter opens up newer thoughts and newer future directions in the field of carbon nanotubes.[8] This review leads to a newer knowledge dimension related to general overview, types, preparation methods, and application avenues of carbon nanotubes.[8]

Danish Environmental Protection Agency Report[9] discussed with deep and cogent insight carbon nanotubes and its types, products, markets, and provisional assessment of the associated risks to man and the environment. This environmental project discusses types and characteristics of nano-tubes, main current, and future applications of carbon nanotubes, exposure to man and the environment, human health effects of carbon nanotubes, ecotoxicology, and provisional integrated risk assessment.[9] The aims of this report are to give a detailed introduction into the physico-chemical complexity of carbon nanotubes, to give an overview of the production volumes, to provide an overview of the current and near market carbon nanotubes based down-stream and consumer products, and to perform a preliminary risk assessment based on current knowledge on carbon nanotubes.[9] The report deeply comprehends structure and physico-chemical characteristics of carbon nanotubes, synthesis, purification, coating, and functionalization of carbon nanotubes, characteristics of commercial carbon nanotubes and the vast application areas of carbon nanotubes. The other pillars of this treatise are the human environmental exposure of the carbon nanotubes.[9] Environmental engineering science and environmental protection science are in the path of newer scientific regeneration and vision today. Carbon nanotubes can solve immense and intricate problems in water treatment. This report is a watershed text in the field of applications of carbon nanotubes in industrial wastewater treatment and opens newer visionary thoughts in decades to come.[9]

Das et al.[10] discussed and described with immense vision and lucidity the present, past, and future of multifunctional carbon nanotubes in water treatment. Human scientific vision, the vast technological challenges and the futuristic vision of nanotechnology will surely contribute immensely toward the scientific success and the scientific truth of environmental protection.[10] The availability of safe and clean drinking water is decreasing day by day, which is expected to escalate in the coming decades. In order to address these scientific issues, various water purification technologies are adopted. Amongst the water treatment tools, carbon nanotube-based water treatment technologies are immensely promising and visionary.[10] The main reason behind the application areas of carbon nanotubes is the large surface area, high aspect ratio, greater chemical reactivity, lower cost and energy, less chemical mass, and veritable impact on the environment. This review deeply discussed most of the effective carbon nanotube based water purification technologies such as adsorption, hybrid catalysis, desalination, disinfection, sensing, and monitoring of three major classes of industrial pollutants such as organic, inorganic, and biological pollutants.[10] Today human factor engineering and environmental engineering are bound together with immense scientific vision and might. This chapter deeply reviews the scientific needs of application of carbon nanotubes in industrial wastewater treatment.[10]

The challenges of commercialization of carbon nanotubes in water treatment are immense and visionary today. The industrial applications depend on the raw material's costs and quality. Today water purification, drinking water treatment, and industrial wastewater treatment stands in the crucial juncture of vision, might, and scientific determination. Technological and scientific validations are the utmost needs of the hour. This chapter depicts profoundly the needs of science and technology, the futuristic vision, and the immense scientific ingenuity in carbon nanotubes applications in modern science.

9.8.1 INTEGRATED WATER RESOURCE MANAGEMENT, HUMAN FACTOR ENGINEERING, AND NANOTECHNOLOGY

The world of science and engineering are witnessing immense scientific challenges and unending barriers. Integrated water resource management and human factor engineering will be the two opposite sides of the visionary scientific coin. In this chapter, the author stresses and reiterates the contribution of human factor engineering to human society and modern science. Provision of clean drinking water is at a dismal state. Water engineering and water resource management are today in the path of newer vision and regeneration. The needs of pure drinking water in developing countries around the world are immense today. Thus, there is need of integrated water resource management and the vast domain of human factor engineering. In the similar vein, civil society participation and the contribution of governments are the necessities of modern science and human scientific progress. Nanotechnology and its applications will surely help in greater scientific understanding and larger scientific redeeming of modern science and technology.

9.9 APPLICATION OF NANOTECHNOLOGY IN WATER PURIFICATION, DRINKING WATER TREATMENT, AND INDUSTRIAL WASTEWATER TREATMENT

Diverse applications of nanotechnology in water purification and drinking water treatment are challenging the vast global scientific fabric. Scientific redeeming, scientific provenance and vast scientific vision are the pillars of research pursuit in water purification today. In this chapter, the author deeply elucidates the immense scientific success, the vast scientific ingenuity and the technological vision behind environmental engineering processes and techniques. Industrial wastewater treatment and the tools of environmental protection in the similar vein need to be re-envisioned and re-envisaged. The success of human civilization and human scientific progress in the similar vein lies in the hands of scientists and engineers globally. This chapter widens the scientific thoughts and the scientific vision in the field of nanofiltration applications, desalination applications, and environmental protection tools. The world of science and engineering today stands in the midst of vision and scientific forbearance. Drinking water treatment is the utmost need of the hour. The author with immense scientific conscience depicts the necessity of traditional as well as non-traditional environmental engineering techniques. Human scientific vision and endeavor are highly stressed today. Provision of clean drinking water stands as an important scientific vision and deep scientific forbearance. Globally, the divide between advantaged and the disadvantaged is extremely high. The visionary definition of "sustainability" by Dr. Gro Harlem Brundtland, former Prime Minister of Norway needs to be re-envisaged and restructured with the passage of human history and time. Nanotechnology is the only answer to the ever-growing concerns for groundwater remediation. Technological redefining, scientific profundity, and the futuristic vision of engineering will thus lead civilization to a more visionary era.

9.10 THE VISION AND THE CHALLENGES OF HEAVY METAL GROUNDWATER REMEDIATION

Groundwater remediation is the necessity of human civilization and human scientific research pursuit today. Membrane science, forward osmosis, and nanofiltration are today challenging the vast scientific fabric of environmental engineering and chemical engineering in particular. In developed and developing nations around the world, the concerns of drinking water heavy metal contamination are veritably challenging the scientific landscape and the vast scientific vision of engineering science. Human scientific regeneration, the vast scientific divination and the futuristic vision of drinking water treatment will surely be the torchbearers toward a newer era in the field of environmental engineering and chemical process engineering. Heavy metal groundwater remediation today stands in the midst of vision and scientific redemption. Nanotechnology and membrane science are the needs of scientific vision in the field of environmental engineering science today. In the Asian countries of India and Bangladesh, arsenic groundwater pollution is changing the vast scientific landscape. Millions of citizens around the world are facing disastrous health effects due to water shortage and groundwater contamination, thus the need for a well-researched treatise in the field of membrane science and drinking water treatment. The authors in this chapter deeply comprehend the scientific necessity, the scientific needs, and the definite vision in the field of drinking water treatment and environmental engineering techniques. Human scientific vision thus needs to be re-envisaged and re-organized with the passage of scientific history and time. This chapter will surely open up newer scientific vision and newer scientific ingenuity in the field of environmental protection in years to come.

9.11   ARSENIC GROUNDWATER REMEDIATION AND THE VISION FOR THE FUTURE

Water pollution control and water purification are the needs of human civilization and human scientific progress today. Arsenic and heavy metal groundwater contamination are a veritable burden and a bane to human civilization today. The vision for the future needs to be reframed in the field of drinking water treatment and industrial wastewater treatment as science, engineering, and technology surges forward. The vision and the challenges of human scientific progress are immense and path-breaking today. Provision of pure drinking water is a veritable need of human civilization today. In South Asia, particularly India and Bangladesh, the water crisis is ever-growing and needs immense attention. Global water shortage is today wrecking the futuristic vision of global chemical process engineering scenario. Thus, the immediate need of novel separation processes, membrane science, and other traditional and non-traditional environmental engineering separation processes. Human scientific thoughts, human scientific rigor, and human academic rigor are today in the state of immense disaster as provision of pure drinking water stands in the midst of deep scientific comprehension. Human Mankind's immense scientific prowess, the scientific rejuvenation, and the vast technological challenges will all lead a long and visionary way in the true emancipation of environmental engineering science today. The world, today, stands transfixed and mesmerized as regards environmental engineering applications to human society and growing concerns of global warming and drinking water shortage. Here comes the importance of a deep scientific deliberation and scientific discussion in the field of chemical process engineering tools such as membrane science. This chapter will surely open up newer visionary avenues in the field of water purification, drinking water treatment, and industrial wastewater treatment. The challenges and the vision of engineering science applications to human society are deeply deliberated with vision and scientific might in this treatise. Human scientific challenges will thus surely be re-envisioned and re-framed as regards protection of environment in near future.

9.12   FUTURE RESEARCH TRENDS AND FUTURE FLOW OF SCIENTIFIC THOUGHTS

Membrane science and industrial wastewater treatment are today linked together by an unsevered umbilical cord. Environmental and energy sustainability are the pivots of human scientific progress today. Technology and engineering science have few answers to the ever-growing concerns of arsenic and heavy metal groundwater contamination. The status of environmental engineering science is extremely dismal today as science and engineering move forward. Future research trends should be targeted toward more visionary scientific innovations and research and development initiatives. Future scientific thoughts in the similar manner need to re-envisioned toward more innovations in membrane science, drinking water treatment, and industrial wastewater treatment. Today science and engineering are surpassing every scientific frontier. The provision of basic human needs such as water, food, shelter, and electricity are highly challenged today and needs to be reframed and re-organized as science and engineering move forward toward newer knowledge dimensions. Energy sustainability, the application of renewable energy initiatives, and the vast vision of energy security are the other future trends of research today. This chapter opens up newer future thoughts and future vision in the field of renewable energy research and development initiatives and also research forays in water process engineering and novel separation processes. Human civilization today stands in the midst of deep vision and scientific contemplation. The necessity of research forays in this critical juncture of human history and time is vision, scientific introspection, and scientific divination. Thus, technology will usher in a new era in the field of chemical process engineering, membrane science, and water technology. Human scientific and academic rigor in the similar vein will also open up newer windows in the field of science of environmental sustainability. The targets of future endeavor should be toward more scientific emancipation in the field of traditional and nontraditional environmental engineering tools.

9.13   CONCLUSION AND VAST SCIENTIFIC PERSPECTIVES

The world of environmental engineering science and human scientific endeavor today stands in the midst of deep scientific vision and contemplation. Water science and technology in the similar vein are in the midst of deep scientific catastrophe. Heavy metal groundwater remediation is today challenging the vast scientific firmament of might, determination, and girth. Scientific perspectives in the field of water purification need to be re-envisioned and reframed with the passage of scientific history and time. The challenges and vision of water treatment and environmental engineering tools thus need to be re-envisaged as science and engineering surges forward toward a newer visionary era. Today, water process engineering, nanotechnology, and novel separation processes are revolutionizing the scientific landscape. In the similar vein, environmental sustainability needs to be targeted with the active participation of civil society, environmental engineers, and scientists. The world today stands mesmerized with the research and development forays in the field of nuclear science and space technology. The vast scientific perspectives in research and development initiatives in environmental engineering thus need to be changed and re-organized. This is the need of environmental sustainability and research initiatives in traditional and non-traditional areas of environmental protection. The authors with immense vision and scientific conscience elucidate on the scientific success and the deep scientific ingenuity in the field of membrane science applications in environmental protection. The needs, the vision, and the scientific profundity behind forward osmosis and nanofiltration applications are re-envisioned in this chapter in minute details. Thus, the success and vision of water science and water process engineering can be emancipated in deep details. Human civilization will thus be re-envisioned and reorganized as regards sustainability.

KEYWORDS

- **environment**
- **forward**
- **osmosis**
- **nanofiltration**
- **vision**
- **nanotechnology**

REFERENCES

1. Akther, N.; Sodiq, A.; Giwa, A.; Daer, S.; Arafat, H. A.; Hasan, S. W. Recent Advancements in Forward Osmosis Desalination: A Review. *Chem. Eng. J.* **2015,** *281,* 502–522.
2. Alsvik, I. L.; Hagg, M. B. Pressure Retarded Osmosis and Forward Osmosis Membranes: Materials and Methods. *Polymers* **2013,** *5,* 303–327.
3. McCutcheon, J.; Huang, L. Forward Osmosis, Encyclopedia of Membrane Science and Technology; Erik, M.; Hoek, V.; Tarabara, V. V., Eds.; John Wiley and Sons, Inc.: USA, 2013.
4. Kochanov, R. Forward Osmosis for Seawater Desalination: Membrane Development and Process Engineering, Doctor of Philosophy Thesis, Imperial College, United Kingdom, 2014.
5. Hilal, N.; Wright, C. J. Exploring the Current State of Play for Cost-effective Water Treatment by Membranes *Nat. Publ. J. (Clean Water)*, **2018,** 1–8.
6. Chorawala, K. K.; Mehta, M. J. Applications of Nanotechnology in Wastewater Treatment. *Int. J. Innov. Emerg. Res. Eng.* **2015,** *2* (1), 21–26.
7. Kunduru, K. R.; Nazarkovsky, M.; Farah, S.; Pawar, R. P.; Basu, A.; Domb, A. J. Nanotechnology for Water Purification: Applications of Nanotechnology Methods in Wastewater Treatment, (Chapter 2), In *Water Purification*; Grumezescu, A., Ed.; Elsevier/Academic Press: USA, 2017; pp 33–74.
8. Pandey, P.; Dahiya, M. Carbon Nanotubes: Types, Methods of Preparation and Applications. *Int. J. Pharm. Sci. Res.* **2006,** *1* (4), 15–21.
9. The Danish Environmental Protection Agency Report. Carbon Nanotubes: Types, Products, Market and Provisional Assessment of the Associated Risks to Man and the Environment, Environmental Project No. 1805, 2015.
10. Das, R.; Hamid, S. B. A.; Ali, M. E.; Ismail, A. F.; Annuar, M. S. M.; Ramakrishna, S. Multifunctional Carbon Nanotubes in Water Treatment: The Present, Past and Future. *Desalination* **2014,** *354,* 160–179.

CHAPTER 10

RESEARCH PROGRESS IN SYNTHESIZATION, COATING, AND CHARACTERIZATION OF MAGNETIC NANOPARTICLES

LAVANYA TANDON and POONAM KHULLAR[*]

Department of Chemistry, BBK DAV College for Women, Amritsar 143001, Punjab, India

[*]*Corresponding author. E-mail: virgo16sep2005@gmail.com*

ABSTRACT

Water contamination is a persistent problem for public health. Protection against pathogens requires disinfectants; however, the combination of increased disinfectant doses and multiple disinfectants leads to the production of harmful byproducts. Due to prevalent use of antibacterial agents, antibiotic-resistant bacteria have emerged; thus, antibacterial agents and therapeutics treat the bacterial infections. The impact of nanoparticles on bacteria helps us to gain an understanding of the impact of nanoparticles on the mechanistic understanding of toxicity and can guide us to design the rules to create a safe environment. Magnetic particles are used in a vast variety of biomedical and engineering disciplines. But there are limitations pertaining to tunable shapes and sizes. An important impact in many different fields is due to plasmonic hybrid nanostructures.

10.1   INTRODUCTION

The research dealing with bionanotechnology targets on functional structures which synergistically combines a wide range of nanotechnology with macromolecules, cells, or multicellular assemblies. The use of cultured microorganisms has been expanded as these provide micrometer-sized cells with tiny nanodevices. For materials designing, a nanoarchitectural strategy which has been inspired with biomimetic has been introduced by Toyoki Kunitake in his work on lipid tubes and proteins architecture and this has directed the study in biological cell nanoencapsulation.[1,2] The nanomodification of living cells has opened the new pathway to control the properties of cells. Cells which are in size of few micrometers are large as compared to those with nanometer particles. Efforts are being made to develop "smart dress" for cells so that cells are protected from the external environment and it also serves as the useful kit which microorganisms can use for their benefit, for example, magnetic field allows cell manipulations. The coatings of "smart dress" are used for determining the fundamental biological properties, like enzyme activity, membrane permeability,[3,4] and viability preservation in the cells.[5] These approaches are also used manufacturing of advanced materials, like fabrication of microcapsules carriers[6,7] or smart polymer nanocomposites.[8] The functionalization of biological cells through direct deposition of nanoparticles as artificial shells which have predesigned and controllable architecture follows the chemical methodology known as the cell surface engineering.[5,9] Lipid bilayer membranes membrane covers the microbial cells and the protective coating called cell wall is developed. It enforces the membrane and the cell becomes resistant to external impacts. The goal of the cell surface engineering is to develop nanocoating on the surface of the cell through chemical engineering nano-assembly and without modifying the genome of the cells. Earlier, the nano-layers were formed over the sacrificial cells through the use of, the layer-by-layer (LbL) deposition technique. But a number of other approaches emerged which offered new avenues in the cell surface engineering. Through surface engineering, the cell wall gets functionalizes with polymer layers and/or nanosized particles and are used to modify the intrinsic properties of microbial cells.[10–14] The live microbial cells can be fabricated with magnetic nanoparticles (MNPs) onto cell walls through the cell encapsulation and it mimics the natural magnetotactic bacteria.[15] Artificial magnetic cells are being used in electrochemical[18] and microfluidic biosensors,[16] and also used for various purposes dealing with selective concentration and isolation of microbial cells.[17] The hydrocarbonoclastic bacteria treated with artificially added magnetic functionality possess a great practical use. The oil spills that resulted during the production of petroleum from natural deposits should be promptly eliminated so as to save the environment. The hydrocarbon degrading microorganisms can be used to eliminate the marine oil spills.[22] Many hydrocarbonoclastic microbial species have been identified as bioremediation means for environmental oil spills.[18,19] The microorganism, like alcanivorax and borkumensis marine bacteria, exhibits good crude oil degradation rates.[20–38] This microorganism has been studied in detail, the genome has been sequenced,[29] the description of proteomic profiles have been provided, and also it possessed the ability to stabilize oil droplets in oil-in-water emulsions which are due to its interaction with surfactant decorated oil–water interface.[30]

Polycyclic aromatic hydrocarbon (PAH) stabilized magnetic nanoparticles (PAH-MNPs) were prepared and were coated onto the *A. borkumensis*. The deposition is shown in the given Figure 10.1.

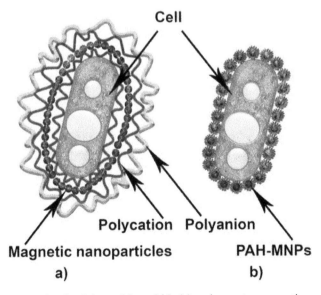

FIGURE 10.1   A sketch of the multistep LbL (a) and one step magnetic nanoparticles complexed with polyallylamine (PAH-MNP) deposition on the bacterial cell (b).
Source: Reproduced with permission from Ref. [31]. © 2016 American Chemical Society.

The cell surface engineering with polycation-coated MNPs is a fast process in which the positively charged iron oxide nanoparticles are deposited onto the surface of microbial cells during the incubation in the presence of excessive concentrations of nanoparticles. The cationic PAH-stabilized superparamagnetic nanoparticles are used to cover the suspended *A. borkumensis* cells. The cells have exhibited a negative ζ-potential of -16 mV, due to which the cationic magnetic complexes get deposited fast on bacterial cell walls. After coating the ζ-potential of the bacteria was found to be 39 mV. After the modification, the cells were made to wash with artificial seawater media so that the loosely attached particles could be removed and the osmotic rupture of cell membranes could be prevented.

It can be seen in Figure 10.2 that the original smooth cell surface has been covered with a uniform layer of aggregated nanoparticles. Under

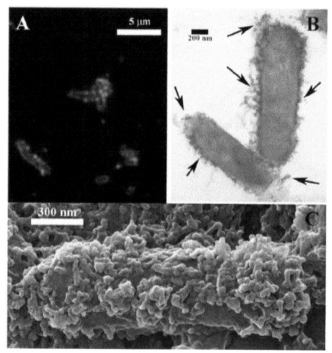

FIGURE 10.2 Magnetically modified *A. borkumensis* bacteria imaged by dark-field (whole-cells in aqueous media) (A), transmission electron (ultrathin resin-embedded slices) (B), and scanning electron wholecells (C) microscopies. Arrows (B) indicate the aggregated polymercoacervates of MNPs.
Source: Reproduced with permission from Ref. [31]. © 2016 American Chemical Society.

native conditions, through dark-field microscopy, the nanoparticles on cell walls could be visualized as bright spots covering the cells. The thin slices of resin-embedded magnetically modified *A. borkumensis* bacteria have been imaged using TEM. It has been cleared from TEM that the cell wall structures were not pierced by the nanoparticles and it traveled into the cytoplasm. In general, bacteria do not employ endocytosis during uptake the external particles. It has been represented that the iron-oxide (20 nm) nanoparticles used in this study coacervated with PAH positively charged hydrogel that has readily assembled on the oppositely charged cells walls. SEM study has confirmed that the coating has been very dense. By reducing the concentration of nanoparticles in coating suspension, the density of coating can be regulated.[31] This is a fast approach and does not require the use of biofilms to deposit the nanoparticles into the matrix.

10.2 FUNCTIONALIZATION OF NANOPARTICLES

Laboratory experiments have a goal in which nanotechnology is to be introduced at the high school in which students should be asked to prepare projects in which synthesis of nanoparticles, functionalization of the nanoparticles, and nanoparticles impact on the environment should be included into the curriculum.

Nanotechnology is one of the multidisciplinary fields or we can say the subject in which the manipulation of the matter is involved at the Nanoscale level and also it encompasses the field of engineering, physics, biology, and chemistry.[32] Nano-sized objects exploration has an excellent potential in generating the enthusiasm of students. The various size regimes like millimeter-sized sample areas, micro-organisms; nanoparticles can be integrated within the project that has been engaged. Through this project, experiments can be used to understand the scale and also provides the opportunity so that the awareness of the technology can be raised in the field of magnetic memory and biomedical applications.[33] The experimental protocol that has been developed, and is being used in laboratory experiments and is dealing in interdisciplinary fields involves students of high school setting and university research laboratories. Sometimes the incorporation of higher research topics into the material education becomes difficult. In this report, the outlined project gives exciting and attention seeking methods through which the subjects can be introduced into the lessons which focus on the micro-organisms and the technology.

Nanomaterials deals with the particles and the matter type whose dimensions are in the range of 1–100 nm. These can be founded in everyday items like strain free clothing, strained glasses make-up and medicines.[34,35] The revolution in nanotechnology has occurred because of the different physical properties of the materials which are due to their high surface to volume ratio and small size.

Examples may include, nanosilver or the properties of superparamagnetism that is, special magnetic properties of the MNPs of the oxides of the iron. A body of research has been explored where MNPs are used as a component of the carriers for drug delivery, biosensors as contrasting agents. When these particles are used as the contrasting agents, then the iron oxide particles are coated with organic molecules or polymers and are regarded as biocompatible. Mostly most of the materials of nanoparticles of magnet which are used in the biomedical application that is, in the field of theranostics have been coated for the easy functionalization and biocompatibility. MNPs possess the properties of the magnet and these possess such abilities that the particles can be activated using an external oscillating magnetic field. Due to this, the MNPs generate the heat and the payload is released at the destination of the target within the body of the human. The particular form of the nanotechnology has been tested in the field of biomedical and scientists in the industry of oil have sought engineered nanorobots which are extremely small to be undetected and load can be carried to the desired location.[36] Nanorobots are being developed and are modeled for the means of energy consumption, mobility, and size.

The micro-organisms can perform the task of nanorobots by using the remote control and for this, the subjects are to be manipulated without resulting in the deaths and injury. For years, eukaryotic organisms are being used in research laboratories for studying bioaccumulation, photodynamic therapies,[37] environmental impact, and toxicology. Previous research deals with the uptake of silver or gold nanoparticles, for example, sore lacking consists of iron oxide particles.

Paramecia have the ability to phagocyte the particulate matter and it becomes a good target for the study. The three species of the paramecium, like *Paramecium aurelia*, *Paramecium caudatum*, and *Paramecium lursaria* are being used in the educational laboratories due to the following reasons:

(1) Through the use of an optical microscope, these can be easily visualized.
(2) These are easy to maintain and handle.
(3) These are inexpensive.

The synthesis of nanoparticles, investigation of polymer coatings, methods of the manipulation of paramecium are included in the longer-scale project and this project is to be carried by mixing the candidates of high school and graduate students. This has been designed in such a way that the students can perform them in regular labs in the setting of high school. In the experiments dealing with two days, the students can functionalize the synthesized polymers, that is, they can be coated with the polymers and then these are added in the tubs in the distilled water. Coated and the uncoated nanoparticles of magnet ingestion are observed using microscopes of different magnification. The microscope can also observe the movement of cilia paramecium, organelles and the MNPs which are ingested. In the *P. aurelia* and *P. caudatum*, the students manipulate the magnetized movement of the micro-organisms by using the bar magnet. It has been found that magnetic properties of the *P. aurelia* can be maintained for one week without cell death. The goal of the experiment deals with the introduction of the school students in the field of nanotechnology, the synthesis, and the coating of the nanoparticles and also with the uptake of paramecium. It will be of great interest to the students. Students learned about the non-aqueous solvents and molecules for the coating of the MNPs and the optical microscope is also used for the uptake. Through laboratories, the following things are learned by the students:

(1) read scientific literature and way to conduct research,
(2) usage of sonicator and centrifuge in the process of washing,
(3) methods and materials used to maintain and manipulate paramecium.

The descriptive, experimental, and comparative methods are used to investigate the synthesis and the coating of the MNPs and also the uptake through paramecia. In the laboratories, scientific reasoning which leads to the scientific explanation and applies the information that has been extracted from various sources and are published in various journal articles and the observations and conclusions are communicated in the reports of laboratories, This is the case dealing with two-day experiments but these are communicated through posters in the case of longer projects.

Chemistry has been published in the National Science Education Standard (CNSES)[38] by the American Chemical Society. It also provides the models in chemistry classrooms of the high school for the learning. The experiments dealing with the laboratories that have been reported here provide a good example for using the scientific inquiry and research in the field of chemistry, biology, physics, and the environmental science. The main drawback dealing with the experiments is that the chemicals used were irritants and should be handled with the gloves, lab aprons, and goggles. For waste solutions, labeled waste containers have been used. For the manipulation of the MNPs, bar magnet is used and should be handled with care so as to avoid the pinching. The solution of micro-organism should be disposed of properly and the contact surface should be washed with bleach so that the illness could be prevented which is caused due to the micro-organisms.

The SEM and TEM images of the MNPs are shown in the given Figure 10.3.

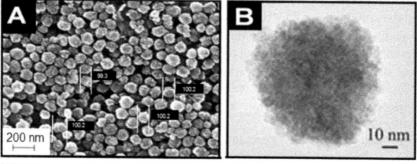

FIGURE 10.3   SEM (A) and TEM (B) images of Fe_3O_4 nanoparticles.
Source: Reproduced with permission from Ref. [40]. © 2017 American Chemical Society.

To ensure the colloidal stability and biocompatibility coating of the nanoparticles is required. The commonly used non-ionic polymer for this purpose is the PVP. Studies have been done to show that the coating on iron oxide nanoparticles has occurred through the interaction between the oxygen atoms of the PVP and iron oxide atoms in the MNP. The PVP presence prevents the aggregation which have aroused through the Van der-Waal interaction occurring between the nanoparticles of the magnet.[39,40]

The PVP coating process is very simple, and thus this experiment is used in the lab experiments.

All protist including paramecia are the single-celled microorganisms. Every paramecium species ate the bacteria, algae, and smaller protozoa and moves with the cilia of *P. caudatum*, *P. aurelia* which are spindle shaped and *P. bursaria* are green in color as it consists of the symbiotic green algae in them.

The optical image of *P. aurelia* after the internalization of the function-alized MNPs is shown in Figure 10.4.

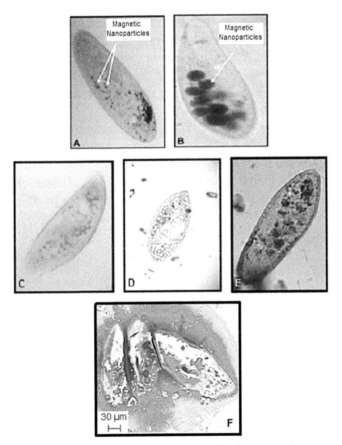

FIGURE 10.4 Viewed with 400× power on an optical microscope after exposure to and internalization of Fe_3O_4 MNPs: (A) *P. aurelia* at 12 h, (B) *P. aurelia* at 4 days, (C) *P. aurelia* at 18 h, (D) *P. bursaria* at 18 h, and (E) *P. caudatum* at 18 h. (F) A scanning electron microscopic (SEM) image of *P. aurelia* with ingested MNPs.

Source: Reproduced with permission from Ref. [40]. © 2017 American Chemical Society.

10.3   METAL OXIDE NANOPARTICLES

Nanoparticles synthesis has been motivated due to interest and applications. The magnetic metal oxide nanoparticles possess application in a large number of fields.[41–45] One of the important oxides of iron is the hematite. It is non-toxic, stable, and resistant to corrosion. It possesses a wide band gap of 2.1 eV and is an *n*-type semiconductor. It is used in the non-linear optics, gas sensors, and catalyst. The properties of the nanoparticles intensely affect the morphology and structure. Hematite nanoparticle fabrication under controlled shape and size is the goal of the scientist. Hematite possess the trigonal crystal system and possess pseudo-cubic morphology and corundum type structure.[41] it has been reported that cubic like nanoparticles of hematite can be synthesized in different ways. Pseudocubic particles are also obtained through the sol–gel method as reported by Sugimoto et al. and also the synthetic parameters influencing the shape have been studied extensively.[42]

   Hematite nanocubes have been synthesized uniformly, which has been reported by Yu et al., through the hydrothermal reaction but not the particles bounded by the plane have been found. Iron particles that have been enclosed by the six planes have been prepared in which the morphology uniformity could be improved. The XRD pattern of the prepared nanocubes is shown in the given Figure 10.5.

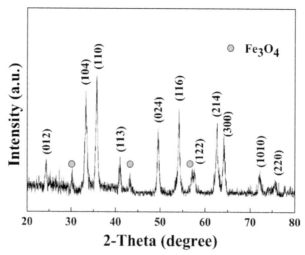

FIGURE 10.5   XRD diffraction spectra of the as-synthesized sample.
Source: Reproduced with permission from Ref. [42].

Within the range from 20°C to 70°C, the two theta value of the major peak is located. It has revealed the diffraction of the rhombohedral iron particles. The two peaks corresponding to the 30° and 43° reveals the diffraction of iron oxide particles.

The TEM images of the obtained nanocrystals are shown in the given Figure 10.6.

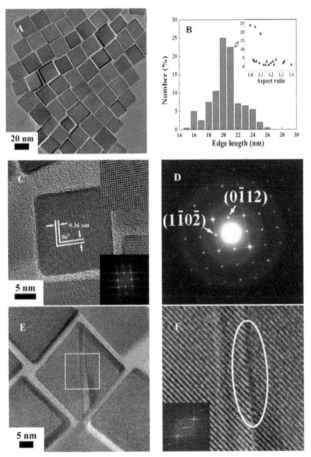

FIGURE 10.6 (A) Low-magnification TEM image; (B) a statistical size distribution histogram; (C) HRTEM images, the bottom inset is the FFT pattern, and the upper inset is the corresponding inverse FFT pattern; (D) SAED pattern; (E) HRTEM image of a nanoparticle with dislocation; (F) the FFT pattern (bottom inset), and the corresponding inverse FFT pattern of the selected area, marked by white box in (E).

Source: Reproduced with permission from Ref. [42]. © 2017 American Chemical Society.

These particles are found to be pseudocubic and possess the narrow size distribution. The size of the nanoparticle are average in 25 nm. The edges of the nanocrystals are found to be clear and the spacing lattice corresponds to the angle of approximately 86 degrees and is similar to the 0.3628 nm. It has been shown through the fast Fourier transform (FFT) that in the no bottom inset, nanoparticles occurs as single crystalline. In the corresponding FFT pattern clear lattice have been discovered with no defects. SAED pattern has also been shown in the above figure. It exhibited the spots of fast diffraction. It has been concluded that nanoparticles are bound by the short {012} plane.

Dislocation has been formed and has been marked in the white box. The corresponding inverse FFT image of the areas selected have been employed, which clearly exhibited the dislocation. The dislocation diagonally crosses the particles and no change in lattice distance has been observed along with the dislocation.

10.4   IRON OXIDE NANOPARTICLES

Iron oxide nanoparticles are used in various oxidation state and occurs in abundance in polymorphism and polymorphus changes in nanophase and are also used in various applications like sensors, catalyst, clinical uses, and high density magnetic recording media. For the iron oxide particles to be used in clinical application, the particles should be superparamagnetic and discrete. Their size should be less than 20 nm and have narrow size distribution for uniform chemical and physical properties. Maghemite and magnetite nanoparticles are used for the chemical applications. Iron oxide nanoparticles are very stable as compared to those of the pure iron metal.

The iron particles prepared through the classical methods are not used for the clinical applications. The iron precursor used in the thermal decomposition is used in the synthesis of good quality of nanoparticles. The decomposition of the $FeCup_3$ precursor leads to the maghemite nanoparticles.

A high temperature is required for the oxidation of magnetite to maghemite. The recent advances used in the synthesis of the nanoparticles have used the pure organometallic precursor and the inert conditions are precisely controlled. Easy and economic syntheses are required for industrial and clinical applications.

The iron oxide nanoparticles are prepared from low-grade reagents.[43] TEM images of nanoparticles are shown in given Figure 10.7.

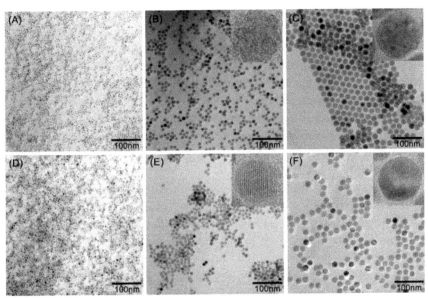

FIGURE 10.7 TEM images of respective intermediate and aerated iron oxide nanoparticles: (A) and (D) 5 nm, (B) and (E) 11 nm, and (C) and (F) 19 nm; insets are HRTEM images.
Source: Reproduced with permission from Ref. [43]. © 2004 American Chemical Society.

The TEM and HRTEM of the intermediate Fe_3O_4 nanoparticles have shown the scattering contrast changes occurring within the particles. One of the major factors which contributed to the scattering contrast changes is the mixed phase of iron oxides. After aeration, the drastic changes in the scattering contrast have been observed.

It has been found that the XRD pattern of the 19 nm has been found to be close to that of magnetic than the maghemite. High intensity has been shown for the larger nanoparticles and the width expected is narrow.

10.5   SYNTHESIS OF IRON OXIDE NANOPARTICLES

To increase the performance of iron oxide synthesized particles in the existing applications, an extensive study of iron oxide nanoparticles have been explored, due to its properties that are shape and size dependent, hydrothermal method has been modified by the use of polyisobutylene bisuccinimmide as a soft template so as to prepare the nanotubes and nanorods of the hematite and in this the shape dependent magnetic properties have been observed.[44] In the hematite nanorods, particles have large size and porosity in the size-dependent electrochemical property and magnetic properties have been revealed, through the method of hydrothermal by following the high concentration of the inorganic salts and high temperature.[45] By using the polyvinylpyrolidine (PVP) single crystalline quasicubic nanoparticle of hematite have been made by Zheng et al. These are found to be superior catalyst that nano- and micro-sized hematite catalyst in terms of conversion efficiency, temperature of activation, and thermal stability in catalyst reaction.[46]

By using CTAB (hexadecyltrimethylammonium bromide) method of hydrothermal synthesis, the hollow sphere of hematite crystals possessing high photocatalytic properties has been synthesized. Coating of silica has been done on the gamma iron oxide particles so that their dispersion or stability could be increased in the wider Ph range in organic media and aqueous solution.[47] Iron oxide particles have been coated with mesoporous silica for the drug delivery and the separation of the multiphase.[48] One of the most popular methods for the iron oxide synthesis with different shapes like rods, ellipsoids, cubes, and rods have been formed by the forced hydrolysis of the ferric salts. Various varieties of additives like PVP, span, phosphate, and CTAB have been used to obtain the size and shape in the controlled manner. The final product that has been formed remains contaminated by the residual activities. An open forced method of hydrolysis has been developed because the opening of the reaction facilitates the byproduct volatization of the HCl and the hydrolysis is not accelerated and the reaction time is reduced and the product shape is tuned.

10.6   COATING OF IRON OXIDE NANOPARTICLES

Iron oxide nanoparticles can be coated with silica and are being used in the delivery of drug, high stability, and the multiphase separation. Through aggregation, these methods are limited and the surface modification process suppresses the formation of new nuclei. The limitations can be breached and monodispersed hematite core–shell silica particles can be obtained. The hematite surface is coated with the silica and a two-phase system has been developed.

The following methods have been used:

(1) By PVP, the water can be bound and the concentration of PVP should be high and in this water is not added as the solvent, therefore water is referred to as the "bound water."[49] Therefore the silica precursor hydrolysis with the PVP can be bounded with water in the WPN system.

(2) In the nanotechnology, PVP has been used as the stabilizer and is used as an affinity agent because of its excellent adsorption ability.

Example silica particles can be coated with the metal and the agglomeration is prevented by the use of PVP.

The WPN system is shown in the given Figure 10.8.

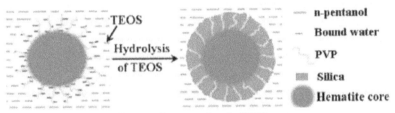

FIGURE 10.8   Proposed silica coating procedure in WPN system.
Source: Reproduced with permission from Ref. [49]. © 2010 American Chemical Society.

The PVP molecules adsorbed on the surface binds water on the particles of hematite and particles are protected from the agglomeration.

By modifying the concentration of the ferric chloride, a direct short-range order assembly of hemispheres has been used in the synthesis. Trial experiments have also been made to assemble the nanoparticles of hematite and coated them with silica so that the ordered structures like that of dimers, trimmers in the process of coating are obtained. The prepared monodispersed silica hematite core–shell spheres are assembled into the ordered long-range order with the photonic bandgap by the use of pressure assembly through the assembled colloids.[50]

Nanoparticles self-assembly has the applications in the field of large varieties of applications like paints, assays, biochemical sensors, photonic crystals and so on.

SEM and TEM images of hematite nanoparticles are shown below in Figure 10.9.

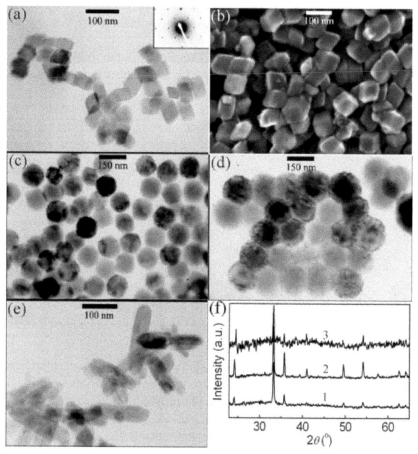

FIGURE 10.9 TEM (a, c–e) and SEM (b) images of hematite nanoparticles made using 0.005 (a), 0.01   (b), 0.02 (c), 0.03 (d), and 0.045 M (e) FeCl$_3$. Inset of (a) shows the corresponding SAED pattern. (f) The corresponding XRD patterns for the samples shown in a (1), c (2), and e (3).

Source: Reproduced with permission from Ref. [49]. © 2010 American Chemical Society.

Figure 10.9a gives the product image of the monodisperse rhombohedral particles which has an average length of about 45 nm. This image corresponds to the SAED pattern. This image has been obtained from the fine single particle which lies on the film support and the electron beam falls perpendicular on the particles that are rhombohedral in shape. As the concentration of FeCl$_3$ is increased, the shape of the product remains rhombohedral and the average edge length increases from 45 to 70 nm.

In Figure 10.9b, the SEM image of the rhombohedral shape can be seen and the particles consist of six rhombic facets. The rhombohedral shape is compressed as an increase in concentration of FeCl$_3$ takes place. And this results in the formation of the spherical particles. Figure 10.9c shows the image of purely monodispersed particles of semispherical in shape. These particles are of the diameter 120 nm in size. The spherical shape of the particle is improved on increasing the concentration of FeCl$_3$ and the average diameter also increases to 150 nm. Figure 10.9d shows the semi spheres short ranged order which is arranged in a hexagonal closed way packing structure. The XRD pattern of the samples with the rhombohedral shape, rod-like, and semispherical shape is shown in the above Figure 10.9.

It has been found that the hematite particles are uniformly found to be coated with silica particles. The size of silica can be tuned between 10 to 200 nm by modifying the amount of TEOS solution or hematite. By increasing the shell thickness of silica, the core–shell improves the spherical shape and the monodispersity. The hematite samples are also coated with short-range ordered shape which has been shown in Figure 10.1d,e shows rod-like structure.

10.7 TARGET-SPECIFIC BINDING OF MNPs

A critical impact on global human health and economies has been due to the emergence of infectious pathogens and subsequent proliferation.[51–54] A deadly pathogen, salmonella, occurs in the contamination of processed and raw food for a short interval of time. It has been reported by Hoelzer et al. that annually about 155,000 deaths occur due to Salmonella infection.[55] Mainly foodborne illnesses are mainly due to the nontyphoid Salmonella which leads to hospitalization and then death.[56] Salmonella cells reproduce asexually and in every 20 min, they double if the environmental conditions are right.[57] There is need for effective, generic, and practical materials which allows capture and separation of pathogenic bacteria. Various number of diagnostic methods which vary in terms of reliability, specificity, and sensitivity have been developed. These include microculture arrays, methods based on polymerase chain reaction (PCR), immunoassays utilizing fluorescence and electrochemical detection.[58–62] But these methods are not considered for regular use as these requires laborious pretreatments and also harmful reagents are used, therefore there is a need for fast and reliable sensing method for Salmonella detection. For cell separation, drug delivery and bioimaging, the Target-specific binding of MNPs are being used advantageously. The choice of target is an important step in the magnetic separation of pathogens.[63–69] A study has been done on the bioconjugated MNPs with the carbohydrate, DNA, antibiotics, and antibody. The magnetic glyco-nanoparticles, vancomycin conjugated MNPs; magneto-DNA nanoparticle can be separated easily by external magnetic fields. Salmonella can be characterized taxonomically into many stereotypes which is based on two structures the flagellin protein (Hantigen), and Opolysaccharide (O-antigen). One of the major components of the outer membrane of Gram-negative bacteria is the O-antigen which is the repetitive glycan polymer The H-antigen is the antigenic type of bacterial flagella comprising proteins. Therefore, we have decorated MNP clusters with O- or H-antibodies for the separation of Salmonella, and compared their effectiveness for cell separation. Although theoretically, both antibodies should work for the Salmonella separation, a surprising difference in cell separation efficiency was observed in the experimental results. The capture efficiency of O-antibodymodified magnetic nanoparticle clusters (OMNC) was ~99%, while the efficiency of H-antibody-modified magnetic nanoparticle clusters (HMNC) was considerably lower (~57%). This report will describe and explain these disparate antibody-dependent magnetic separation results, which can be extended to other important pathogens. The schematic illustration of the preparation of bioconjugated magnetic nanoclusters is provided in Figure 10.10.

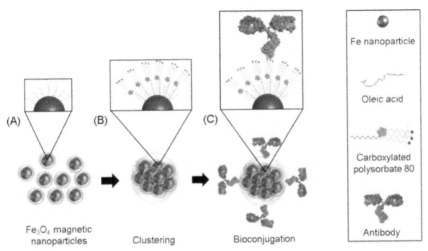

FIGURE 10.10 Schematic illustration of the bioconjugated magnetic nanoclusters. (A) Monodispersed MNPs coated with hydrophobic oleic acid. (B) Colloidal cluster of nanoparticles developed using a microemulsion method. (C) PCMNCs functionalized as target antibodies using a bioconjugation process.

Source: Reproduced with permission from Ref. [63]. © 2016 American Chemical Society.

The oleic acid-coated superparamagnetic nanoparticles (NPs) were prepared by thermal decomposition of iron(III) acetylacetonate as a precursor and oleic acid as primary ligand in benzyl ether.[70] These MNPs are typically coated with hydrophobic ligands. Therefore, it was necessary to modify the surface coating to make the MNPs more water-soluble and biocompatible. Polysorbate 80 is a biocompatible compound commonly used for biochemical applications and in the food production industry due to its nontoxic and biodegradable properties.[71] Also, other important features of polysorbate 80 are that it is commonly used as protein stabilizer and blocking agent for nonspecific binding.[72] Polysorbate compounds are composed of three functional groups: (1) the aliphatic ester chains, which prevent nonspecific binding of biomolecules, (2) hydrophilic three-terminal hydroxyl groups, which can be modified for further applications, and (3) an aliphatic chain that can easily adsorb on a hydrophobic surface of oleic-coated MNPs via noncovalent interactions (Fig. 10.S1).[73,74] Prior to the modification of the oleic-coated MNP surface, the terminal hydroxyl groups of polysorbate 80 are replaced with carboxyl groups using succinic anhydride.[75] The carboxyl group can be readily conjugated with amine-functionalized targeting moieties, such as antibodies. Hydrophilic

PCMNCs were synthesized using a water-in-oil microemulsion technique (Figs. 10.1B and 10.S2).[76] The size increments of MNP clusters are induced using polysorbate 80, which have a low Critical Micelle Concentration (CMC), and an Evaporation Induced Self Assembly (EISA) method. In the last step of the process, the carboxyl group of polysorbate 80 on O-antibody-coated polysorbate 80-coated magnetic nanoclusters (PCMNCs) is conjugated with the amine group of the antibodies.

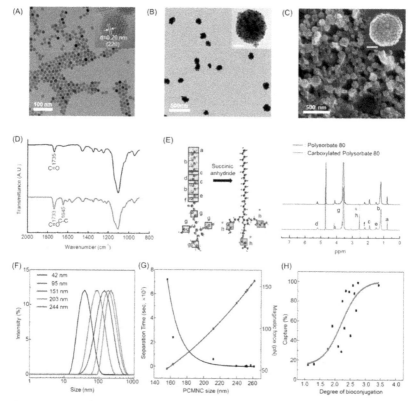

FIGURE 10.11 (See color insert.) TEM images of (A) monodispersed MNPs with oleic acid (inset shows a HRTEM image) and (B) magnetic nanoclusters with polysorbate 80. (C) SEM images of magnetic nanoclusters. Insets show TEM and SEM images at high magnification. Inset scale bar is 100 nm. (D) FT-IR spectra of oleic acid-coated MNPs (black line) and polysorbate 80-coated MNCs (red line). (E) 1H NMR spectrum of carboxylated polysorbate 80. (F) Size distributions of MNCs analyzed by DLS. (G) Relationship between magnetic separation time (black line) and magnetic force under specific field gradients (blue line). (H) Capture efficacy corresponds to the ELISA values for degree of bioconjugation.

Source: Reproduced with permission from Ref. [63]. © 2016 American Chemical Society.

Figure 10.11 shows that the resulting monodisperse MNPs were composed of highly crystalline nanocrystals ~10 nm in size. The lattice fringe pattern with the spacing of 0.29 nm can be identified as the (220) planes of Fe_3O_4 (inset in Fig. 10.11A). Figure 10.11B and C shows the TEM and SEM images of the PCMNCs with size of about 220 nm. The magnetic hysteresis loop of MNPs and PCMNCs were analyzed using a vibration sample magnetometer at 298 K. The magnetic saturation values at 15T are measured to be 67.7 and 68.3 emu g^{-1} for MNPs and PCMNCs, respectively. The magnetic saturation value is increased with increasing cluster size effects. Both FTIR and NMR spectra show that corresponding results of the modification of PCMNCs. The original form of polysorbate 80 contains a C=O stretching vibration band at 1735 cm^{-1}, which is slightly shifted to 1733 cm^{-1} for the modified form; this is due to the presence of negatively charged tricarboxylate groups on the PCMNCs (Fig. 10.11D). Consistently, modified structures of PCMNC were further supported by a characteristic peak observed at 2.6 ppm for 1H NMR spectra (Fig. 10.11E). The hydrodynamic sizes of synthesized spheroidal PCMNCs range from 40 to 250 nm, as confirmed by DLS analyses (Fig. 10.11F). The PCMNCs are densely self-assembled single MNPs, as shown in Figure 10.11G as the relationships between PCMNC size and separation time, as well as between size and theoretical magnetic field gradient. Separation times increased exponentially with a decrease in PCMNC size due to a reduced magnetic force. For PCMNCs with a hydrodynamic size greater than 200 nm (as per DLS analysis), separation times were below 10 min. In order to achieve rapid separations of PCMNCs, we controlled the relative amount of MNP and polysorbate 80 using a drift-diffusion equation. In biofunctional MNPs, the control of separation time has significant advantages, including that (1) time and cost of the experimental process are dramatically reduced and (2) rapid separations can reduce the deformation of biomaterials. The model of the drift diffusion equation provides insight into the physics of particle transport and is useful for optimizing the performance in the separation system for antibodies. In this model, the magnetic force can be applied to a single particle with a cluster.

Infectious bacterial diseases are the leading cause of disability and death,[77] the majority of these infections are more prevalent in developing countries. Every year, approximately 17 million deaths are due to these infectious diseases and this represents one-third of global mortality. Infections due to gram-positive bacteria is the major cause of morbidity and mortality in humans.[78,79] Such diseases can be treated by the ability to rapidly diagnose Gram-positive pathogens. Staphylococci and Streptococci are the most common Gram-positive pathogenic bacteria in human-sarecocci (sphere-shapedbacteria). Many antibiotics are being developed to treat Gram-positive infections. These antibiotics work by inhibiting cell wall synthesis or by blocking transcription/translation processes. Vancomycin is a glycopeptide which is composed of peptide sequence and glycosides. It is a commonly used antibiotic which results in inhibition of cell wall synthesis. The binding with the target bacteria takes place through the peptide. Vancomycin forms hydrogen bonds with the terminal D-alanyl-D-alanine (D-Ala-D-Ala) moieties of the N-acetylmuramic acid (NAM) and N-acetylglucosamine (NAG) peptide subunits[80,81] and exerts its antibacterial activity. Due to this binding, subunits cannot be incorporated into a major structural component of gram-positive cell walls and thus inhibition of cell wall takes place. New generation antibiotics that is, daptomycin, linezolide, and pristinamycin. Through hydrophobic tail, Daptomycin binds to the cell wall of gram-positive bacteria and results in perturbation and depolarization of the cell membrane.

Both vancomycin and daptomycin bind to the bacterial cell wall of gram-positive bacteria, and it is reasoned that bioorthogonal derivatives of these antibiotics can be used for the rapid detection and capture of gram-positive bacteria. Vancomycin-resistant bacterial strains are emerged due to extensive use of vanomycin.

Vancomycin binding to Gram-positive bacteria is shown in Figure 10.12.

Figure 10.12A illustrates the mechanism of labeling using vancomycin-TCO (vanc-TCO) in which vanomycin act as the targeting ligand.[5] The binding is due to five-point hydrogen bond interaction between the drug and the D-Ala-D-Ala moieties of the NAM/NAG peptides. HPLC, ESI-MS, and HRMS were used to determine the purity and identity of the synthesized product. Through bioorthogonal labeling approach, bright fluorescence on the surface and outer layer of the bacterial cells is targeted and has been shown by the confocal microscopy and transmission electron microscopy (TEM) show the presence of nanoparticles, coated evenly across the surface of bacterial cells (Fig. 10.12B).

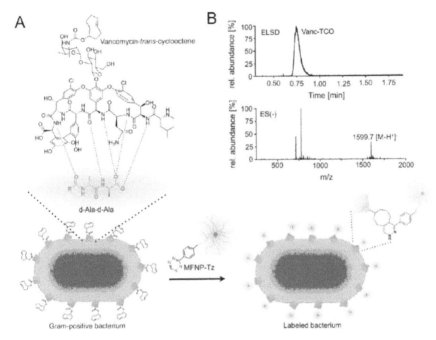

FIGURE 10.12 (See color insert.) Chemistry underlying bioorthogonal magnetofluorescent nanoparticle (MFNP) labeling. (A) Vancomycin-transcyclooctene (vanc-TCO) targets gram-positive bacteria by binding onto their membrane subunits. Following incubation with MFNP-Tz, bacteria are labeled and can be detected via fluorescent or magnetic sensors. (B) HPLC (top) and ESI-MS (bottom) traces of vanc-TCO verifying its identity and purity.

Source: Reproduced with permission from Ref. [77]. © 2011 American Chemical Society.

Infectious diseases are caused by bacterial infections, which lead to sepsis, bacteremia, pneumonia, endocarditis, or even death.[82–84] The involvement of high amount of bacteria lead to serious diseases. Antibiotics are being developed to treat bacterial infections. Resistance has been developed against many pathogenic bacteria due to the extensive use of antibacterial agents.[85,86] To solve the emerging problem new types of antibiotics and effective therapeutic strategies should be developed (Fig. 10.13).

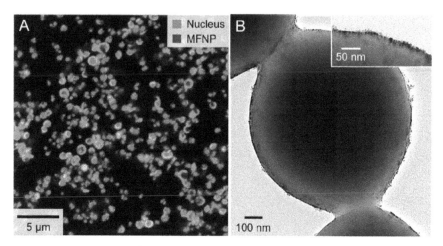

FIGURE 10.13   (See color insert.) Bacterial labeling. (A) Confocal microscopy and (B) transmission electron microscopy of *S. aureus* labeled with vanc-TCO and MFNP-Tz. Inset in top right of (B) shows labeling at a higher magnification.
Source: Reproduced with permission from Ref. [77]. © 2011 American Chemical Society.

Pathogenic bacteria, such as Gram-positive, Gram-negative, and vancomycin-resistant bacterial strains can be targeted using vancomycin-immobilized nanoparticles (VanNPs).[87–94] Pathogenic bacteria can be probed using NPs with different shapes and compositions, such as spherical gold (Au) NPs, magnetic NPs,[10,11] polygonal Au NPs,[12] Fe_3O_4@ Au nanoeggs, silica beads, and spherical Ag NPs, immobilized with vancomycin. To maintain the antibiotic activity of vancomycin on the surface of Van-NPs is being the most challenging part in the generation of effective Van-NPs against bacteria. The activity of vanomycin relies on the orientation of the vancomycin structure anchored on the surface of the Van-NPsthat is, the active binding site of vancomycin specific for bacteria should not be blocked during the immobization on the surface of NPs, antibiotic activity is lost during the generation of Van-NPs. Using simple immobilization procedures reduce the probability of losing the antibacterial activity during the preparation of Van-NPs β-lactam antibiotic can be used as a reducing agent and protective group to generate Au NPs. Glucose is used to reduce Au ions for generation of Au NPs with addition of aqueous NaOH at high pH[30–32] vancomycin can be used to reduce Au ions and be directly capped on the generated Au NPs as vanomycis is composed of glycosides.

The mechanism proposed for vancomycin-directed synthesis of Au NPs is given in Figure 10.14.

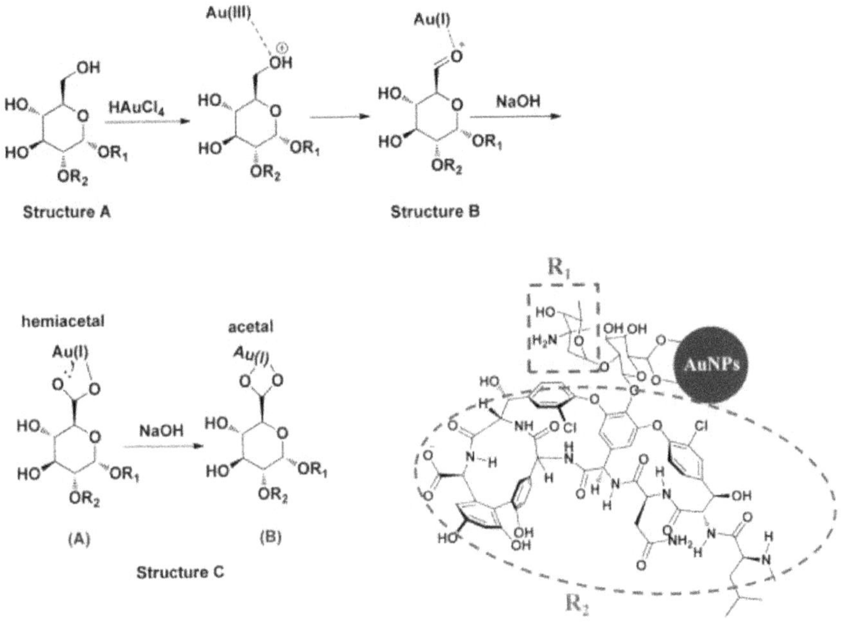

FIGURE 10.14   Proposed reaction of vancomycin-directed synthesis of Au NPs.
Source: Reproduced with permission from Ref. [86]. © 2015 American Chemical Society.

In vancomycin, the C−O group of glycoside was being involved in the binding between vancomycin and Au NPs.

TEM images were obtained by incubating different bacteria with the Van-Au NPs (Fig. 10.15).

A facile approach to generate Van-Au NPs from one-pot reaction has been developed successfully. The cell growth of pathogenic bacteria is inhibited through Van-Au NPs, and these nanoparticles have been generated in a straightforward manner. The Van-Au NPs act as drug carriers and as effective antibiotics, and also Van-Au NPs are effective against antibioticresistant bacterial strains, which are engulfed by macrophages this approach leads to new therapeutics for treating bacterial infections.

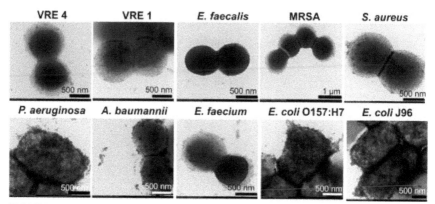

FIGURE 10.15 TEM images obtained by incubating different bacteria with the Van-Au NPs. All scale bars on the images are equal to 500 nm except the scale bar in the image of MRSA is 1 μm.

Source: Reproduced with permission from Ref. [86]. © 2015 American Chemical Society.

10.8 ISOLATION OF BACTERIA

The efficient and fast method for elimination or isolation of bacteria, for example, for water disinfection and decontamination of food and biotechnological applications which involves bacteria, for example, during the production of recombinant drug-protein or therapeutic treatments or clinical diagnostics is a challenging task today. There are three major classes of materials for targeting of bacteria which can be differentiated, antibacterial coatings and antibacterial surfaces for the protection against growth of bacteria, for example, in the clinical settings or household settings antibacterial filters through the unspecific size discrimination, for the removal of bacteria from air and solution. The last system includes strong and selective binding of bacteria to the particles; these can then be isolated from the sample. The technique of magnetic separation based on beads of magnet allows the rapid and easy removal of the bounded bacteria from the different media through the contactless and harmless external magnet. The commercially available systems for this method are Dynabeads. These are the polystyrene beads of 2−5 μm. The polystyrene-based particles are non-biocompatible and are prone to unspecific interactions which limit their applicability. A more advanced particle system for isolation of bacteria for the biomedical and biotechnological applications will combine high loading capacity biocompatibility and reduced unspecific interactions. A new microparticle system has been developed which is based on sugar-functionalized (MaPoS) poly(ethylene glycol) (PEG) microgels, are porous and magnetic for the magnetic removal and the selective binding of bacteria from solution.

Three features are combined by the MaPoS particles:

(1) A macroporous PEG hydrogel matrix, it offers a large surface area for binding of bacteria effectively.
(2) Covalent attachment of sugar ligands to PEG network for the bacteria targeting.
(3) Loading of superparamagnetic iron oxide NPs in the particles for the removal of bacteria after it has bound.

The PEG hydrogel matrix is a hydrophilic and biocompatible polymer through which unspecific interactions are prevented and it is thus regarded as the commonly used polymer for biomedical and biotechnological applications.[30] Different chemical functionalities can be introduced to PEG particles,[31] which covalently links the cationic moieties for the electrostatic loading of iron oxide NPs and sugar ligands for targeting of the receptor. Efforts are being made to employ high loading capacities and fast binding of bacteria through application of porous particles which presents pores of the bacteria size. These particles present the high loading capacity and allow high yield of binding of bacteria due to their increased surface area. The macroporous microparticles which are composed of PEG32 will be used and will be functionalized to prepare MaPoS particles. The targeting systems that have been developed carries the antibodies which bind specifically to antigens on the bacterial surface. Targeting and binding are achieved through carbohydrate–protein interactions. These bacteria carry carbohydrate-binding proteins and are thus responsible for the microbial adhesion to cells and infections, example the bacteria *Escherichia coli* is involved in food contamination and is used for biotechnological applications and it presents proteins at the surface of the organelles known as fimbriae. On comparing with the antibodies, ligands of carbohydrates are found to be less prone to denaturation and presented specificity of border interaction. The carbohydrate ligands are used for the detection of new bacteria strains. The magnetic beams combine the biocompatibility and minimize the unspecific interactions and also avoids hydrophobic interactions and thus specific ligand modification, and porosity is introduced for the isolation of bacteria. The MaPoS particles are synthesized and characterized. In order to obtain particles (MaPoS) with large surface areas so that they select and bind bacteria through the specific ligand–receptor interactions, the synthesis, and functionalization of porous PEG hydrogels is done and the bacteria can be removed from the solution through the simple use of magnet. It is represented in given Figure 10.16.

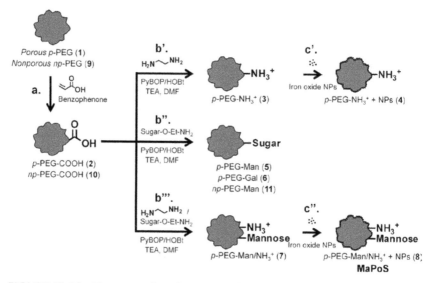

FIGURE 10.16   The preparation of MaPoS particles.
Source: Reproduced with permission from Ref. [90]. © 2013 American Chemical Society.

Figure 10.16a shows the principle of the isolation of bacteria using MaPoS microparticles. During incubation, bacteria bind to the particles via carbohydrate–protein interactions. Then, MaPoS particles and the bacteria bound to them can be isolated with the help of a magnet. (b) Synthesis of MaPoS particles: porous PEG microgels were first synthesized by hard templating with CaCO$_3$ porous microparticles (1) and functionalized to present both amine groups and mannose moieties (2). The microgels were finally made magnetic by loading negatively charged iron oxide NPs into the particles via electrostatic interactions with NH$_3^+$ groups (3).

The scheme demonstrates the preparation of MaPoS particles and it comprises the following:

(1) Casting of the PEG microgels, which are porous from the $CaCO_3$ template microparticles.

(2) Through the radical benzophenone chemistry, activation of the PEG backbone which is followed by the functionalization with sugar ligands and amine groups.

(3) Loading of the superparamagnetic iron oxide NPs with the PEG microgels.

The concentration of NPs during the loading step is an important measure to control the amount of loaded iron oxide NPs. Magnetic PEG microgels are obtained even at the low concentration of 0.2 mg·mL^{-1} of citrate-coated iron oxide NPs. It has been found that the below this concentration, the obtained microgels were not magnetic. In the PBS buffer (pH 7.4), the loading has been found to be stable and all microgels have been found to retain their magnetic properties and even no leaking of nanoparticles has been observed. These are stable in the medium of high ionic strength which is because of electrostatic interactions and also due to the formation of coordination bonds between amines of iron of NPs and those of PEG particles. The cryo-scanning electron microscopy (cryo-SEM) has been used to study the presence of NPs on the surface of PEG microgels. The study of TEM has depicted that in the pores of the micro-gels the NPs have been homogeneously loaded. This data is also useful for observing the core of the porous PEG microgels. It is quite difficult to clearly visualize pure PEG microgels but these can be enhanced with the physisorbed iron oxide NPs. It has been found that nanoparticles are being bound to the PEG hydrogel and cannot be diffused into the hydrogel network because of their large size.[55] The internal morphology of PEG microgels have been revealed, and the NPs are used to detect and visualize the interface between PEG hydrogels and the pores.

Sugar ligands can be introduced for targeting of receptors and binding of bacteria. PEG microgels can be functionalized by the use of peptide coupling chemistry and also by use of high excess of a protected mannose ligand with a short amine-linker at the anomeric position ((2-aminoethyl)-2,3,4,6-tetra-O-acetyl-α-D-mannopyranoside hydrochloride). Galactose functionalized particles (PEG-Gal, 6) can also be synthesized as a negative control for ligand−receptor binding. The carbohydrates were deprotected after coupling with sodium methoxide (MeONa) and the final PEG-sugar microgels can be redispersed in water readily. The particles which have been obtained have shown an increase in ζ-potential from -50.8 ± 5.9 mV for the PEG-COO$^-$ particles to -19.4 ± 4.7 mV for the PEG-mannose particles (PEG-Man, 5). This has indicated successful functionalization of carboxylate groups with sugar ligands. The sugar ligands have the ability to bind specifically the protein receptors and this distribution of sugar ligands can be visualized by incubation with fluorescently labeled Concanavalin A (ConA), which is a mannose-binding protein. The uniform fluorescence of PEG-Man microgels have been revealed by the confocal microscopy and it also depicted the binding of FITC-ConA and in this, no fluorescence has been observed with nonfunctionalized PEG, PEG-Gal, and PEGCOOH particles. It is depicted in Figure 10.17.

An inhibition experiment has been performed with methyl α-D-mannopyranoside so as to prove that the binding of FITC-ConA to PEG-Man is specific. it has been found that the competitor to the mannose moieties on PEG particles is the Methyl α-D-mannopyranoside. After incubation with methyl α-D-mannopyranoside, particles were not found to be fluorescent. The binding of FITC-ConA is found to be specific because of sugar–protein interactions. Sugar ligands are used to discriminate between strains of bacteria and present different receptors and bind selectively bacteria which present a large number of sugar receptors. The sugar ligands present the specificity of interactions in comparison with the antibodies and allow the detection of unanticipated or new bacteria strains. The features of magnetic properties and sugar ligands used for binding of bacteria have been combined so as to obtain magnetic, porous, sugar-functionalized (MaPoS) PEG microgels (8). PEG-COOH microgels have been functionalized with mannose moieties and amine groups in a one-step procedure through the addition of stoichiometric amounts of ethylenediamine and mannose ligand to the carboxylate groups which are activated.

The final MaPoS particles (8) have found to be magnetic and are able to bind to the bacteria and are also found to be used for the bacterial removal from contaminated solutions. The number of bacteria is found to be bound per particle. a strong dependency of bacteria bound per particle on the concentration of bacteria in solution has been found at the lower ratios of bacteria to particle. This is because of the slow diffusion of bacteria in diluted samples which limits the interaction between the particles and bacteria and thus the probability of a binding event at a lower bacteria/ particle ratio is reduced. At high bacterial concentration, the possible binding sites of MaPoS particles bind maximum bacteria per particle. On increasing the further bacterial concentration, the number of bound bacteria was not found to be increased, this is because the maximum binding capacity was reached already. In the yield of removal of bacteria, the average efficiency in bacteria removed from solution comprising of 70% for bacteria/particle ratios is below 30/1. It can be increased to 90% by performing second incubation with particles of MaPoS. The coverage of bacteria per particles becomes maximum at the particle ratios of 30/1 and after this, reduction in bacterial yield is observed. Through this, from a solution of known bacteria concentration, the bacteria can be removed by using the optimal concentration of MaPoS particles.

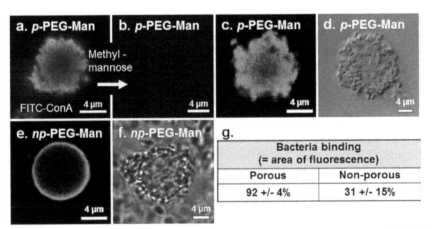

FIGURE 10.17 (a) CLSM image of porous PEG-Man (5) after labeling with FITC-ConA; (b) CLSM image of porous PEG-Man (5) after first labeling with FITC-ConA and then addition of a 2 M solution of methyl α-D-mannopyranoside; (c) CLSM and (d) optical microscopy image (differential interference contrast DIC mode) of porous PEG-Man (5) after 1 h incubation with *E. coli* bacteria strain ORN178 and staining with DAPI; (e) CLSM and (f) optical microscopy image of nonporous PEG-Man (11) after 1 h incubation with the same bacteria and staining; (g) table presenting the efficiency of binding via area of fluorescence due to the bound, labeled bacteria for porous and nonporous PEG-Man particles.
Source: Reproduced with permission from Ref. [90]. © 2013 American Chemical Society.

10.9 CONCLUSION

Nanoparticles can be biosynthesized by algae, bacteria, and fungi followed by dissolved metal uptake. "Greener" nanoparticle syntheses requires the optimization of these biosynthesis processes for gold recovery from aqueous solutions metal recovery biosorbants require less energy, toxic reagents, or labor than carbon sorption, electro deposition, solvent extraction, ion exchange, and metal nanoparticles are a byproduct of this process. In nanoparticle biosynthesis, dissolved gold ions (Au^{3+}) are reduced to Au^0 intracellular or extracellular without using acidic or caustic reagents. Using single microbial species with the growth condition modification (i.e., pH, initial Au concentration, and temperature) AuNP bioproduction is attractive as a "green" synthesis. FT-IR spectroscopy, X-ray photoelectron spectroscopy, zeta potential measurements, enzymatic activity assays, and TEM−electron spectroscopic imaging are used to study the mechanisms of nanoparticle.[95–98]

There is a tremendous variety of biomolecules which are capable of reducing Au^{3+} to Au^0, due to organism specific biochemistries. It has been suggested by one theory that negatively charged cell wall electrostatically attracts Au^{3+} or enzymes at the cell wall and reduces Au^{3+} to Au^0 and follows nucleation that nanoparticles are transported through the cell wall. Other pathway includes bioreduction of Au^{3+} by bacteria and fungi involving thylakoid membranes, reductase enzymes, and cytochromes. At present, we have limited information regarding biomolecules associated with intracellular nanoparticle surfaces for intact cells. Algal studies dealing with intracellular AuNP biosynthesis focus solely on metal uptake for recovery from waste streams or omit discussion of the AuNP reduction mechanism.

KEYWORDS

- **coated nanoparticles**
- **magnetic nanoparticles**
- **self-propulsion**
- **bactericidal**
- **magnetic control**

REFERENCES

1. Nakashima, N.; Asakuma, S.; Kunitake, T. Optical Microscopic Study of Helical Superstructures of Chiral Bilayer Membranes. *J. Am. Chem. Soc.* **1985,** *107,* 509−510.

2. Lvov, Y.; Ariga, K.; Ichinose, I.; Kunitake, T. Assembly of Multicomponent Protein Films by Means of Electrostatic Layer-by-layer Adsorption. *J. Am. Chem. Soc.* **1995,** *117*, 6117−6122.

3. Georgieva, R.; Moya, S.; Donath, E.; Bäumler, H. Permeability and Conductivity of Red Blood Cell Templated Polyelectrolyte Capsules Coated with Supplementary Layers. *Langmuir* **2004,** *20*, 1895−1900.

4. Delcea, M.; Möhwald, H.; Skirtach, A. G. Stimuli-responsive LbL Capsules and Nanoshells for Drug Delivery. *Adv. Drug Delivery Rev.* **2011,** *63*, 730−747.

5. Fakhrullin, R. F.; Lvov, Y. M. Face-Lifting" and "Make-Up" for Microorganisms: Layer-by-Layer Polyelectrolyte Nanocoating. *ACS Nano* **2012,** *6* (6), 4557−4564.

6. Song, W.; He, Q.; Möhwald, H.; Yang, Y.; Li, J. Smart Polyelectrolyte Microcapsules as Carriers for Water-Soluble Molecular Drug. *J. Contr. Release* **2009,** *139*, 160−166.

7. Qi, W.; Yan, X.; Fei, J.; Wang, A.; Cui, Y.; Li, J. Triggered Release of Insulin from Glucose-Sensitive Enzyme Multilayer Shells. *Biomaterials* **2009,** *30*, 2799−2806.

8. Li, D.; He, Q.; Li, J. Smart Core/shell Nanocomposites: Intelligent Polymers Modified Gold Nanoparticles. *Adv. Colloid Interface Sci.* **2009,** *149*, 28−38.

9. Hong, D.; Park, M.; Yang, S. H.; Lee, J.; Kim, Y.-G.; Choi, I. S. Artificial Spores: Cytoprotective Nanoencapsulation of Living Cells. *Trends Biotechnol.* **2013,** *31* (8), 442−447.

10. Park, J. H.; Hong, D.; Lee, J.; Choi, I. S. Cell-in-shell Hybrids: Chemical Nanoencapsulation of Individual Cells. *Acc. Chem. Res.* **2016,** *49*, 792.

11. Granicka, L. H. Nanoencapsulation of Cells Within Multilayer Shells for Biomedical Applications. *J. Nanosci. Nanotechnol.* **2014,** *14* (1), 705−716.

12. Catania, C.; Thomas, A. W.; Bazan, G. C. Tuning Cell Surface Charge in *E. coli* with Conjugated Oligoelectrolytes. *Chem. Sci.* **2016,** *7*, 2023−2029.

13. Yang, S. H.; Ko, E. H.; Jung, Y. H.; Choi, I. S. Bioinspired Functionalization of Silica-Encapsulated Yeast Cells. *Angew. Chem. Int. Ed.* **2011,** *50*, 6115−6118.

14. Yang, S. H.; Lee, K.-B.; Kong, B.; Kim, J.-H.; Kim, H.-S.; Choi, I. S. Biomimetic Encapsulation of Individual Cells with Silica. *Angew. Chem. Int. Ed.* **2009,** *48*, 9160−9163.

15. Yan, L.; Zhang, S.; Chen, P.; Liu, H.; Yin, H.; Li, H. Magnetotactic Bacteria, Magnetosomes and their Application. *Microbiol. Res.* **2012,** *167* (9), 507−519.

16. García-Alonso, J.; Fakhrullin, R. F.; Paunov, V. N.; Shen, Z.; Hardege, J. D.; Pamme, N.; Haswell, S. J.; Greenway, G. M. Microscreening Toxicity System Based on Living Magnetic Yeast and Gradient Chips. *Anal. Bioanal. Chem.* **2011,** *400*, 1009−1013.

17. Zhang, D.; Fakhrullin, R. F.; Özmen, M.; Wang, H.; Wang, J.; Paunov, V. N.; Li, G.; Huang, W. E. Functionalization of Whole-Cell Bacterial Reporters with Magnetic Nanoparticles. *Microb. Biotechnol.* **2011,** *4*, 89−97.

18. Atlas, R. M. Petroleum Biodegradation and Oil Spill Bioremediation. *Mar. Pollut. Bull.* **1995,** *31*, 178−182.

19. Kadali, K. K.; Simons, K. L.; Skuza, P. P.; Moore, R. B.; Ball, A. S. A Complementary Approach to Identifying and Assessing the Remediation Potential of Hydrocarbonoclastic Bacteria. *J. Microbiol. Methods* **2012,** *88* (3), 348−55.

20. Bertrand, J.-C.; Bianchi, M.; Mallah, M. A.; Acquaviva, M.; Mille, G. Hydrocarbon Biodegradation and Hydrocarbonoclastic Bacterial Communities Composition

Grown in Seawater as a Function of Sodium Chloride Concentration. *J. Exp. Mar. Biol. Ecol.* **1993**, *168*, 125–138.

21. Yakimov, M. M.; Golyshin, P. N.; Lang, S.; Moore, E. R. B.; Abraham, W.-R.; Lunsdorf, H.; Timmis, K. N. Alcanivorax borkurnensis gen. now, sp. nov., a New, Hydrocarbon-degrading and Surfactant Producing Marine Bacterium. *Int. J. Syst. Bacteriol.* **1998**, *48*, 339–348.

22. Lange, A. B.; Tenberge, K. B.; Robenek, H.; Steinbüchel, A. Cell Surface Analysis of the Lipid-Discharging Obligate Hydrocarbonoclastic Species of the Genus Alcanivorax. *Eur. J. Lipid Sci. Technol.* **2010**, *112*, 681–691.

23. Cappello, S.; Denaro, R.; Genovese, M.; Giuliano, L.; Yakimov, M. M. Predominant Growth of Alcanivorax During Experiments on "Oil Spill Bioremediation" in Mesocosms. *Microbiol. Res.* **2007**, *162* (2), 185–190.

24. Hara, A.; Syutsubo, K.; Harayama, S. Alcanivorax which Prevails in Oil-Contaminated Seawater Exhibits Broad Substrate Specificity for Alkane Degradation. *Environ. Microbiol.* **2003**, *5* (9), 746–753.

25. Schneiker, S.; Martins dos Santos, V. A. P.; Bartels, D.; Bekel, T.; Brecht, M.; Buhrmester, J.; Chernikova, T. N.; Denaro, R.; Ferrer, M.; Gertler, C.; Goesmann, A.; Golyshina, O. V.; Kaminski, F.; Khachane, A. N.; Lang, S.; Linke, B.; McHardy, A. C.; Meyer, F.; Nechitaylo, T.; Pühler, A.; Regenhardt, D.; Rupp, O.; Sabirova, J. S.; Selbitschka, W.; Yakimov, M. M.; Timmis, K. N.; Vorhölter, F.-J.; Weidner, S.; Kaiser, O.; Golyshin, P. N. Genome Sequence of the Ubiquitous Hydrocarbon-Degrading Marine Bacterium Alcanivorax borkumensis. *Nat. Biotechnol.* **2006**, *24*, 997–1004.

26. Kalscheuer, R.; Stöveken, T.; Malkus, U.; Reichelt, R.; Golyshin, P. N.; Sabirova, J. S.; Ferrer, M.; Timmis, K. N.; Steinbüchel, A. Analysis of Storage Lipid Accumulation in Alcanivorax borkumensis: Evidence for Alternative Triacylglycerol Biosynthesis Routes in Bacteria. *J. Bacteriol.* **2007**, *189* (3), 918–928.

27. Naether, D. J.; Slawtschew, S.; Stasik, S.; Engel, M.; Olzog, M.; Wick, L. Y.; Timmis, K. N.; Heipieper, H. J. Adaptation of the Hydrocarbonoclastic Bacterium Alcanivorax borkumensis SK2 to Alkanes and Toxic Organic Compounds: A Physiological and Transcriptomic Approach. *Appl. Environ. Microbiol.* **2013**, *79* (14), 4282–4293.

28. Miri, M.; Bambai, B.; Tabandeh, F.; Sadeghizadeh, M.; Kamali, N. Production of a Recombinant Alkane Hydroxylase (AlkB2) from Alcanivorax borkumensis. *Biotechnol. Lett.* **2010**, *32* (4), 497–502.

29. Golyshin, P. N.; Martins Dos Santos, V. A.; Kaiser, O.; Ferrer, M.; Sabirova, Y. S.; Lünsdorf, H.; Chernikova, T. N.; Golyshina, O. V.; Yakimov, M. M.; Pühler, A.; Timmis, K. N. Genome Sequence Completed of Alcanivorax borkumensis, a Hydrocarbon-degrading Bacterium that Plays a Global Role in Oil Removal from Marine Systems. *J. Biotechnol.* **2003**, *106* (2–3), 215–220.

30. Bookstaver, M.; Bose, A.; Tripathi, A. Interaction of *Alcanivorax borkumensis* with a Surfactant Decorated Oil–Water Interface Micelle. *Langmuir* **2015**, *31* (21), 5875–5881.

31. Svetlana, A.; Konnova, Yuri, M. Lvov; Rawil, F. Fakhrullin. Nanoshell Assembly for Magnet-Responsive Oil-Degrading Bacteria. *Langmuir* **2016**, *32*, 12552–12558.

32. Mahmoudi, M.; Azadmanesh, K.; Shokrgozar, M. A.; Journeay, W. S.; Laurent, S. Effect of Nanoparticles on the Cell Life Cycle. *Chem. Rev.* **2011**, *111*, 3407–3432.

33. Kolhatkar, A. G.; Jamison, A. C.; Litvinov, D.; Willson, R. C.; Lee, T. R. Tuning the Magnetic Properties of Nanoparticles. *Int. J. Mol. Sci.* **2013**, *14*, 15977−16009.

34. Ochsenkuhn, M. A.; Jess, P. R. T.; Stoquert, H.; Dholakia, K.; Campbell, C. J. Nanoshells for Surface-Enhanced Raman Spectroscopy in Eurkaryotic Cells: Cellular Response and Sensor Development. *ACS Nano* **2009**, *3*, 3613−3621.

35. Park, H. -H.; Jamison, A. C.; Lee, T. R. Rise of the Nanomachine: The Evolution of a Revolution in Medicine. *Nanomedicine* **2007**, *2*, 425−439.

36. Sanni, M. L.; Kamal, R. A.; Kanj, M. Y. Reservoir Nanorobots. *Saudi Aramco J. Technol.* **2008**, No. Spring, 44−52.

37. Detty, M. R.; Gibson, S. L.; Wagner, S. J. Current Clinical and Preclinical Photosensitizers for Use in Photodynamic Therapy. *J. Med. Chem.* **2004**, *47*, 3897−3915.

38. Collins, K. D.; Taylor, T. M. The Changing Landscape of Teaching High School Chemistry. *J. Chem. Educ.* **2009**, *86*, 21−22.

39. Chen, C. -W.; Tano, D.; Akashi, M. Colloidal Platinum Nanoparticles Stabilized by Vinyl Polymers with Amide Side Chains. Dispersion Stability and Catalytic Activity in Aqueous Electrolyte Solutions. *J. Colloid Interface Sci.* 2000, *225*, 349−358.

40. Lynn, M.; Tarkington, William W. Bryan et al. Magnetic Microorganisms: Using Chemically Functionalized Magnetic Nanoparticles to Observe and Control Paramecia. *J. Chem. Educ.* **2017**, *94*, 85−90.

41. Gurlo, A.; Lauterbach, S.; Miehe, G.; Kleebe, H. -J.; Riedel, R. *J. Phys. Chem. C* **2008**, *112*, 9209.

42. Lili, Wang; Lian, Gao. Morphology-controlled Synthesis and Magnetic Property of Pseudocubic Iron Oxide Nanoparticles. *J. Phys. Chem. C* **2009**, *113*, 15914−15920.

43. Kyoungja, Woo; Jangwon, Hong; Sungmoon, Choi; Hae-Weon, Lee et al. Easy Synthesis and Magnetic Properties of Iron Oxide Nanoparticles. *Chem. Mater.* **2004**, *16*, 2814-2818.

44. Liu, L.; Kou, H. Z.; Mo, W. L.; Liu, H. J.; Wang, Y. Q. *J. Phys. Chem. B* **2006**, *110*, 15218.

45. Kiwi, J.; C€artzel, M. *J. Chem. Soc., Faraday Trans.* **1987**, *83*, 1101.

46. Zheng, Y.; Cheng, Y.; Wang, Y.; Bao, F.; Zhou, L.; Wei, X.; Zhang, Y.; Zheng, Q. *J. Phys. Chem B* **2006**, *110*, 3093.

47. Klotz, M.; Ayral, A.; Guizard, C.; Menager, C.; Cabuil, V. *J. Colloid Interface Sci.* **1999**, *220*, 357.

48. Zhao, W.; Gu, J.; Zhang, L.; Chen, H.; Shi, J. *J. Am. Chem. Soc.* **2005**, *127*, 8916.

49. Zhang, J.; Thurber, A.; Hanna, C.; Punnoose, A. Highly Shape-selective Synthesis, Silica Coating, Self-assembly, and Magnetic Hydrogen Sensing of Hematite Nanoparticles. *Langmuir* **2010**, *26* (7), 5273−5278.

50. Zhang, J.; Liu, H.; Wang, Z.; Ming, N. B. *J. Appl. Phys.* **2008**, *103*, 013517.

51. Morens, D. M.; Folkers, G. K.; Fauci, A. S. The Challenge of Emerging and Re-emerging Infectious Diseases. *Nature* **2004**, *430*, 242−249.

52. Binder, S.; Levitt, A. M.; Sacks, J. J.; Hughes, J. M. Emerging Infectious Diseases: Public Health Issues for the 21st Century. *Science* **1999**, *284*, 1311−1313.

53. Jones, K. E.; Patel, N. G.; Levy, M. A.; Storeygard, A.; Balk, D.; Gittleman, J. L.; Daszak, P. Global Trends in Emerging Infectious Diseases. *Nature* **2008**, *451*, 990−993.

54. O'Connor, S. M.; Taylor, C. E.; Hughes, J. M. Emerging Infectious Determinants of Chronic Diseases. *Emerging Infect. Dis.* **2006,** *12,* 1051–1057.

55. Hoelzer, K.; Switt, A. I. M.; Wiedmann, M. Animal Contact as a Source of Human Non-typhoidal Salmonellosis. *Vet. Res.* **2011,** *42,* 34.

56. Scallan, E.; Griffin, P. M.; Angulo, F. J.; Tauxe, R. V.; Hoekstra, R. M. Foodborne Illness Acquired in the United States-unspecified Agents. *Emerging Infect. Dis.* **2011,** *17,* 16–22.

57. Winter, S. E.; Baumler, A. J. Salmonella Exploits Suicidal Behavior of Epithelial Cells. *Front. Microbiol.* **2011,** 2, DOI: 10.3389/ fmicb.2011.00048

58. Kwon, D.; Joo, J.; Lee, J.; Park, K. H.; Jeon, S. Magnetophoretic Chromatography for the Detection of Pathogenic Bacteria with the Naked Eye. *Anal. Chem.* **2013,** *85,* 7594–7598.

59. Lazcka, O.; Del Campo, F. J.; Munoz, F. X. Pathogen Detection: A Perspective of Traditional Methods and Biosensors. *Biosens. Bioelectron.* **2007,** *22,* 1205–1217.

60. Heo, J.; Hua, S. Z. An Overview of Recent Strategies in Pathogen Sensing. *Sensors* **2009,** *9,* 4483–4502.

61. Lim, D. V.; Simpson, J. M.; Kearns, E. A.; Kramer, M. F. Current and Developing Technologies for Monitoring Agents of Bioterrorism and Biowarfare. *Clin. Microbiol. Rev.* **2005,** *18,* 583–607.

62. Fura, J. M.; Sabulski, M. J.; Pires, M. M. d-Amino Acid Mediated Recruitment of Endogenous Antibodies to Bacterial Surfaces. *ACS Chem. Biol.* **2014,** *9,* 1480–1489.

63. Yong-Tae, Kim; Kook-Han, Kim; Eun, Sung Kang, et al. Synergistic Effect of Detection and Separation for Pathogen Using Magnetic Clusters. *Bioconjugate Chem.* **2016,** *27,* 59–65.

64. El-Boubbou, K.; Gruden, C.; Huang, X. Magnetic Glyco-nanoparticles: A Unique Tool for Rapid Pathogen Detection, Decontamination, and Strain Differentiation. *J. Am. Chem. Soc.* **2007,** *129,* 13392–13393.

65. Chung, H. J.; Castro, C. M.; Im, H.; Lee, H.; Weissleder, R. A Magneto-DNA Nanoparticle System for Rapid Detection and Phenotyping of Bacteria. *Nat. Nanotechnol.* **2013,** *8,* 369–375.

66. Lim, E. K.; Kim, T.; Paik, S.; Haam, S.; Huh, Y. M.; Lee, K. Nanomaterials for Theranostics: Recent Advances and Future Challenges. *Chem. Rev.* **2015,** *115,* 327–394.

67. Ling, D. S.; Hackett, M. J.; Hyeon, T. Surface Ligands in Synthesis, Modification, Assembly and Biomedical Applications of Nanoparticles. *Nano Today* **2014,** *9,* 457–477.

68. Huh, Y. M.; Jun, Y. W.; Song, H. T.; Kim, S.; Choi, J. S.; Lee, J. H.; Yoon, S.; Kim, K. S.; Shin, J. S.; Suh, J. S., et al. In vivo Magnetic Resonance Detection of Cancer by Using Multifunctional Magnetic Nanocrystals. *J. Am. Chem. Soc.* **2005,** *127,* 12387–12391.

69. Sun, S.; Anders, S.; Hamann, H. F.; Thiele, J. -U.; Baglin, J. E. E.; Thomson, T.; Fullerton, E. E.; Murray, C. B.; Terris, B. D. Polymer Mediated Self-Assembly of Magnetic Nanoparticles. *J. Am. Chem. Soc.* **2002,** *124,* 2884–2885.

70. Cheng, C. J.; Lin, C. C.; Chiang, R. K.; Lin, C. R.; Lyubutin, I. S.; Alkaev, E. A.; Lai, H. Y. Synthesis of Monodisperse Magnetic Iron Oxide Nanoparticles from Submicrometer Hematite Powders. *Cryst. Growth Des.* **2008,** *8,* 877–883.

71. Wu, H.; Zhu, H.; Zhuang, J.; Yang, S.; Liu, C.; Cao, Y. C. Water-soluble Nanocrystals Through Dual-interaction Ligands. *Angew. Chem. Int. Ed.* **2008**, *47*, 3730–3734.

72. Burnette, W. N. Western Blotting": Electrophoretic Transfer of Proteins from Sodium Dodecyl Sulfate-Polyacrylamide Gels to Unmodified Nitrocellulose and Radiographic Detection with Antibody and Radioiodinated Protein A. *Anal. Biochem.* **1981**, *112*, 195–203.

73. Park, S.; Mohanty, N.; Suk, J. W.; Nagaraja, A.; An, J. H.; Piner, R. D.; Cai, W. W.; Dreyer, D. R.; Berry, V.; Ruoff, R. S. Biocompatible, Robust Free-Standing Paper Composed of a TWEEN/Graphene Composite. *Adv. Mater.* **2010**, *22*, 1736–1740.

74. Ren, W. L.; Tian, G.; Jian, S.; Gu, Z. J.; Zhou, L. J.; Yan, L.; Jin, S.; Yin, W. Y.; Zhao, Y. L. TWEEN Coated NaYF4:Yb,Er/ NaYF4 Core/shell Upconversion Nanoparticles for Bioimaging and Drug Delivery. *RSC Adv.* **2012**, *2*, 7037–7041.

75. Cho, E. J.; Yang, J.; Mohamedali, K. A.; Lim, E. K.; Kim, E. J.; Farhangfar, C. J.; Suh, J. S.; Haam, S.; Rosenblum, M. G.; Huh, Y. M. Sensitive Angiogenesis Imaging of Orthotopic Bladder Tumors in Mice Using a Selective Magnetic Resonance Imaging Contrast Agent Containing VEGF121/rGel. *Invest. Radiol.* **2011**, *46*, 441–449.

76. Bai, F.; Wang, D. S.; Huo, Z. Y.; Chen, W.; Liu, L. P.; Liang, X.; Chen, C.; Wang, X.; Peng, Q.; Li, Y. D. A Versatile Bottomup Assembly Approach to Colloidal Spheres from Nanocrystals. *Angew. Chem. Int. Ed.* **2007**, *46*, 6650–6653.

77. Hyun Jung, Chung; Thomas, Reiner; Ghyslain, Budin, et al. Ubiquitous Detection of Gram-Positive Bacteria with Bioorthogonal Magnetofluorescent Nanoparticles. *Am. Chem. Soc.* **2011**, *5* (11), 8834–8841.

78. Menichetti, F. Current and Emerging Serious Gram-Positive Infections. *Clin. Microbiol. Infect.* **2005**, *11*, 22–28.

79. Woodford, N.; Livermore, D. Infections Caused by Gram-Positive Bacteria: A Review of the Global Challenge. *J. Infect.* **2009**, *59*, S4–16.

80. Nagarajan, R. Antibacterial Activities and Modes of Action of Vancomycin and Related Glycopeptides. *Antimicrob. Agents Chemother.* **1991**, *35*, 605–609.

81. Reynolds, P. Structure, Biochemistry and Mechanism of Action of Glycopeptide Antibiotics. *Eur. J. Clin. Microbiol. Infect. Dis.* **1989**, *8*, 943–950.

82. Hoffman, L. R.; D'Argenio, D. A.; MacCoss, M. J.; Zhang, Z.; Jones, R. A.; Miller, S. I. Aminoglycoside Antibiotics Induce Bacterial Biofilm Formation. *Nature* **2005**, *436*, 1171–1175.

83. Furno, F.; Morley, K. S.; Wong, B.; Sharp, B. L.; Arnold, P. L.; Howdle, S. M.; Bayston, R.; Brown, P. D.; Winship, P. D.; Reid, H. J. Silver Nanoparticles and Polymeric Medical Devices: A New Approach to Prevention of Infection. *J. Antimicrob. Chemother.* **2004**, *54*, 1019–1024.

84. Klevens, R. M.; Edwards, J. R.; Richards, C. L.; Horan, T. C.; Gaynes, R. P.; Pollock, D. A.; Cardo, D. M. Estimating Health Care-Associated Infections and Deaths in US Hospitals, 2002. *Public Health Rep.* **2007**, *122*, 160–166.

85. Tamar, L.; Jay, G.; Frank, R. E.; Liliana, G. Pediatric Vancomycin Use in 421 Hospitals in the United States, 2008. *PLoS One* **2012**, *7*, e43258.

86. Hong-Zheng, Lai; Wei-Yu, Chen; Ching-Yi, Wu; Yu-Chie, Chen. Potent Antibacterial Nanoparticles for Pathogenic Bacteria. *ACS Appl. Mater. Interfaces* **2015**, *7*, 2046–2054.

87. Gu, H.; Ho, P.; Tong, E.; Wang, L.; Xu, B. Presenting Vancomycin on Nanoparticles to Enhance Antimicrobial Activities. *Nano Lett.* **2003**, *3*, 1261–1263.

88. Kell, A. J.; Stewart, G.; Ryan, S.; Peytavi, R.; Boissinot, M.; Huletsky, A.; Bergeron, M. G.; Simard, B. Vancomycin-Modified Nanoparticles for Efficient Targeting and Preconcentration of Gram- Positive and Gram-Negative Bacteria. *ACS Nano* **2008**, *2*, 1777–1788.

89. Gu, H.; Ho, P.-L.; Tsang, K. W.; Wang, L.; Xu, B. Using Biofunctional Magnetic Nanoparticles to Capture Vancomycin- Resistant Enterococci and Other Gram-Positive Bacteria at Ultralow Concentration. *J. Am. Chem. Soc.* **2003**, *125*, 15702–15703.

90. Muriel, Behra; Nahid, Azzouz; Stephan, Schmidt; Dmitry V., Volodkin, et al., Magnetic Porous Sugar-Functionalized PEG Microgels for Efficient Isolation and Removal of Bacteria from Solution. *Biomacromolecules* **2013**, *14*, 1927–1935.

91. Huang, W.-C.; Tsai, P.-J.; Chen, Y.-C. Functional Gold Nanoparticles as Photothermal Agents for Selective-Killing of Pathogenic Bacteria. *Nanomedicine* **2007**, *2*, 777–787.

92. Huang, W.-C.; Tsai, P.-J.; Chen, Y.-C. Multifunctional Fe_3O_4@Au Nanoeggs as Photothermal Agents for Selective Killing of Nosocomial and Antibiotic-Resistant Bacteria. *Small* **2009**, *5*, 51–56.

93. Qi, G.; Li, L.; Yu, F.; Wang, H. Vancomycin-Modified Mesoporous Silica Nanoparticles for Selective Recognition and Killing of Pathogenic Gram-Positive Bacteria Over Macrophage-Like Cells. *ACS Appl. Mater. Int.* **2013**, *5*, 10874–10881.

94. Yu, T.-J.; Li, P.-H.; Tseng, T.-W.; Chen, Y.-C. Multifunctional Fe_3O_4/Alumina Core/Shell MNPs as Photothermal Agents for Targeted Hyperthermia of Nosocomial and Antibiotic-Resistant Bacteria. *Nanomedicine* **2011**, *6*, 1353–1363.

95. Durán, N.; Marcato, P. D.; Durán, M.; Yadav, A.; Gade, A.; Rai, M. Mechanistic Aspects in the Biogenic Synthesis of Extracellular Metal Nanoparticles by Peptides, Bacteria, Fungi, and Plants. *Appl. Microbiol. Biotechnol.* **2011**, *90*, 1609–1624.

96. Rai, M. *Metal Nanoparticles in Microbiology*; Springer: Berlin, 2011.

97. Das, S. K.; Liang, J.; Schmidt, M.; Laffir, F.; Marsili, E. Biomineralization mechanism of gold by zygomycete fungi Rhizopous oryzae. *ACS Nano* **2012**, *6* (7), 6165–6173.

98. Quester, K.; Avalos-Borja, M.; Castro-Longoria, E. Biosynthesis and Microscopic Study of Metallic Nanoparticles. *Micron* **2013**, *54–55*, 1–27.

CHAPTER 11

RESEARCH PROGRESS AND NEW INSIGHTS IN BIOSYNTHESIS OF SILVER NANOPARTICLES WITH PARTICULAR APPLICATIONS

DEBARSHI KAR MAHAPATRA[1,*] and SANJAY KUMAR BHARTI[2]

[1]*Department of Pharmaceutical Chemistry, Dadasaheb Balpande College of Pharmacy, Nagpur 440037, Maharashtra, India*

[2]*Institute of Pharmaceutical Sciences, Guru Ghasidas Vishwavidyalaya (A Central University), Bilaspur 495009, Chhattisgarh, India*

Corresponding author. E-mail: mahapatradebarshi@gmail.com

ABSTRACT

The plant mediated biosynthesis of silver nanoparticles (AgNPs) is an attractive method for commercial fabrication of nanomaterials with indispensable environment protection attributes. The central theme of this article is to provide an approach for replacing the influence of certain harmful chemicals like sodium borohydride, dimethyl formamide, hydrazine hydrate, ethylene glycol, etc. in the fabrication of AgNPs, with plant extracts for overcoming the drawbacks. The present chapter comprehensively focuses on the common techniques and principles for the fabrication of AgNPs by the utilization of some common plant biomass. The content focuses on the information of specific plant species, plant part employed, size of produced nanoparticles, etc. In addition to it, the nanomedicinal applications of AgNPs as anti-cancer (in vivo and cell-line), anti-diabetic, anti-parasite (malaria, dengue, and filariasis), anti-inflammatory, anti-platelet, theranostic utility, and anti-microbial (bacteria and fungi) as well as their impact on environment have been highlighted exclusively. The probable mechanisms of actions of AgNP-based products (Acti-coat®, SilvaSorb®, Silverline®, ON-Q-SilverSoaker®) exerted via various biochemical pathways as well as the plausible mechanism of formation of reduced silver forms from plant extracts are also shown exhaustively.

11.1 INTRODUCTION

The nanotechnology in biomedical research and clinical practice emerged as nanomedicine that could potentially make a major impact on human health.[213] Nanotechnology has revolutionized all fields across the globe and opened up several frontiers in nanobiotechnology,[348] nanopharmaco-therapeutics,[178] and material and applied sciences.[257] The role of silver (Ag) materials as bactericidal agents is known for ages and many archives have highlighted the role and utilization of Ag-fabricated products in their daily life. The synthesis of silver nanoparticles (AgNPs) is a dynamic field in academics and applied research. AgNPs are increasingly used in therapy, diagnostics, imaging, and targeted drug delivery due to their specific func-tion at the cellular, atomic and molecular levels.[212] Recently, AgNPs gained key importance in biomedical sciences,[433] drug delivery,[369] catalysis,[269] optics,[133] and photoelectrochemistry.[367] Nanomedicine has potential to integrate diagnostics/imaging with therapeutics and facilitates the develop-ment of theranostics for personalized medicine.[283] Generally, the processes employed for the synthesis of nanoparticles are broadly classified into two types: the "top-down" process and "bottom-up" process (Fig. 11.1). In "top-down" approach, the bulk material breaks down into particles at the nanoscale with various lithographic procedures like grinding, milling, etc., whereas in "bottom-up" approach, the atoms bring together to form a nucleus which grows eventually into a particle of nanosize.[397]

A variety of chemical and physical processes have been reported for the synthesis of metallic nanoparticles. The use of biological resources in synthesis of nanomaterials is rapidly developing due to their growing success, ease of formation of nanoparticles, economic, and ecofriendly in nature.[196] Typically, a plant-mediated bioreduction process occurs when the aqueous extract of plant reacts with an aqueous solution of the metal salt (here, $AgNO_3$) (Fig. 11.2). The complete reaction occurs at room temperature within a few minutes. Due to presence of a wide variety of chemicals, the bioreduction process is relatively complex. Biological

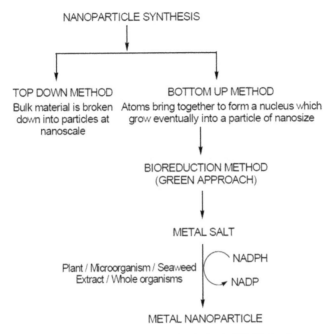

FIGURE 11.1   Top-down and bottom-up methods for AgNP fabrication.

synthesis of nanoparticles using plant extracts are relatively scalable and less expensive compared with microbial processes.[235] The source of the plant extract also influences the characteristics of nanoparticles because of varying concentrations and combinations of organic reducing agents.[92] Thus, there is a need to explore ecofriendly green synthetic protocols for the synthesis of AgNPs.

11.2   NANOMEDICINE PERSPECTIVES OF AgNPs

11.2.1   *AgNPs AS ANTI-DIABETIC AGENTS*

Silver nanomaterials obtained from the *Tephrosia tinctoria* aqueous extract demonstrated increased glucose uptake in muscular cells and noteworthy α-amylase and α-glucosidase inhibitory potential with inhibition percentages of 94.76 and 83.52, respectively. The mediated pharmacological activity may be due to the antioxidant potential of the AgNPs.[308]

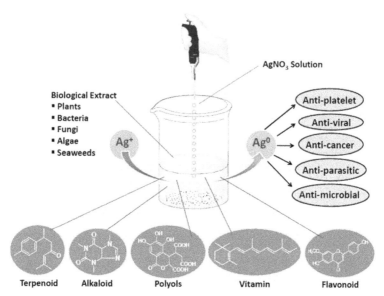

FIGURE 11.2   Diagrammatic representation of common technique for the fabrication of AgNPs.

11.2.2 AgNPs AS THERANOSTIC AGENT

Although at present in a very stage for use in mammals, the biosynthesized AgNPs find applications as a future generation nanomedicine for the advances in cancer theranostic agents (imaging, sensing, and therapy) owing to their physicochemical attributes. The safety and effectiveness of fabricated AgNPs have not yet established very well. The translation process of the existing knowledge and the emerging technologies into a few application-oriented commercial products for the simultaneous imaging and pharmacotherapeutics by overcoming the main challenges and issues will be the future perspective.[246,283]

11.2.3 AgNPs AS ANTI-INFLAMMATORY AGENTS

Noteworthy, anti-inflammatory perspective has been demonstrated by the biosynthesized AgNPs without any specific toxic effects. The in vivo experimental findings revealed that the AgNPs expressed a significant decrease in inflammatory responses in the peritoneal adhesions and their clinical application provides directions toward the postoperative adhesion prevention. The study presented   substantiation for comprehending the mechanism of anti-inflammatory action(s).[438]

11.2.4   AgNPs AS ANTI-PLATELET AGENTS

Considerable in vitro integrin-mediated platelet responses suppression has been exhibited by the plant extract-mediated synthesized AgNPs of size 15–30 nm in a dose-dependent manner. On focusing the mechanism of action of the AgNPs, it was seen that the compound accrues inside the platelet and diminishes the interplatelet proximity. AgNPs do not have any lytic effect on the human platelets which opens new avenues of new-generation antithrombotic agents.[368] In a similar manner, AgNPs of size range 10–50 nm procured from *Brevibacterium casei* have remarkable anticoagulant potentials where AgNPs adheres with the immobilized collagen, thereby, preventing the platelet aggregation and formation of fibrin clot along with inhibiting the integrin-mediated platelet function in a dose-dependent manner. The nanoparticles are fabricated from the proteins present in the bacterium which acts as possible bioreductants and capping agent, thereby marked stability to the nanoparticles.[163]

11.2.5   AgNPs AS ANTI-MICROBIAL AGENTS

Nanoparticles of Ag find importance as bactericidal from millenniums. Even in the old literature, Ag in ionic form or in any other forms has received adequate attention. At present, AgNPs have been reported to exert bactericidal effect on a large number of severely pathogenic bacterial species such as *Aeromonas hydrophilla, Bacillus cereus, Corney bacterium, Enterobacter cloacae, Escherichia coli, Klebsiella pneumoniae, Micrococcus Luteus, Proteus mirabilis, Proteus vulgaris, Pseudomonas aeruginosa, Salmonella enterica, Salmonella paratyphi, Shigella dysenteriae, Shigella sonnei, Staphylococcus aureus, Streptococcus pyogens, Staphylococcus saprophyticus,* and *Vibrio cholera.*[102] The ionized Ag forms distinctly interact with the thiol part of proteins, bacterial plasmid, DNA bases, peptidoglycan structures, plasma membrane, release lipid peroxide, and bacterial protein assembly (ribosomes) that cause inactivation of the cellular enzymes, generate reactive oxygen species (by strongly reacting with the electron-donating groups), increase permeation of the cell membrane and mitochondrial systems, ceases DNA replication process, detaches respiratory electron transport from the oxidative phosphorylation, protein synthesis inhibition, causes DNA condensation, alters membrane permeability, deformation of microcellular constitution, therefore, ultimately leading to complete bacterial growth inhibition and inactivation.[65] The bactericidal mechanism in *E. coli* by AgNPs was comprehensively studied by Li et al.[205] where the nanomaterial at concentration of 100 mg/mL forms a pit in the cell membrane causing oozing of the proteins and reducing sugars from the cytoplasm (Fig. 11.3).

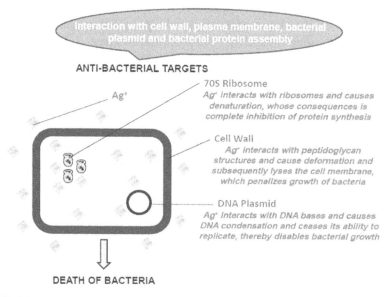

FIGURE 11.3   Anti-bacterial mechanism of action(s) of AgNPs.

11.2.6   *AgNPs AS ANTIPARASITIC AGENTS*

The plant extract-mediated AgNPs have been reported to exhibit noteworthy larvicidal and pupicidal activity. Malaria, dengue, and filariasis are the chief diseases transmitted by the malaria vector (by bite of infected mosquitoes) in the tropical and subtropical regions like Africa.[437] Biomass-mediated synthesized AgNPs play pivotal activity against *Aedes aegypti, Anopheles stephensi, Anopheles subpictus, Culex gelidus, Culex quinquefasciatus, Culex tritaeniorhynchus, Haemaphysalis bispinosa, Hippobosca maculate, Hyalomma anatolicumanatolicum, Hyalomma marginatumisaaci, Pediculus humanus capitis*, etc.

AgNPs synthesized from the leaf extract of *Andrographis paniculata* have been reported to express antiplasmodial activity against *Plasmodium falciparum* with IC_{50} values of 83% at 100 µg/mL.[273] A significant *Anopheles* larvicidal activity has been displayed by the AgNPs produced from the plant biomass. The larvicidal activity of *Gmelina asiatica* aqueous leaf extract,[251] aqueous leaf extracts of *Nelumbo nucifera*,[346] leaf extract of *Euphorbia hirta*,[301] floral extract-mediated biosynthesized AgNPs from *Chrysanthemum indicum* ,[22] aqueous leaf extracts of *Musa paradisiaca* L.,[151] and biosynthesized AgNPs using *Murraya koenigii* leaf extract[384] against late third instar larvae of *Anopheles stephensi*. Considerable larvicidal activity against *Aedes aegypti* have been observed by biosynthesized AgNPs from aqueous extract from *Eclipta prostrate*,[306] plant latex of *Plumeria rubra*,[281] and leaf extract of *Mukia maderaspatana*.[68] Remarkable larvicidal potentials of plant extract-mediated AgNPs have been noticed extensively. Bark aqueous extract of *Ficus racemosa*,[420] *Couroupita guianensis* leaf and fruit extracts[429] against *Culex quinquefasciatus* and *Culex gelidus*.

Similarly, authors also reported the pediculocidal and larvicidal activities of biosynthesized AgNPs from aqueous leaf extract of *Tinospora cordifolia* against the head louse *Pediculus humanus capitis* and fourth instar larvae of malaria vector, *Anopheles subpictus* and filariasis vector, *Culex quinquefasciatus*.[150] Aqueous leaf extract of *Ocimum canum* mediated synthesis have shown antiparasitic activities against the larvae of *Hyalomma anatolicum* (a.) *anatolicum* and *Hyalomma marginatum* (m.) *isaaci*.[151]

Plant-mediated synthesized AgNPs of size 70–140 nm using aqueous leaf extract of *Manilkara zapota* to control *Rhipicephalus (Boophilus) microplus* was reported with LC_{50} values of 16.72 and 3.44 mg/L, respectively.[306]

11.2.7 AgNPs AS ANTICANCER AGENTS

Cytotoxicity and genotoxicity remained the mainstream biological activity of the biofabricated AgNPs. At low concentration, AgNPs on contact with the lysosome under acidic environment exert pronounced cytotoxicity by directly liberating reactive oxygen species (ROS) production.[449] The ionized Ag escapes from the lysosomes and supplementary tempt for the intensification of the intracellular ROS (like hydroxyl radicals, superoxide anions, and hydrogen peroxide) and subsequently diffuse into the nucleus via nuclear pore complexes causing chromosomal abnormalities and produce DNA damage.[24] The ionized Ag interacts with the cellular components and release lipid peroxide which leads to cytoplasmic contents leakage, resulting in the rupture of lysosomal membranes, activation of lysosome-mediated apoptosis, mitochondria injury which often give rise to necrosis and impairs electron transfer system (Fig. 11.4). In addition to it, in vivo Ag^+ release occur in contact with H_2O_2 which results in metabolism upregulation and activation of genes essentially linked with the oxidative stress, thus, greatly elevates the ROS production.[435]

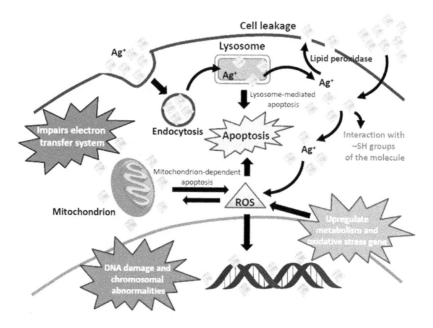

FIGURE 11.4   Anti-cancer mechanism of action(s) of AgNPs.

Metals and metal containing compounds possess a tremendous high nucleolytic potential and exhibit cytotoxicity.[56] In recent times, the cytotoxic potentials of AgNPs on numerous human cell lines like A549, BRL3A, G292, H157, HaCaT, HeLa, HEp-2, HepG2, HL-60, HSC-3, HT144, IMR-90, KB, Lu, MCF-7, U251, Vero, etc. have been reported by various researchers. In vitro cytotoxic potential of AgNPs have been reported against HepG2, KB, Lu, and MCF-7 with IC_{50} range of 3.74–6.09

μg/mL.[402] The potential of AgNPs on the skin were evaluated against HaCaT cell line with IC_{50} value of 6.8 μM after 7 days of contact which concluded a long-lasting cell growth inhibition.[444] Leaf extract of *Piper longum* mediated fabrication of AgNPs of size range 17–41 nm expressed striking cytotoxic effect against Hep-2 cell line.[144] A noticeable antiproliferative activity of AgNPs fabricated from *Citrullus colocynthis* calli extract have been reported where the prepared nanomaterials significantly reduced the HEp2 viability to 50% of the initial level via apoptosis process.[355] Biocompatible AgNPs of size range 10–20 nm fabricated from table sugar exhibited excellent antitumor perspective against both HT144 and H157 cell lines.[264] In an interesting study, fabricated stable AgNPs from *Melia azedarach* displayed dose-dependent cytotoxic action against HeLa cells with LD_{50} value of 300 μg/mL.[385]

In search for chemonanotherapeutics, Nano-Ag biosynthesized from the *Syzygium cumini* fruit extract expressed potential anticancer activity against dalton lymphoma cell line.[234] A dose-dependent cytotoxic action against HeLa cell line have been demonstrated by AgNPs biofabricated from *Calotropis gigantean* latex extract with IC_{50} value of 91.3 μg/mL.[313] Nanoparticles of Ag biosynthesized from the *Eucalyptus chapmaniana* leaf extract wielded a dose-dependent cytotoxic activity against HL-60 cell line.[386] In a similar research, screening of *Eucalyptus chapmaniana* mediated fabricated AgNPs against HL-60 cell line presented an effective dose-dependent decline in cellular viability.[386] In the direction of exploring the potentials of biofabricated AgNPs, *Erythrina indica* root extract mediated synthesized Ag nanomaterial demonstrated noteworthy effect against MCF-7 and HepG2 cell lines with IC_{50} values of 23.89 and 13.86 μg/mL, respectively.[383] The screening of a marketed product (Nanocid®) of mean particle size 7–20 nm against osteoblast (G292) cell line revealed that the product exerted concentration-dependent toxicity with IC_{50} of 3.42 μg/mL.[238]

11.3 NANOSILVER-BASED PRODUCTS IN MARKET

In the recent years, a few innovative products have come to the market for clinical utility. The nanocrystalline AgNP product Acticoat® as wound dressing in burns to improve wound healing.[90] SilvaSorb® is a product of Medline composed of AgNP-based hand gel used for personal and medical uses.[63] Recently, Ag-coated catheters Silverline® and ON-Q-SilverSoaker® finds application for the drainage of fluids and drug delivery.[65]

11.4 INDIAN PLANTS AS BIOMASS FOR THE REDUCTION OF SILVER NITRATE AND THEIR PROBABLE MECHANISMS

India is privileged for its wide biodiversity. Several plants grow in this subcontinent due to different climatic conditions. Commonly, *Allium sativum, Camellia sinensis, Terminalia catechu, Mentha piperita, Parthenium hysterophorus, Carica papaya, Citrus limon, Datura metel, Piper longum, Swietenia mahogany, Glycyrrhiza glabra, Andrographis paniculata, Astragalus gummifer, Piper betle, Piper nigrum, Azadirachta indica, Allium cepa,* etc. are the main plants present. Several researchers have reported the fabrication of AgNPs from these biomasses (Table 11.1).

TABLE 11.1   Some Common Indian Plants Used for Biosynthesis of AgNPs.

S. no.	Plant	Plant part	Size of AgNPs (nm)	References
	Abutilon indicum	Leaf	5–25	[228]
	Abutilon indicum	Leaf	7–17	[33]
	Acalypha indica	Leaf	20–30	[187]
	Allium cepa	Whole plant	33.6	[359]
	Allium sativum	Whole plant	3–6	[436]
	Andrographis paniculata	Leaf	13–27	[392]
	Andrographis paniculata	Leaf	35–55	[273]
	Andrographis paniculata	Leaf	40–60	[377]
	Astragalus gummifer	Gum	12–14	[184]
	Azadirachta indica	Leaf	5–35	[366]
	Azadirachta indica	Leaf	15–20	[404]
	Azadirachta indica	Leaf	NA*	[201]
	Azadirachta indica	Leaf	20–50	[229]
	Azadirachta indica	Leaf	NA	[330]
	Bacopa monniera	Leaf	10–30	[215]

TABLE 11.1   *(Continued)*

S. no.	Plant	Plant part	Size of AgNPs (nm)	References
	Camellia sinensis	Leaf	40	[428]
	Camellia sinensis	Leaf	4	[207]
	Carica papaya	Fruit	15	[147]
	Carica papaya	Callus	60–80	[243]
	Citrus limon	Fruit	2–10	[360]
	Citrus limon	Fruit	50	[298]
	Citrus limon	Fruit	75	[326]
	Citrus sinensis	Peel	35	[172]
	Citrus sinensis	Peel	91	[36]
	Datura metel	Leaf	16–40	[174]
	Datura metel	Flower	20–30	[265]
	Datura metel	Leaf	20–36	[398]
	Datura metel	Leaf	50–100	[270]
	Glycyrrhiza glabra	Root	20–30	[83]
	Murraya koenigii	Leaf	10	[287]
	Murraya koenigii	Leaf	10–25	[71]
	Murraya koenigii	Leaf	20–35	[384]
	Murraya koenigii	Leaf	35.43	[52]
	Ocimum sanctum	Leaf	10–20	[287]
	Ocimum sanctum	Leaf	42	[325]
	Ocimum sanctum	Root and stem	8–12 (root) 4–6 (stem)	[4]
	Ocimum sanctum	Leaf	3–20	[219]
	Ocimum sanctum	Leaf	18	[318]
	Ocimum sanctum	Leaf	40–50	[33]
	Ocimum sanctum	Leaf	4–30	[376]
	Piper betle	Leaf	5–30	[319]
	Piper longum	Leaf	17–41	[144]
	Piper nigrum	Flower	5–50	[111]
	Swietenia mahogany	Leaf	10–25	[240]
	Zingiber officinale	Leaf	10–30	[373]

NA = not available

The mechanism of AgNPs formation by the bioreduction process is quite interesting. The constituents present in plant extracts like vitamins, polyphenols, enzymes, alkaloids, proteins, terpenoids, amino acids, flavonoid, polysaccharides, etc. and other chemically complex environmentally benevolent substances are supposed to play imperative role in the reduction of Ag^+ ion Ag^0. White II et al.[436] synthesized AgNPs of size 3–6 nm using *Allium sativum* extract. The reduction of metal salts into metal nanoparticles takes place by sugar components like sucrose, fructose, and other reducing sugars. Daniel et al.[74] fabricated AgNPs using *Achyranthus aspera* extract containing various polyol principles which account for exhibiting bioreduction of Ag. Similarly, polyols present in extracts of *Anacardium occidentale, Mirabilis jalapa*, etc. are reported to exhibit bioreduction of metal. Polyphenols or the "tannins" present in various plants is one of the prime principles for nano-Ag biofabrication. Various plant extracts like *Camellia sinensis, Dioscorea bulbifera, Dioscorea oppositifolia, Leonuri herba, Terminalia catechu*, etc. exhibited bioreduction by their polyphenolic components. Rajaram et al.[308] investigated and characterized the rich phenolic and flavonoids groups of compounds in *Tephrosia tinctoria* using phytochemical quantification and FT-IR. Authors revealed that during the green synthesis, the isoflavonoids and phenolic groups were attached with the Ag to form AgNPs. Thus, concluded the role of biochemical components in fabrication of nanoparticles. Some of the individual components present in plant extracts are splendidly known to exhibit bioreduction of Ag metal salts. Few phytopharmaceuticals like plastohydroquinone (in *Datura metel*),[174] menthol (in *Mentha piperita*),[9] oxalic acid (in *Chenopodium album*) ,[92] ascorbic acid (in *Allium cepa*),[276] neem principles (salanin, nimbin, azadirone, azadirachtin),[345] citric acid (in *Citrus limon*),[298] piperine (in *Piper longum*),[144] eugenol (in *Szyygium aromaticum*),[371] sorbic acid (in *Sorbus aucuparia*),[87] verbascoside, isoverbascoside, luteolin, and chrysoeriol-7-*O*-diglucoronide (in *Lippia citriodora*),[73] curacycline (in *Jatropha curcas*),[46] catechin (in *Parthenium hysterophorus*),[275] thiamine (in *Glycyrrhiza glabra*) ,[83] hydroxylimonoids (in *Swietenia mahogany*),[240] etc. are chief components for biosynthesis of AgNPs. Proteins have also been identified as potential constituent for exhibiting bioreduction of silver metal salt into its corresponding nanoparticle. Excellent fabrications of AgNPs are reported from extracts of *Astragalus gummifer*, *Hydrilla verticilata*, *Pedilanthus tithymaloides*, *Piper betle*, *Piper nigrum*, *Plumeria rubra*, etc. containing various types of proteins. A large number of antioxidant components gifted by nature serve excellent reducing and capping agents for production of silver nanomaterials. Phytoconstituents like flavonoids, terpenoids, hydroxy-flavones, flavones, glycoside, limonoids, etc. present in plant extracts like *Andrographis paniculata*, *Carica papaya*, *Coleus aromaticus*, *Glycyrrhiza glabra*, *Lantana camara*, *Morinda pubescens*, *Parthenium hysterophorus*, *Swietenia mahogany*, *Trianthema decandra*, etc. have demonstrated rapid reduction of silver ion into silver metal and it gets oxidized simultaneously.[170]

11.5 IMPACT OF AgNPs ON ENVIRONMENT

Although AgNPs have wide range of applications, due to their small size and high surface area, they tend to react very swiftfully to produce several adverse effects. These nanomaterials in the size range of 10–500 nm are quite toxic to the aquatic life and fish often bioconcentrates the aquatic contaminants which in the successive food cycles or food chains affect the human health. Several policies and guidelines have been framed by the regulatory bodies for the prevention and control of nanomaterials from mixing, deposition, accumulation of these metallic wastes in fields. To combat such situations, improved analytical techniques are must to detect the exposure of AgNPs and will have attributes for differentiating AgNPs from the ionic silver species in the environment.

11.6 FUTURE DIRECTIONS

Nanotechnology is a subject that cannot be neglected and is the coming future. Till date, nanomaterials of size 1–100 nm have come into varied applications and a number of advanced materials are coming in the near years. None of the fields, whether, it will be therapeutics, diagnostics, theranostics, biomedical research and clinical practices, all utilize nano-materials for the specific function at the cellular, atomic, and molecular levels. But, meeting so much high requirement of nanomaterials since their applications in wide variety of activities like antiparasitic, anticancer, antimicrobial, anti-inflammatory, antiviral, etc. by using the most prefer-able economic wet chemical approach is the most hazardous technique as it severly affects the environment due to toxic, production of harmful byproducts, and flammable nature of chemicals. Recently, the world has witnessed the impact of climatic change and waste disposal problems due to the applications of these harmful approaches over the years. A shift to green approach will revolutionize the modern ways of nanomaterial production. The plant extract mediated ecofriendly, simple, and size/shape-controlled biosynthesis of AgNPs will be an impressive approach to meet the demands of the upcoming generation. The bulk production of nanoparticles by the efficient green chemistry method (biogenic approach) at higher synthesis rate will replace the traditional synthetic methods.

11.7 CONCLUSION

In this decade, global warming, climatic changes, and environment destruction remained the burning issues across the globe. Nature itself is the greatest inspiration to mankind. All hidden secrets are present in the lap of Mother Nature. The chapter will provide an approach for replacing the influence of certain harmful chemicals like sodium borohydride, dimethyl formamide, hydrazine hydrate, ethylene glycol, etc. in the fabri-cation of AgNPs, with plant extracts for overcoming the drawbacks. The extract-mediated synthesis of therapeutic nanostructures has high efficacy, is easily obtainable, economic, environment friendly, and sustainable method. However, few associated disadvantages such as sluggish process, to some extent complicated in attaining the desired size/shape, nonspecific conjugation of phytoconstituents during the synthesis process, and failure to validate the precise way(s) for nanoparticles development remained few challenges in the application of natural bioreductants. In spite of all these issues, the plant-mediated biosynthesis of AgNPs is an attractive method for commercial fabrication of nanomaterials with indispensable environ-ment protection attributes.

KEYWORDS

- nanoparticles
- silver
- fabrication
- nanomedicine
- nanoproducts
- mechanisms
- plant extract

REFERENCES

1. Abdel-Aziz, M. S.; Shaheen, M. S.; El-Nekeety, A. A.; Abdel-Wahhab, M. A. Antioxidant and Antibacterial Activity of Silver Nanoparticles Biosynthesized Using Chenopodium Murale Leaf Extract. *J. Saudi Chem. Soc.* **2014,***18*, 356–363.
2. Ahmad, A.; Mukherjee, P.; Senapati, S.; Mandal, D.; Khan, M. I.; Kumar, R.; Sastry, M. Extracellular Biosynthesis of Silver Nanoparticles Using the Fungus Fusarium Oxysporum. *Colloids Surf. B Biointerfaces* **2003,** *28*, 313–318.
3. Ahmad, N.; Sharma, S. Green Synthesis of Silver Nanoparticles Using Extracts of *Ananas comosus. Green Sustainable Chem.* **2012,** *2*, 141–147.
4. Ahmad, N.; Sharma, S.; Alam, M. K.; Singh, V. N.; Shamsi, S. F.; Mehta, B. R.; Fatma, A. Rapid Synthesis of Silver Nanoparticles Using Dried Medicinal Plant of Basil. *Colloids Surf. B Biointerfaces* **2010,** *81*, 81–86.
5. Ahmad, N.; Sharma, S.; Rai, R. Rapid Green Synthesis of Silver and Gold Nanoparticles Using Peels of *Punica granatum. Adv. Mater. Lett.* **2012,** *3* (5), 376–380.
6. Ahmed, S.; Ahmad, M.; Swami, B. L.; Ikram, S. A Review on Plants Extract Mediated Synthesis of Silver Nanoparticles for Antimicrobial Applications: A Green Expertise. *J. Adv. Res.* **2015,** *7* (1), 17–28. http://dx.doi.org/10.1016/j.jare.2015.02.007.
7. Ajitha, B. Y.; Reddy, A. K.; Sreedhara, P. R. Biosynthesis of Silver Nanoparticles Using Plectranthus Amboinicus Leaf Extract and Its Antimicrobial Activity. *Spectrochim. Acta A Mol. Biomol. Spectrosc.* **2014,** *128*, 257–262.
8. Ali, D. M.; Thajuddin, N.; Jeganathan, K.; Gunasekaran, M. Plant Extract Mediated Synthesis of Silver and Gold Nanoparticles and Its Antibacterial Activity Against Clinically Isolated Pathogens. *Colloids Surf B Biointerfaces* **2011,** *85*, 360–365.
9. Ali, D. M.; Sasikala, M.; Gunasekaran, M.; Thajuddin, N. Biosynthesis and Characterization of Silver Nanoparticles Using Marine Cyanobacterium, *Oscillatoria Willei* NTDM01. *Digest J. Nanomaterials Biostruct.* **2011,** *6* (2), 385–390.
10. Amaladhas, T. P.; Sivagami, S.; Akkini, T. D.; Ananthi, N.; Velammal, S. P. Biogenic Synthesis of Silver Nanoparticles by Leaf Extract of *Cassia angustifolia. Adv. Nat. Sci.: Nanosci. Nanotechnol.* **2012,** *3*, 045006.

11. Amaladhas, T. P.; Usha, M.; Naveen, S. Sunlight Induced Rapid Synthesis and Kinetics of Silver Nanoparticles Using Leaf Extract of *Achyranthus Aspera* and Their Antimicrobial Applications. *Adv. Mater. Lett.* **2013,** *4* (10), 779–785.

12. Amaladhas, T. P.; Sivagami, S.; Akkini, D. T.; Ananthiand, N.; Velammal, S. P. Biogenic Synthesis of Silver Nanoparticles By Leaf Extract of *Cassia angustifolia. Adv. Nat. Sci. Nanosci. Nanotechnol.* **2012,** *3,* 045006.

13. Amin, M.; Anwar, F.; Janjua, M. R. S. A.; Iqbal, M. A.; Rashid, U. Green Synthesis of Silver Nanoparticles Through Reduction with *Solanum xanthocarpum* L. Berry Extract: Characterization, Antimicrobial and Urease Inhibitory Activities Against *Helicobacter pylori. Int. J. Mol. Sci.* **2012,** *13,* 9923–9941.

14. Anastas, P.; Eghbali, N. Green Chemistry: Principles and Practice. *Chem. Soc. Rev.* **2010,** *39,* 301–312.

15. Ankamwar, B.; Damle, C.; Absar, A.; Sastry, M. Biosynthesis of Gold and Silver Nanoparticles Using Emblica Officinalis Fruit Extract, Their Phase Transfer and Transmetallation in an Organic Solution. *J. Nanosci. Nanotech.* **2005,** *5* (10), 1665–1671.

16. Ankanna, S.; Savithramma, N. Biological Synthesis of Silver Nanoparticles by Using Stem of *Shorea Tumbuggaia* Roxb. and Its Antimicrobial Efficacy. *Asian J. Pharm. Clin. Res.* **2011,** *4* (2), 137–141.

17. Ankanna, S.; Prasad, T. N. V. K. V.; Elumalai, E. K.; Savithramma, N. Production of Biogenic Silver Nanoparticles Using *Boswellia Ovalifoliolata* Stem Bark. *Digest J. Nanomaterials Biostruct.* **2010,** *5* (2), 369–372.

18. Annamalai, A.; Babu, S. T.; Jose, N. A.; Sudha, D.; Lyza, C. V. Biosynthesis and Characterization of Silver and Gold Nanoparticles Using Aqueous Leaf Extraction of *Phyllanthus amarus* Schum and Thonn. *World App. Sci. J.* **2011,** *13* (8), 1833–1840.

19. Annamalai, A.; Christina, V. L. P.; Sudha, D.; Kalpana, M.; Lakshmi, P. T. V. Green Synthesis, Characterization and Antimicrobial Activity of Au NPs Using *euphorbia hirta* l. Leaf Extract. *Colloids Surfaces B Biointerfaces* **2013,** *108,* 60–65.

20. Antony, J. J.; Sivalingam, P.; Siva, D.; Kamalakkannan, S.; Anbarasu, K.; Sukirtha, R.; Krishnan, M.; Achiraman, S. Comparative Evaluation of Antibacterial Activity of Silver Nanoparticles Synthesized Using Rhizophora Apiculata and Glucose. *Colloids Surfaces B Biointerfaces* **2011,** *88,* 134–140.

21. Anuradha, G.; Sundar, B. S.; Kumar, J. S.; Ramana, M. V. Synthesis and Characterization of Silver Nanoparticles from *Ocimum basilicum* L. Var Thyrsiflorum. *Eur. J. Acad. Essay* **2014,** *1* (5), 5–9.

22. Arokiyaraj, S. Arasu, M. V.; Vincent, S.; Prakash, N. U.; Choi, S. H.; Oh, Y. –K.; Choi, K. C.; Kim, K. H. Rapid Green Synthesis of Silver Nanoparticles from Chrysanthemum Indicum L and Its Antibacterial and Cytotoxic Effects: An In Vitro Study. *Int. J. Nanomed.* **2014,** *9,* 379–388.

23. Arokiyaraj, S.; Kumar, V. D.; Elakya, V.; Kamala, T.; Park, S. K.; Ragam, M.; Saravanan, M.; Bououdina, M.; Arasu, M. V.; Kovendan, K.; Vincent, S. Biosynthesized Silver Nanoparticles Using Floral Extract of *Chrysanthemum Indicum* L.: Potential for Malaria Vector Control. *Environ. Sci. Pollut. Res.* **2015.** DOI 10.1007/s11356-015-4148-9.

24. Arora, J. J.; Rajwade, J. M.; Paknikar, K. M. Cellular Responses Induced by Silver Nanoparticles: *In Vitro* Studies. *Toxicol. Lett.* **2008,** *179* (2), 93–100.

25. Arun, S.; Saraswathi, U.; Singaravelu. Green Synthesis of Silver Nanoparticles Using Mangrove *Excoecaria agallocha*. *Int. J. Pharm. Sci. Invent.* **2014**, *3* (10), 54–57.

26. Arunachalam, K. D.; Arun, L. B.; Annamalai, S. K.; Arunachalam, A. M. Potential Anticancer Properties of Bioactive Compounds of *Gymnema Sylvestre* and Its Biofunctionalized Silver Nanoparticles. *Int. J. Nanomed.* **2015**, *10*, 31–41.

27. Arunachalam, K. D.; Annamalai, S. K. *Chrysopogon zizanioides* Aqueous Extract Mediated Synthesis, Characterization of Crystalline Silver and Gold Nanoparticles for Biomedical Applications. *Int. J. Nanomed.* **2013**, *8*, 2375–2384.

28. Arunachalam, K. D.; Annamalai, S. K.; Hari, S. One-step Green Synthesis and Characterization of Leaf Extract-mediated Biocompatible Silver and Gold Nanoparticles from *Memecylon umbellatum*. *Int. J. Nanomed.* **2013**, *8*, 1307–1315.

29. Arunachalam, R.; Dhanasingha, S.; Kalimuthua, B.; Uthirappana, M.; Rosea, C.; Mandal, A. B. Phytosynthesis of Silver Nanoparticles Using Coccinia Grandis Leaf Extract and Its Application in the Photocatalytic Degradation. *Colloids Surf. B Biointerfaces* **2012**, *94*, 226–230.

30. Arvizo, R. R.; Bhattacharyya, S.; Rachel, A. K.; Giri, K.; Bhattacharya, R.; Mukherjee, P. Intrinsic Therapeutic Applications of Noble Metal Nanoparticles: Past, Present and Future. *Chem. Soc. Rev.* **2012**, *41*, 2943–2970.

31. AshaRani, P. V.; Mun, G. L. K.; Hande, M. P.; Valiyaveettil, S. Cytotoxicity and Genotoxicity of Silver Nanoparticles in Human Cells. *ACS Nano* **2009**, *3* (2), 279–290.

32. AshaRani, P. V.; Hande, M. P.; Valiyaveettil, S. Anti-proliferative Activity of Silver Nanoparticles. *BMC Cell Biol.* **2009**, *10*, 65.

33. Ashokkumar, S.; Ravi, S.; Kathiravan, V.; Velmurugan, S. Synthesis of Silver Nanoparticles Using *A. Indicum* Leaf Extract and Their Antibacterial Activity. *Spectrochim. Acta Part A Mol. Biomol. Spectrosc.* **2015**, *134*, 34–39.

34. Ashokkumar, S.; Ravi, S.; Velmurugan, S. Green Synthesis of Silver Nanoparticles from *Gloriosa Superba* L. Leaf Extract and Their Catalytic Activity. *Spectrochim. Acta Part A Mol. Biomol. Spectrosc.* **2013**, *115*, 388–392.

35. Austin, L. A.; Kang, B.; Yen, C.; El-Sayed, M. A. Nuclear Targeted Silver Nanospheres Perturb the Cancer Cell Cycle Differently than Those of Nanogold. *Bioconjugate Chem.* **2011**, *22*, 2324–2331.

36. Awad, M. A.; Hendi, A. A.; Ortashi, K. M. O.; Elradi, D. F. A.; Eisa, N. E.; Al-lahieb, L. A.; Al-Otiby, S. M.; Merghani, N. M.; Awad, A. A. G. Silver Nanoparticles Biogenic Synthesized Using an Orange Peel Extract and Their Use as an Anti-bacterial Agent. *Int. J. Phys. Sci.* **2014**, *9* (3), 34–40.

37. Awwad, A. M.; Salem, N. M.; Abdeen, A. O. Green Synthesis of Silver Nanoparticles Using Carob Leaf Extract and Its Antibacterial Activity. *Int. J. Ind. Chem.* **2013**, *4*, 29.

38. Awwad, A. M.; Salem, N. M.; Abdeen, A. Biosynthesis of Silver Nanoparticles Using *Olea Europaea* Leaves Extract and Its Antibacterial Activity. *Nanosci. Nanotechnol.* **2012**, *2*, 164–170.

39. Babu, S. A.; Prabu, H. G. Synthesis of AgNPs Using the Extract of *Calotropis Procera* Flower at Room Temperature. *Mater. Lett.* **2011**, *65*, 1675–1677.

40. Baeriswyl, V.; Christofori, G. The Angiogenic Switch in Carcinogenesis. *Sem. Cancer Biol.* **2009**, *19* (5), 329–337.

41. Bai, H.; Yang, B.; Chai, C.; Yang, G.; Jia, W.; Yi, Z. Green Synthesis of Silver Nanoparticles Using Rhodobacter Sphaeroides. *World J. Microbiol. Biotechnol.* **2011,** 27, 2723–2728.

42. Balaji, D. S.; Basavaraja, S.; Deshpande, R.; Mahesh, D. B.; Prabhakara, B. K.; Venkataraman, A. Extracellular Biosynthesis of Functionalized Silver Nanoparticles by Strains of *Cladosporium cladosporioides* Fungus. *Colloids Surf. B Biointerfaces* **2009,** 68, 88–92.

43. Balasubramanian, S.;, Jeyapaul, U.; Bosco, A. J.; Kala, S. M. J. Green Synthesis of Silver Nanoparticles Using *Cressa Cretica* Leaf Extract and Its Antibacterial Efficacy. *Int. J. Adv. Chem. Sci. Appl.* **2015,** 3 (1), 65–71.

44. Banerjee, P.; Satapathy, M.; Mukhopahayay, A.; Das, P. Leaf Extract Mediated Green Synthesis of Silver Nanoparticles from Widely Available Indian Plants: Synthesis, Characterization, Antimicrobial Property and Toxicity Analysis. *Bioresour. Bioprocess.* **2014,** 1, 3.

45. Bankar, A.; Joshi, B.; Kumar, A. R.; Zinjarde, S. Banana Peel Extract Mediated Novel Route for the Synthesis of Silver Nanoparticles. *Colloids Surf. A Physicochem. Eng. Aspects* **2010,** 368, 58–63.

46. Bar, H.; Bhui, D. K.; Sahoo, G. P.;, Sarkar, P.; De, S. P.; Misra, A. Green Synthesis Of Silver Nanoparticles Using Latex of *Jatropha curcas. Colloids Surf. A Physicochem. Eng. Aspects* **2009,** 339, 134–139.

47. Barud, H. S.; Barrios, C.; Regiani, T.; Marques, R. F. C.; Verelst, M.; Dexpert-Ghys, J. Self-supported Silver Nanoparticles Containing Bacterial Cellulose Membrane. *Mat. Sci. Eng. C* **2008,** 28, 515.

48. Basarkar, U. G.; Nikumbh, P. S.; Thakur, H. A. Biosynthesis of Silver Nanoparticles Using *Euphorbia hirta* Leaves Extract and Evaluation of Their Antimicrobial Activity. *Int. J. Life Sci.* **2014,** A2, 1–5.

49. Basavaraja, S.; Balaji, S. D.; Lagashetty, A.; Rajasab, A. H.; Venkataraman, A. Extracellular Biosynthesis of Silver Nanoparticles Using the Fungus Fusarium Semitectum. *Mater. Res. Bull.* **2008,** 43, 1164–1170.

50. Basavegowda, N.; Idhayadhulla, A.; Lee, Y. R. Preparation of Au and Ag Nanoparticles Using Artemisia Annua and Their In Vitro Antibacterial and Tyrosinase Inhibitory Activities. *Mater. Sci. Eng. C Mater. Biol. Appl.* **2014,** 43, 58–64.

51. Basu, S.; Maji, P.; Ganguly, J. Rapid Green Synthesis of Silver Nanoparticles by Aqueous Extract of Seeds of *Nyctanthes Arbor-tristis. Appl. Nanosci.* **2016,** 6 (1), 1–5. DOI 10.1007/s13204-015-0407-9.

52. Behera, S.; Rout, J.; Nayak, P. L. Antimicrobial Activity of Green Synthesized and Characterized Silver Nano Particles from *Murraya koenigii. Appl. Sci. Adv. Mater. Int.* **2014,** 1 (2), 46–49.

53. Bergers, G.; Benjamin, L. E. Tumorigenesis and the Angiogenic Switch. *Nat. Rev. Cancer* **2003,** 3, 401–410.

54. Bhainsa, K. C.; D'Souza, S. F. Extracellular Biosynthesis of Silver Nanoparticles Using the Fungus *Aspergillus fumigates. Colloids Surf. B Biointerfaces* **2006,** 47, 160–164.

55. Bharti, S. K.; Singh, S. K. Metal Based Drugs: Current Use and Future Potential. *Der Pharmacia Lettre* **2009,** 1 (2), 39–51.

56. Bharti, S. K.; Singh, S. K. Recent Developments in the Field of Anticancer Metallopharmaceuticals. *Int. J. Pharm. Tech. Res.* **2009,** *1* (4), 1406–1420.

57. Bhati-Kushwaha, H.; Malik, C. P. Biosynthesis of Silver Nanoparticles Using Fresh Extracts of *Tridax procumbens* Linn. *Ind. J. Exp. Biol.* **2014,** *52*, 359–368.

58. Bindhu, M. R.; Umadevi, M. Synthesis of Monodispersed Silver Nanoparticles Using *Hibiscus Cannabinus* Leaf Extract and Its Antimicrobial Activity. *Spectrochim. Acta Part A Mol. Biomol. Spectrosc.* **2013,** *101*, 184–190.

59. Birla, S. S.; Tiwari, V. V.; Gade, A. K.; Ingle, A. P.; Yadav, A. P.; Rai, M. K. Fabrication of Silver Nanoparticles by *Phoma glomerata* and Its Combined Effect Against *Escherichia coli, Pseudomonas aeruginosa* and *Staphylococcus aureus. Lett. Appl. Microbiol.* **2009,** *48*, 173.

60. Bonde, S. A Biogenic Approach for Green Synthesis of Silver Nanoparticles Using Extract of *Foeniculum vulgare* and Its Activity Against *Staphylococcus aureus* and *Escherichia coli. Bioscience* **2011,** *3* (2), 59–63.

61. Borase, H. P.; Patil, C. D.; Suryawanshi, R. K.; Patil, S. V. *Ficus carica* Latex-Mediated Synthesis of Silver Nanoparticles and Its Application as a Chemophotoprotective Agent. *Appl. Biochem. Biotechnol.* **2013,** *171*, 676–688.

62. Borase, H. P.; Salunkhe, R. B.; Patil, C. D.; Suryawanshi, R. K.; Salunke, B. K.; Patil, S. V. *Biotechnol. Appl. Biochem.* **2015.** doi: 10.1002/bab.1341.

63. Castellano, J. J.; Shafii, S. M.; Ko, F.; Donate, G.; Wright, T. E.; Mannari, R. J.; Payne, W. G.; Smith, D. J.; Robson, M. C. Comparative Evaluation of Silver-containing Antimicrobial Dressings and Drugs. *Int. Wound J.* **2007,** *4* (2), 114–122.

64. Castro-Longoria, E.; Vilchis-Nestor, A. R.; Avalos-Borja, M. Biosynthesis of Silver, Gold and Bimetallic Nanoparticles Using the Filamentous Fungus *Neurospora crassa. Colloids Surfaces B Biointerfaces* **2011,** *83*, 42–48.

65. Chaloupka, K.; Malam, Y.; Seifalian, A. M. Nanosilver as a New Generation of Nanoproduct in Biomedical Applications. *Trends Biotechnol.* **2010,** *28* (11), 580–88.

66. Chandran, S. P.; Chaudhary, M.; Pasricha, R.; Ahmad, A.; Sastry, M. Synthesis of Gold Nanotriangles and Silver Nanoparticles Using *Aloe vera* Plant Extract. *Biotechnol. Prog.* **2006,** *22*, 577–583.

67. Chen, N.; Zheng, Y.; Yin, J.; Li, X.; Zheng, C. Inhibitory Effects of Silver Nanoparticles Against Adenovirus Type 3 In Vitro. *J. Virol. Methods* **2013,** *193*, 470–477.

68. Chitra, G.; Balasubramani, G.; Ramkumar, R.; Sowmiya, R.; Perumal. P. *Mukia maderaspatana* (Cucurbitaceae) Extract-mediated Synthesis of Silver Nanoparticles to Control Culex Quinquefasciatus and Aedes Aegypti (Diptera: Culicidae). *Parasitol. Res.* **2015,** *114*, 1407–1415.

69. Choudhary, R. S.; Bhamare, N. B.; Mahure, B. V. Bioreduction of Silver Nanoparticles Using Different Plant Extracts and Its Bioactivity Against *E. coli* and *A. Niger. IOSR J. Agric. Vet. Sci.* **2014,** *7* (7), 7–11.

70. Chowdhury, I. H.; Ghosh, S.; Roy, M.; Naskar, M. K. Green Synthesis of Water-dispersible Silver Nanoparticles at Room Temperature Using Green Carambola (Star Fruit) Extract. *J. Sol-Gel Sci. Technol.* **2015,** *73*, 199–207.

71. Christensen, L.; Vivekanandhan, S.; Misra, M.; Mohanty, A. K. Biosynthesis of Silver Nanoparticles Using *Murraya Koenigii* (Curry Leaf): An Investigation on the Effect of Broth Concentration in Reduction Mechanism and Particle Size. *Adv. Mater. Lett.* **2011,** *2* (6), 429–434.

72. Clement, J. L.; Jarrett, P. S. Antibacterial Silver. *Met. Based Drugs* **1994**, *1* (5–6), 467–482.

73. Cruz, D.; Fale, P. L.; Mourato, B. A.; Vaz, P. D.; Serralheiro, M. L.; Lino, A. R. L. Preparation and Physicochemical Characterization of Ag Nanoparticles Biosynthesized by *Lippia Citriodora* (Lemon Verbena). *Colloids Surf. B Biointerfaces* **2010**, *81* (1), 67–73.

74. Daniel, S. C. G. K.; Ayyappan, S.; Philiphan, N. J. P.; Sivakumar, M.; Menaga, G.; Sironmani, T. A. Green Synthesis and Transfer of Silver Nanoparticles in a Food Chain Through Chiranamous Larva to Zebra Fish: A New Approach for Therapeutics. *Int. J. NanoSci. Nanotechnol.* **2011**, *2* (3), 159–169.

75. Dar, M. A.; Ingle, A.; Rai, M. Enhanced Antimicrobial Activity of Silver Nanoparticles Synthesized by Cryphonectria Sp. Evaluated Singly and in Combination with Antibiotics. *Nanomed.: Nanotechnol. Biol. Med.* **2013**, *9*, 105–110.

76. Das, J.; Das, M. P.; Velusamy, P. *Sesbania grandiflora* Leaf Extract Mediated Green Synthesis of Antibacterial Silver Nanoparticles Against Selected Human Pathogens. *Spectrochim. Acta Part A Mol. Biomol. Spectrosc.* **2013**, *104*, 265–270.

77. de Lima, R.; Seabra, A. B.; Duran, N. Silver Nanoparticles: A Brief Review of Cytotoxicity and Genotoxicity of Chemically and Biogenically Synthesized Nanoparticles. *J. Appl. Toxicol.* **2012**, *32* (11), 867–879.

78. Devi, J. S.; Bhimba, B. V.; Ratnam, K. *In Vitro* Anticancer Activity of Silver Nanoparticles Synthesized Using the Extract of *Gelidiella Sp. Int. J. Pharm. Pharm. Sci.* **2012**, *4* (4), 710–715.

79. Devi, J. S.; Bhimba, B. V. Antimicrobial Potential of Silver Nanoparticles Synthesized Using *Ulva Reticulata. Asian J. Pharm. Clin. Res.* **2014**, *7* (2), 82–85.

80. Dhal, S.; Panda, S. S.; Rout, N. C.; Dhal, N. K. Biosynthesis of Silver Nanoparticles Using *Cordia macleodii* (Griff.) Hook. F & Thomas and Its Antibacterial Activity. *World J. Pharm. Sci.* **2014**, *2* (9), 1051–1057.

81. Dhanalakshmi, T.; Rajendran, S. Synthesis of Silver Nanoparticles Using *Tridax Procumbens* and Its Antimicrobial Activity. *Arch. Appl. Sci. Res.* **2012**, *4* (3), 1289–1293.

82. Dinesh, D.; Murugan, K.; Madhiyazhagan, P.; Panneerselvam, C.; Kumar, P.; Nicoletti, M.; Jiang, W.; Benelli, G.; Chandramohan, B.; Suresh, U. Mosquitocidal and Antibacterial Activity of Green-synthesized Silver Nanoparticles from Aloe Vera Extracts: Towards an Effective Tool Against the Malaria Vector Anopheles Stephensi? *Parasitol. Res.* **2015**, *114*, 1519–1529.

83. Dinesh, S.; Karthikeyan, S.; Arumugam, P. Biosynthesis of Silver Nanoparticles from *Glycyrrhiza Glabra* Root Extract. *Arch. Appl. Sci. Res.* **2012**, *4* (1), 178–187.

84. Donda, M. R.; Kudle, K. R.; Alwala, J.; Miryala, A.; Sreedhar, B.; Rudra, M. P. P. Synthesis of Silver Nanoparticles Using Extracts of *Securinega Leucopyrus* and Evaluation of Its Antibacterial Activity. *Int. J. Curr. Sci.* **2013**, *7*, E1–E8.

85. Dubey, M.; Bhadauria, S.; Kushwah, B. S. Green Synthesis of Nanosilver Particles from Extract of *Eucalyptus Hybrida* (Safeda) Leaf. *Digest J. Nanomaterials Biostruct.* **2009**, *4* (3), 537–543.

86. Dubey, S. P.; Lahtinen, M.; Sarkka, H.; Sillanpaa, M. Bioprospective of Sorbus Aucuparia Leaf Extract in Development of Silver and Gold Nanocolloids. *Colloids Surfaces B Biointerfaces* **2010**, *80*, 26–33.

87. Dubey, S. P.; Lahtinen, M.; Sillanpää, M. Green Synthesis and Characterizations of Silver and Gold Nanoparticles Using Leaf Extract of Rosa Rugosa. *Colloids and Surfaces A: Physicochem. Eng. Aspects* **2010**, *364*, 34–41.

88. Dubey, S. P.; Lahtinen, M.; Sillanpaa, M. Tansy Fruit Mediated Greener Synthesis of Silver and Gold Nanoparticles. *Proc. Biochem.* **2010**, *45*, 1065–1071.

89. Durán, N.; Marcato, P. D.; Alves, O. L.; de Souza, G. I. H.; Esposito, E. Mechanistic Aspects of Biosynthesis of Silver Nanoparticles by Several *Fusarium Oxysporum* Strains. *J. Nanobiotechnol.* **2005**, *3*, 8.

90. Dunn, K.; Edwards-Jones, V. The Role of Acticoat™ with Nanocrystalline Silver in the Management of Burns. *Burns* **2004**, *30*, Sl–S9.

91. Dwivedi, A. D.; Gopal, K. Biosynthesis of Silver and Gold Nanoparticles Using *Chenopodium Album* Leaf Extract. *Colloids Surf. A Physicochemical Eng. Aspects* **2010**, *369* (1), 27–33.

92. Dwivedi, A. D.; Gopal, K. Plant Mediated Biosynthesis of Silver and Gold Nanoparticle. *J. Biomed. Nanotech.* **2011**, *7* (1), 163–164.

93. Edison, T. J. I.; Sethuraman, M. G. Instant Green Synthesis of Silver Nanoparticles Using *Terminalia Chebula* Fruit Extract and Evaluation of Their Catalytic Activity on Reduction of Methylene Blue. *Proc. Biochem.* **2012**, *47*, 1351–1357.

94. Eftekhari, K.; Pasha, K. M.; Tarigopula, S. P.; Sura, M.; Daddam, J. R. Biosynthesis and Characterization of Silver and Iron Nanoparticles from *Spinacia Oleraceae* and their Antimicrobial Studies. *Int. J. Plant Animal Environ. Sci.* **2015**, *5* (1), 116–122.

95. Elavazhagan, T.; Arunachalam, K. D. *Memecylon edule* Leaf Extract Mediated Green Synthesis of Silver and Gold Nanoparticles. *Int. J. Nanomed.* **2011**, *6*, 1265–1278.

96. Elumalai, E. K.; Prasad, T. N. V. K. V.; Hemachandran, J.; Therasa, S. V.; Thirumalai, T.; David, E. Extracellular Synthesis of Silver Nanoparticles Using Leaves of *Euphorbia Hirta* and Their Antibacterial Activities. *J. Pharm. Sci. Res.* **2010**, *2* (9), 549–554.

97. Elumalai, E. K.; Prasad, T. N. V. K. V.; Kambala, V.; Nagajyothi, P. C.; David, E. Green Synthesis of Silver Nanoparticle Using *Euphorbia Hirta* L and Their Antifungal Activities. *Arch. Appl. Sci. Res.* **2010**, *2* (6), 76–81.

98. Faedmaleki, F.; Shirazi, F. H.; Salarian, A.; Ashtiani, H. A.; Rastegar, H. Toxicity Effect of Silver Nanoparticles on Mice Liver Primary Cell Culture and HepG2 Cell Line. *Iranian J. Pharm. Res.* **2014**, *13* (1), 235–242.

99. Farooqui, M. A.; Chauhan, P. S.; Krishnamoorthy, P.; Shaik, J. Extraction of Silver Nanoparticles from the Leaf Extracts of Clerodendrum Inerme. *Digest J. Nanomaterials Biostruct.* **2010**, *5* (1), 43–49.

100. Fatima, F.; Bajpai, P.; Pathak, N.; Singh, S.; Priya, S.; Verma, S. R. Antimicrobial and Immunomodulatory Efficacy of Extracellularly Synthesized Silver and Gold Nanoparticles by a Novel Phosphate Solubilizing Fungus *Bipolaris Tetramera*. *BMC Microbiol.* **2015**, *15*, 52.

101. Fayaz, M.; Balaji, K.; Girilal, M.; Yadav, R.; Kalaichelvan, P. T.; Venketesan, R. Biogenic Synthesis of Silver Nanoparticles and Their Synergistic Effect with Antibiotics: A Study Against Gram-positive and Gram-negative Bacteria. *Nanomed. Nanotechnol. Biol. Med.* **2010**, *6* (1), 103–109.

102. Feng, Q.; Wu, J.; Chen, G.; Cui, F.; Kim, T.; Kim, J. A Mechanistic Study of the Antibacterial Effect of Silver Ions on *Escherichia coli* and *Staphylococcus aureus*. *J. Biomed. Mater. Res.* **2000**, 52, 662–668.

103. Firdhouse, M. J.; Lalitha, P. Green Synthesis of Silver Nanoparticles Using the Aqueous Extract of *Portulaca Oleracea (L.)*. *Asian J. Pharm. Clin. Res.* **2012**, *6* (1), 92–94.

104. Foldbjerg, R.; Dang, D. A.; Autrup, H. Cytotoxicity and Genotoxicity of Silver Nanoparticles in the Human Lung Cancer Cell Line, A549. *Arch. Toxicol.* **2011**, *85*, 743–750.

105. Forough, M.; Farhadi, K. Biological and Green Synthesis of Silver Nanoparticles. *Turkish J. Eng. Env. Sci.* **2010**, *34*, 281287.

106. Franco-Molina, M. A.; Mendoza-Gamboa, E.; Sierra-Rivera, C. A; Gómez-Flores, R. A.; Zapata-Benavides, P.; Castillo-Tello, P.; Alcocer-González, J. M.; Miranda-Hernández, D. F.; Tamez-Guerra, R. S.; Rodríguez-Padilla, C. Antitumor Activity of Colloidal Silver on Mcf-7 Human Breast Cancer Cells. *J. Exp. Clin. Cancer Res.* **2010**, *29*, 148.

107. Gade, A. K.; Bonde, P. P.; Ingle, A. P.; Marcato, P.; Duran, N.; Rai, M. K. Exploitation of *Aspergillus Niger* for Synthesis of Silver Nanoparticles. *J. Biobased Mater. Bioenergy* **2008**, *2*, 1.

108. Gaikwad, S.; Ingle, Avinash; Gade, Aniket; Rai, Mahendra; Falanga, Annarita; Incoronato, Novella; Russo, Luigi; Galdiero, Stefania; Galdiero, Massimilano. Antiviral Activity of Mycosynthesized Silver Nanoparticles Against Herpes Simplex Virus and Human Parainfluenza Virus Type 3. *Int. J. Nanomed.* **2013**, *8*, 4303–4314.

109. Gajbhiye, M.; Kesharwani, J.; Ingle, A.; Gade, A.; Rai, M. Fungus-mediated Synthesis of Silver Nanoparticles and their Activity Against Pathogenic Fungi in Combination with Fluconazole. *Nanomed. Nanotechnol. Biol. Med.* **2009**, *5*, 382–386.

110. Gan, P. P.; Li, S. F. Y. Potential of Plant as a Biological Factory to Synthesize Gold and Silver Nanoparticles and Their Applications. *Rev. Environ. Sci. Biotechnol.* **2012**, *11*, 169–206.

111. Garg, S. Rapid Biogenic Synthesis of Silver Nanoparticles Using Black Pepper (*Piper Nigrum*) Corn Extract. *Int. J. Innov. Biol. Chem. Sci.* **2012**, *3*, 5–10.

112. Gavade, N. L.; Kadam, A. N.; Suwarnkar, M. B.; Ghodake, V. P.; Garadkar, K. M. Biogenic Synthesis of Multi-applicative Silver Nanoparticles by Using *Ziziphus Jujuba* Leaf Extract. *Spectrochim. Acta Part A* **2015**, *136*, 953–960.

113. Gavarkar, P. S.; Adanik, R. S.; Mohite, S. K.; Magdum, C. S. 'Green' Synthesis and Antimicrobial Activity of Silver Nanoparticles (SNP) of *Cucumis Melo* Extract. *Int. J. Uni. Pharm. Bio. Sci.* **2014**, *3* (4), 392–396.

114. Gebru, H.; Taddesse, A.; Kaushal, J.; Yadav, O. P. Green Synthesis of Silver Nanoparticles and Their Antimicrobial Activity. *J. Surf. Sci. Technol.* **2013**, *29* (1–2), 47–66.

115. Geetha, N.; Geetha, T. S.; Manonmani. P.; Thiyagarajan, M. Green Synthesis of Silver Nanoparticles Using *Cymbopogon citratus* (Dc) Stapf. Extract and Its Antibacterial Activity. *Austr. J. Basic Appl. Sci.* **2014**, *8* (3), 324–331.

116. Geethalakshmi, R.; Sarada, D. V. L. Synthesis Of Plant-mediated Silver Nanoparticles Using *Trianthema Decandra* Extract and Evaluation of Their Antimicrobial Activities. *Int. J. Eng. Sci. Technol.* **2010**, *2* (5), 970–975.

117. Geethalakshmi, R.; Sarada, D. V. L. Gold and Silver Nanoparticles from *Trianthema Decandra*: Synthesis, Characterization, and Antimicrobial Properties. *Int. J. Nanomed.* **2012**, *7*, 5375–5384.

118. Gengan, R. M.; Anand, K.; Phulukdaree, A.; Chuturgoon, A. A549 Lung Cell Line Activity of Biosynthesized Silver Nanoparticles Using Albizia Adianthifolia Leaf. *Colloids Surf. B Biointerfaces* **2013**, *105*, 87–91.

119. Ghaffari-Moghaddam, M.; Hadi-Dabanlou, R. Plant Mediated Green Synthesis and Antibacterial Activity of Silver 3 Nanoparticles Using Crataegus Douglasii Fruit Extract. *Ind. Eng. Chem.* **2013.** Http://Dx.Doi.Org/10.1016/ J.Jiec.2013.09.005.

120. Ghosh, S.; Patil, S.; Ahire, M.; Kitture, R.; Kale, S.; Pardesi, K.; Cameotra, S. S.; Bellare, J.; Dhavale, D. D.; Jabgunde, A.; Chopade, B. A. Synthesis of Silver Nanoparticles Using *Dioscorea Bulbifera* Tuber Extract and Evaluation of Its Synergistic Potential in Combination with Antimicrobial Agents. *Int. J. Nanomed.* **2012,** *7*, 483–496.

121. Gnanadesigan, M.; Anand, M.; Ravikumar, S.; Maruthupandy, M.; Vijayakumar, V.; Selvam, S.; Dhineshkumar, M.; Kumaraguru. A. K. Biosynthesis of Silver Nanoparticles by Using Mangrove Plant Extract and Their Potential Mosquito Larvicidal Property. *Asian Pacific J. Tropical Med.* **2011,** 799–803.

122. Gnanadesigan, M.; Anand, M.; Ravikumar, S.; Maruthupandy, M.; M. Ali, M. S.; Vijayakumar, V.; Kumaraguru, A. K. Antibacterial Potential of Biosynthesised Silver Nanoparticles Using Avicennia Marina Mangrove Plant. *Appl. Nanosci.* **2012,** *2*, 143–147.

123. Gnanajobitha, G.; Annadurai, G.; Kannan, C. Green Synthesis of Silver Nanoparticle Using *Elettaria Cardamomom* and Assessment of Its Antimicrobial Activity. *Int. J. Pharm. Sci. Res.* **2012,** *3* (3), 323–330.

124. Gogoi, N.; Babu, P. J.; Mahanta, C.; Bora, U. Green Synthesis and Characterization of Silver Nanoparticles Using Alcoholic Flower Extract of *Nyctanthes Arbor-tristis* and In Vitro Investigation of Their Antibacterial and Cytotoxic Activities. *Mater. Sci. Eng. C* **2015,** *46*, 463–469.

125. Gopinath, V.; Mubarakali, D.; Priyadarshini, S.; Priyadharsshini, N. M.; Thajuddin, N.; Velusamy, P. Biosynthesis of Silver Nanoparticles from Tribulus Terrestris and Its Antimicrobial Activity: A Novel Biological Approach. *Colloids Surf. B Biointerfaces* **2012,** *96*, 69–74.

126. Gopinath, V.; Priyadarshini, S.; Priyadharsshini, N. M.; Pandian, K.; Velusamy, P. Biogenic Synthesis of Antibacterial Silver Chloride Nanoparticles Using Leaf Extracts of *Cissus Quadrangularis* Linn. *Mater. Lett.* **2013,** *91*, 224–227.

127. Govindaraju, K.; Tamilselvan, S.; Kiruthiga, V.; Singaravelu, G. Biogenic Silver Nanoparticles by *Solanum Torvum* and Their Promising Antimicrobial Activity. *J. Biopesticides* **2010,** *3* (1), 394–399.

128. Gude, V.; Upadhyaya, K.; Prasad, M. N. V.; Rao, N. V. S. Green Synthesis of Gold and Silver Nanoparticles Using *Achyranthes Aspera* L. Leaf Extract. *Adv. Sci. Eng. Med.* **2012,** *4*, 1–6.

129. Guo, D.; Zhu, L.; Huang, Z.; Zhou, H.; Ge, Y.; Ma, W.; Wu, J.; Zhang, X.; Zhou, X.; Zhang, Y.; Zhao, Y.; Gu, N. Anti-leukemia Activity of PVP-coated Silver Nanoparticles via Generation of Reactive Oxygen Species and Release of Silver Ions. *Biomaterials* **2013,** *34*, 7884–7894.

130. Gurunathan, S.; Lee, K.; Kalishwaralal, K.; Sheikpranbabu, S.; Vaidyanathan, R.; Eom, S. H. Antiangiogenic Properties of Silver Nanoparticles. *Biomaterials* **2009,** *30*, 6341–6350.

131. Gurunathan, S.; Kalishwaralal, K.; Vaidyanathana, R.; Deepak. V.; Pandian, S. R. K.; Muniyandi, J.; Hariharan, N.; Eom, S. H. Biosynthesis, Purification and Characterization of Silver Nanoparticles Using Escherichia coli. *Colloids Surf. B Biointerfaces* **2009**, *74*, 328–335.

132. Haldar, K. M.; Haldar, B.; Chandra, G. Fabrication, Characterization and Mosquito Larvicidal Bioassay of Silver Nanoparticles Synthesized from Aqueous Fruit Extract of Putranjiva, Drypetes Roxburghii (Wall). *Parasitol. Res.* **2013**, *112*, 1451–1459.

133. Haynes, C. L.; McFarland, A. D.; Zhao, L.; Duyne, R. P. V.; Schatz, G. C. Nanoparticle Optics: The Importance of Radiative Dipole Coupling in Two-Dimensional Nanoparticle Arrays. *J. Phys. Chem. B* **2003**, *107*, 7337–7342.

134. He, Y.; Du, Z.; Lv, H.; Jia, Q.; Tang, Z.; Zheng, X.; Zhang, K.; Zhao, F. Green Synthesis of Silver Nanoparticles by *Chrysanthemum Morifolium* Ramat Extract and Their Application in Clinical Ultrasound Gel. *Int. J. Nanomed.* **2013**, *8*, 1809–1815.

135. Heydari, R.; Rashidipour, M. Green Synthesis of Silver Nanoparticles Using Extract of Oak Fruit Hull (Jaft): Synthesis and In Vitro Cytotoxic Effect on MCF-7 Cells. *Int. J. Breast Cancer* **2015**, 6.

136. Huang, D.; Liao, F.; Molesa, S.; Redinger, D.; Subramanian, V. Plastic-compatible Low Resistance Printable Gold Nanoparticle Conductors for Flexible Electronics. *J. Electrochem. Soc.* **2003**, *150* (7), G412–G417.

137. Huang, J.; Lin, L.; Li, Q.; Sun, D.; Wang, Y.; Lu, Y.; He, N.; Yang, K.; Yang, X.; Wang, H.; Wang, W.; Lin, W. Continuous-flow Biosynthesis of Silver Nanoparticles by Lixivium of Sundried *Cinnamomum camphora* Leaf in Tubular Microreactors. *Ind. Eng. Chem. Res.* **2008**, *47*, 6081–6090.

138. Huang, J.; Zhan, G.; Zheng, B.; Sun, D.; Lu, F.; Lin, Y.; Chen, H.; Zheng, Z.; Zheng, Y.; Li, Q. Biogenic Silver Nanoparticles by *Cacumen Platycladi* Extract: Synthesis, Formation Mechanism, and Antibacterial Activity. *Ind. Eng. Chem. Res.* **2011**, *50*, 9095–9096.

139. Hussain, S. M.; Hess, K. L.; Gearhart, J. M.; Geiss, K. T.; Schlager, J. J. In Vitro Toxicity of Nanoparticles in BRL 3A Rat Liver Cells. *Toxicol. In Vitro* **2005**, *19*, 975–983.

140. Im, A.; Han, L.; Kim, E. R.; Kim, J.; Kim, Y. S.; Park, Y. Enhanced Antibacterial Activities of Leonuri Herba Extracts Containing Silver Nanoparticles. *Phytother. Res.* **2012**, *26*, 1249–1255.

141. Ingle, A.; Gade, A.; Pierrat, S.; Sonnichsen, C.; Rai, M. Mycosynthesis of Silver Nanoparticles Using the Fungus *Fusarium Acuminatum* and Its Activity Against Some Human Pathogenic Bacteria. *Curr. Nanosci.* **2008**, *4*, 141.

142. Ingle, A.; Rai, M.; Gade, A.; Bawaskar, M. Fusarium Solani: A Novel Biological Agent for the Extracellular Synthesis of Silver Nanoparticles. *J. Nanopart. Res.* **2009**, *11*, 2079–2085.

143. Iravani, S.; Zolfaghari, B. Green Synthesis of Silver Nanoparticles Using *Pinus eldarica* Bark Extract. *BioMed Res. Int.* **2013**, 5.

144. Jacob, S. J. P.; Finub, J. S.; Narayanan, A. Synthesis of Silver Nanoparticles Using Piper Longum Leaf Extracts and Its Cytotoxic Activity Against Hep-2 Cell Line. *Colloids Surf. B Biointerfaces* **2012**, *91*, 212–214.

145. Jadhav, K. V.; Dhamecha, D. L.; Dalvi, B. R.; Patil, M. B. Green Synthesis of Silver Nanoparticles Using Salacia chinensis: Characterization and Its Antibacterial Activity. *Particulate Sci. Technol. Int. J.* **2015**. DOI: 10.1080/02726351.2014.1003628.

146. Jagtap, U. B.; Bapat, V. A. Green Synthesis of Silver Nanoparticles Using *Artocarpus heterophyllus* Lam. Seed Extract and Its Antibacterial Activity. *Ind. Crops Products* **2013**, *46*, 132–137.

147. Jain, D.; Daima, H. K.; Kachhwaha, S.; Kothari, S. L. Synthesis of Plant-mediated Silver Nanoparticles using Papaya Fruit Extract and Evaluation of their Anti Microbial Activities. *Digest J. Nanomater. Biostruct.* **2009**, *4* (3), 557–563.

148. Jain, D.; Kachhwaha, S.; Jain, R.; Srivastava, G.; Kothari, S. L. Novel Microbial Route to Synthesize Silver Nanoparticles Using Spore Crystal Mixture of Bacillus Thuringiensis. *Indian J. Exp. Biol.* **2010**, *48*, 1152–1156.

149. Jain, N.; Bhargava, A.; Majumdar, S.; Tarafdar, J. C.; Panwar, J. Extracellular Biosynthesis and Characterization of Silver Nanoparticles Using *Aspergillus flavus* NJP08: A Mechanism Perspective. *Nanoscale* **2011**, *3*, 635–641.

150. Jayaseelan, C.; Rahuman, A. A.; Rajakumar, G.; Kirthi, A. V.; Santhoshkumar, T.; Marimuthu, S.; Bagavan, A.; Kamaraj, C.; Zahir, A. A.; Elango, G. Synthesis of Pediculocidal and Larvicidal Silver Nanoparticles by Leaf Extract from Heartleaf Moonseed Plant, *Tinospora cordifolia* Miers. *Parasitol. Res.* **2011**, *109*, 185–194.

151. Jayaseelan, C.; Rahuman, A. A.; Rajakumar, G.; Santhoshkumar, T.; Kirthi, A. V.; Marimuthu, S.; Bagavan, A.; Kamaraj, C.; Zahir, A. A.; Elango, G.; Velayutham, K.; Rao, K. V. B.; Karthik, L.; Raveendran, S. Efficacy of Plant-mediated Synthesized Silver Nanoparticles Against Hematophagous Parasites. *Parasitol. Res.* **2012**, *111*, 921–933.

152. Jayaseelan, C.; Rahuman, A. A. Acaricidal Efficacy of Synthesized Silver Nanoparticles Using Aqueous Leaf Extract of Ocimum Canum Against Hyalomma Anatolicum and Hyalomma Marginatum Isaaci (Acari: Ixodidae). *Parasitol Res.* **2012**, *111*, 1369–1378.

153. Jeeva, K.; Thiyagarajana, M.; Elangovanb, V.; Geethac, N.; Venkatachalam, P. *Caesalpinia coriaria* Leaf Extracts Mediated Biosynthesis of Metallicsilver Nanoparticles and Their Antibacterial Activity Against Clinically Isolated Pathogens. *Ind. Crops Products* **2014**, *52*, 714–720.

154. Jena, J.; Pradhan, N.; Dash, B. P.; Sukla, L. B.; Panda, P. K. Biosynthesis And Characterization of Silver Nanoparticles Using Microalga *Chlorococcum humicola* and *its* Antibacterial Activity. *Int. J. Nanomater. Biostruct.* **2013**, *3* (1), 1–8.

155. Jha, A. K.; Prasad, K. Green Synthesis of Silver Nanoparticles Using Cycas Leaf. *Int. J. Green Nanotechnol. Phys. Chem.* **2010**, *1* (2), P110–P117.

156. Jha, A. K.; Prasad, K.; Kumar, V.; Prasad, K. Biosynthesis of Silver Nanoparticles Using Eclipta Leaf. *Biotechnol. Prog.* **2009**, *25* (5), 1476–1479.

157. Joseph, S.; Mathew, B. Microwave Assisted Facile Green Synthesis of Silver and Gold Nanocatalysts Using the Leaf Extract of *Aerva Lanata*. *Spectrochim. Acta Part A Mol. Biomol. Spectrosc.* **2015**, *136*, 1371–1379.

158. Juibari, M. M.; Abbasalizadeh, S.; Jouzani, G. S.; Noruzi, M. Intensified Biosynthesis of Silver Nanoparticles Using a Native Extremophilic Ureibacillus thermosphaericus Strain. *Mater. Lett.* **2011**, *65*, 1014–1017.

159. Jun, S. H.; Cha, S.; Kim, J.; Cho, S.; Park, Y. Crystalline Silver Nanoparticles by Using *Polygala Tenuifolia* Root Extract as a Green Reducing Agent. *J. Nanosci. Nanotechnol.* **2015**, *15* (2), 1567–1574.

160. Jun, S. H.; Song-Hyun, C.; Jae-Hyun, K.; Minho, Y.; Seonho, C.; Youmie, P. Silver Nanoparticles Synthesized Using *Caesalpinia sappan* Extract as Potential Novel

Nanoantibiotics Against Methicillin-resistant *Staphylococcus aureus*. *J. Nanosci. Nanotechnol.* **2015**, *15* (8), 5543–5552.

161. Kalainila, P.; Subha, V.; Ravindran, R. S. E.; Renganathan, S. Synthesis and Characterization of Silver Nanoparticle from *Erythrina indica*. *Asian J. Pharm. Clin. Res.* **2014**, *7* (2). 39–43.

162. Kalimuthu, K.; Babu, R. S.; Venkataraman, D.; Bilal, M.; Gurunathan, S. Biosynthesis of Silver Nanocrystals by *Bacillus licheniformis*. *Colloids Surf. B Biointerfaces* **2008**, *65*, 150–153.

163. Kalishwaralal, K.; Deepak, V.; Pandian, S. R. K.; Kottaisamy, M.; BarathManiKanth, S.; Kartikeyan, B.; Gurunathan, S. Biosynthesis of Silver And Gold Nanoparticles Using *Brevibacterium casei*. *Colloids Surf. B Biointerfaces* **2010**, *77*, 257–262.

164. Kannan, N.; Shekhawat, M. S.; Ravindran, C. P.; Manokari, M. Preparation of Journal of Silver Nanoparticles Using Leaf and Fruit Extracts of *Morinda coreia* Buck. Ham.: A Green Approach. *Scientific Innov. Res.* **2014**, *3* (3), 315–318.

165. Kannan, N.; Mukunthan, K. S.; Balaji, S. A Comparative Study of Morphology, Reactivity and Stability of Synthesized Silver Nanoparticles Using *Bacillus subtilis* and *Catharanthus roseus* (L.) G. Don. *Colloids Surf. B Biointerfaces* **2011**, *86*, 378–383.

166. Kannan, R. R. R.; Arumugam, R.; Ramya, D.; Manivannan, K.; Anantharaman, P. Green Synthesis of Silver Nanoparticles Using Marine Macroalga Chaetomorpha linum. *Appl. Nanosci.* **2013**, *3*, 229–233.

167. Kannan, R. R. R.; Stirk, W. A.; Staden, J. V. Synthesis of Silver Nanoparticles Using the Seaweed Codium capitatum P.C. Silva (Chlorophyceae). *South African J. Botany* **2013a**, *86*, 1–4.

168. Karuppiah, M.; Rajmohan, R. Green Synthesis of Silver Nanoparticles Using Ixora Coccinea Leaves Extract. *Mater. Lett.* **2013**, *97*, 141–143.

169. Kathiresan, K.; Manivannan, S.; Nabeel, M. A.; Dhivya, B. Studies on Silver Nanoparticles Synthesized By a Marine Fungus, *Penicillium Fellutanum* Isolated from Coastal Mangrove Sediment. *Colloids Surf. B Biointerfaces* **2009**, *71*, 133–137.

170. Kavitha, K. S.; Baker, S.; Rakshith, D.; Kavitha, H. U.; Rao, Y. H. C.; Harini, B. P.; Satish, S. Plants as Green Source Towards Synthesis of Nanoparticles. *Int. Res. J. Biol. Sci.* **2013**, *2* (6), 66–76.

171. Kaviya, S.; Santhanalakshmi, J.; Viswanathan, B. Green Synthesis of Silver Nanoparticles Using Polyalthia longifolia Leaf Extract along with D-Sorbitol: Study of Antibacterial Activity. *J. Nanotechnol.* **2011a**, 5.

172. Kaviya, S.; Santhanalakshmi, J.; Viswanathan, B.; Muthumary, J.; Srinivasan, K. Biosynthesis of Silver Nanoparticles Using *Citrus Sinensis* Peel Extract and Its Antibacterial Activity. *Spectrochim. Acta Part A* **2011b**, *79*, 594–598.

173. Kaviya, S.; Santhanalakshmi, J.; Viswanathan, B. Biosynthesis of Silver Nano-flakes by *Crossandra infundibuliformis* Leaf Extract. *Mater. Lett.* **2012**, *67*, 64–66.

174. Kesharwani, J.; Yoon, K. Y.; Hwang, J. Phytofabrication of Silver Nanoparticles by Leaf Extract of *Datura metel*: Hypothetical Mechanism Involved in Synthesis. *J. Biosci.* **2009**, *3* (1), 39–44.

175. Khan, M.; Khan, M.; Adil, S. F.; Tahir, M. N.; Tremel, W.; Alkhathlan, H. Z.; Al-Warthan, A.; Rafiq, M.; Siddiqui, H. Green Synthesis of Silver Nanoparticles Mediated by *Pulicaria glutinosa* Extract. *Int. J. Nanomed.* **2013**, *8*, 1507–1516.

176. Khatoon, N.; Ahmad, R.; Sardar, M. Robust and Fluorescent Silver Nanoparticles Using Artemisia Annua: Biosynthesis, Characterization and Antibacterial Activity. *Biochem. Eng. J.* **2015.** http://dx.doi.org/10.1016/j.bej.2015.02.019.

177. Kim, S.; Choi, J. E.; Choi, J.; Chung, K.; Park, K.; Yi, J.; Ryu, D. Oxidative Stress-dependent Toxicity of Silver Nanoparticles in Human Hepatoma Cells. *Toxicol. In Vitro* **2009,** *23,* 1076–1084.

178. Kim, G. J.; Nie, S. Targeted Cancer Nanotherapy. *Mater. Today* **2005,** *8* **(8), 28–33.**

179. Klaus, T.; Joerger, R.; Olsson, E.; Granqvist. C. G. Silver Based Crystalline Nanoparticles, Microbially Fabricated. *Proc. Natl. Acad. Sci. USA* **1999,** *96,* 13611–13614.

180. Kaus-Joerger, T.; Joerger, R.; Olsson, E.; Granqvist, C. Bacteria as Workers in the Living Factory: Metal-accumulating Bacteria and Their Potential for Materials Science. *Trends in Biotech* **2001,** *19* (1), 15–20.

181. Kokila, T.; Ramesh, P. S.; Geetha, D. Biosynthesis of Silver Nanoparticles from Cavendish Banana Peel Extract and Its Antibacterial and Free Radical Scavenging Assay: A Novel Biological Approach. *Appl. Nanosci.* **2015.** DOI 10.1007/s13204-015-0401-2.

182. Kora, A. J.; Sashidharb R, B.; Arunachalam, J. Gum Kondagogu (*Cochlospermum Gossypium*): A Template for the Green Synthesis and Stabilization of Silver Nanoparticles with Antibacterial Application. *Carbohydrate Polym.* **2010,** *82,* 670–679.

183. Kora, A. J.; Sashidharb, R. B.; Arunachalam, J. Aqueous Extract of Gum Olibanum (*Boswellia Serrata*): A Reductant and Stabilizer for the Biosynthesis of Antibacterial Silver Nanoparticles. *Proc. Biochem.* **2012,** *47,* 1516–1520.

184. Kora, A. J.; Arunachalam, J. Green Fabrication of Silver Nanoparticles by Gum Tragacanth (Astragalus Gummifer): A Dual Functional Reductant and Stabilizer. *J. Nanomaterial.* **2012,** 8.

185. Kouvaris, P.; Delimitis, A.; Zaspalis, V.; Papadopoulos, D.; Tsipas, S. A.; Michailidis, N. Green Synthesis and Characterization of Silver Nanoparticles Produced Using *Arbutus Unedo* Leaf Extract. *Mater. Lett.* **2012,** *76,* 18–20.

186. Kowshik, M.; Ashtaputre, S, Kharrazi, S.; Vogel, W.; Urba, J.; Kulkarni, S. K. Paknikar, K. M. Extracellular Synthesis of Silver Nanoparticles by a Silver-tolerant Yeast Strain MKY3. *Nanotechnology* **2003,** *14,* 95–100.

187. Krishnaraj, C.; Jagan, E. G.; Rajasekhar, S.; Selvakumar, P.; Kalaichelvan, P. T.; Mohan, N. Synthesis of Silver Nanoparticles Using *Acalypha Indica* Leaf Extracts and Its Antibacterial Activity Against Water Borne Pathogens. *Colloids Surfaces B Biointerfaces* **2010,** *76,* 50–56.

188. Kudle, K. R.; Donda, M. R.; Alwala, J.; Koyyati, R.; Nagati, V.; Merugu, R.; Prashanthi, Y.; Rudra, M. P. P. Biofabrication of Silver Nanoparticles Using *Cuminum Cyminum* Through Microwave Irradiation. *Int. J. Nanomater. Biostruct.* **2012,** *2* (4), 65–69.

189. Kudle, K. R.; Donda, M. R.; Kudle, M. R.; Merugu, R.; Prashanthi, Y.; Rudra, M. P. P. Fruit (Epicarp And Endocarp) Extract Mediated Synthesis of Silver Nanoparticles from *Sterculia Foetida* Plant and Evaluation of Their Antimicrobial Activity. *Nanosci. Nanotechnol.* **2013,** *3* (3), 56–59.

190. Kumar, C. G.; Mamidyala, S. K. Extracellular Synthesis of Silver Nanoparticles Using Culture Supernatant of *Pseudomonas aeruginosa*. *Colloids Surf. B Biointerfaces* **2011**, *84* (2), 462–466.

191. Kumar, S. A.; Abyaneh, M. K.; Gosavi, S. W.; Kulkarni, S. K.; Pasricha, R. Ahmad, A.; Khan, M. I. Nitrate Reductase-mediated Synthesis of Silver Nanoparticles from AgNO$_3$. *Biotech. Lett.* **2007**, *29* (3), 439–445.

192. Kumar, K. M.; Sinha, M.; Mandal, B. K.; Ghosh, A. R.; Kumar, K. S.; Reddy, P. S. Green Synthesis of Silver Nanoparticles Using Terminalia Chebula Extract at Room Temperature and Their Antimicrobial Studies. *Spectrochim. Acta Part A* **2012**, *91*, 228–233.

193. Kumar, P.; Selvi, S. S.; Prabha, A. L.; Kumar, K. P.; Ganeshkumar, R. S.; Govindaraju, M. Synthesis of Silver Nanoparticles from Sargassum Tenerrimum and Screening Phytochemicals for Its Antibacterial Activity. *Nano Biomed. Eng.* **2012**, *4* (1), 12–16.

194. Kumar, P. P. N. V.; Pammi, S. V. N.; Kollu, P.; Satyanarayana, K. V. V.; Shameem, U. Green Synthesis and Characterization of Silver Nanoparticles Using *Boerhaavia Diffusa* Plant Extract and Their Antibacterial Activity. *Ind. Crops Products* **2014**, *52*, 562–566.

195. Kumar, R.; Roopan, S. M.; Prabhakarn, A.; Khanna, V. G.; Chakroborty, S. Agricultural Waste *Annona Squamosa* Peel Extract: Biosynthesis of Silver Nanoparticles. *Spectrochim. Acta Part A* **2012**, *90*, 173–176.

196. Kumar, V.; Yadav, S. C. Plant Mediated Synthesis of Silver and Gold Nanoparticles and Their Applications. *J. Chem. Technol. Biotechnol.* **2008**, *84* (2), 154–157.

197. Kumar, V.; Yadav, S. C.; Yadav, S. K. *Syzygium cumini* Leaf and Seed Extract Mediated Biosynthesis of Silver Nanoparticles and Their Characterization. *J. Chem. Technol. Biotechnol.* **2010**, *85*, 1301–1309.

198. Kumar, A.; Kaur, K.; Sharma, S. Synthesis, Characterization and Antibacterial Potential of Silver Nanoparticles by Morus Nigra Leaf Extract. *Indian J. Pharm. Biol. Res.* **2013**, *1* (4), 16–24.

199. Kumar, S.; Daimary, R. M.; Swargiary, M.; Brahma, A.; Kumar, S.; Singh, M. *Int. J. Pharm. Bio. Sci.* **2013**, *4* (4), 378–384.

200. Kuppusamy, P.; Ichwan, S. J. A.; Parine, N. R.; Yusoff, M. M.; Maniam, G. P.; Govindan, N. Intracellular Biosynthesis of Au and Ag Nanoparticles Using Ethanolic Extract of *Brassica Oleracea* L. and Studies on Their Physicochemical and Biological Properties. *J. Environ. Sci.* **2015**. http://dx.doi.org/10.1016/j.jes.2014.06.050

201. Lalitha, A.; Subbaiya, R.; Ponmurugan, P. Green Synthesis of Nanoparticles from Leaf Extract *Azadirachta Indica* and To Study Its Anti-bacterial and Anti-oxidant Property. *Int. J. Curr. Microbiol. App. Sci. 2* (6), 228–235.

202. Lara, H. H.; Ayala-Nuñez, N. V.; Ixtepan-Turrent, L.; Rodriguez-Padilla, C. Mode of Antiviral Action of Silver Nanoparticles Against HIV-1. *J. Nanobiotechnol.* **2010**, *8*, 1.

203. Lengke, M. F.; Fleet, M. E.; Southam, G. Biosynthesis of Silver Nanoparticles by Filamentous Cyanobacteria from a Silver(I) Nitrate Complex. *Langmuir* **2007**, *23*, 2694–2699.

204. Li, S.; Shen, Y.; Xie, A.; Yu, X.; Qiu, L.; Zhang, L.; Zhang, Q. Green Synthesis of Silver Nanoparticles Using *Capsicum annuum* L. Extract. *Green Chem.* **2007**, *9*, 852–858.

205. Li, W.; Xie, X. B.; Shi, Q. S.; Zeng, H. Y.; Yng, Y.; Chen, Y. B. Antibacterial Activity and Mechanism of Silver Nanoparticles on *Escherichia coli. Appl. Microbiol. Biotechnol.* **2010,** *85,* 1115–1122.
206. Lin, L.; Wang, W.; Huang, J.; Li, Q.; Sun, D.; Yang, X.; Wang, H.; He, N.; Wang, Y. Nature Factory of Silver Nanowires: Plant-mediated Synthesis Using Broth of *Cassia fistula* Leaf. *Chem. Eng. J.* **2010,** *162,* 852–858.
207. Loo, Y. Y.; Chieng, B. W.; Nishibuchi, M.; Radu, S. Synthesis of Silver Nanoparticles by Using Tea Leaf Extract from *Camellia sinensis. Int. J. Nanomed.* **2012,** *7,* 4263–4267.
208. Lu, L.; Sun, R. W.; Chen, R.; Hui, C. K.; Ho, C.; Luk, J. M.; Lau, G. K. K.; Che, C. Silver Nanoparticles Inhibit Hepatitis B Virus Replication. *Antiviral Ther.* **2014,** *13,* 253–262.
209. Lukman, A. I.; Gong, B.; Marjo, C. E.; Roessner, U.; Harris, A. T. Facile Synthesis, Stabilization, and Anti-bacterial Performance of Discrete Ag Nanoparticles Using *Medicago sativa* Seed Exudates. *J. Colloid Interface Sci.* **2011,** *353,* 433–444.
210. Luna, C.; Chávez, V. H. G.; Barriga-Castro, E. D.; Núñez, N. O.; Mendoza-Reséndez, R. Biosynthesis of Silver Fine Particles and Particles Decorated with Nanoparticles Using the Extract of *Illicium verum* (Star Anise) Seeds. *Spectrochim. Acta Part A Mol. Biomol. Spectr.* **2015,** *141,* 43–50.
211. Luo, X.; Morrin, A.; Killard, A. J.; Smyth, M. R. Application of Nanoparticles in Electrochemical Sensors and Biosensors. *Electroanalysis* 2006, *18* (4), 319–326.
212. Mahapatra, D. K.; Tijare, L. K.; Gundimeda, V.; Mahajan, N. M. Rapid Biosynthesis of Silver Nanoparticles of Flower-like Morphology from the Root Extract of *Saussurea lappa. Res. Rev. J. Pharmacognosy* **2018,** *5* (1), 20–24.
213. Mahapatra, D. K.; Bharti, S. K.; Asati, V. Nature Inspired Green Fabrication Technology for Silver Nanoparticles. *Curr. Nanomed.* **2017,** *7* (1), 5–24.
214. Maheswari, R. U.; Prabha, A. L.; Nandagopalan, V.; Anburaja, V. Green Synthesis of Silver Nanoparticles by Using Rhizome Extract of *Dioscorea oppositifolia* L. and Their Anti Microbial Activity Against Human Pathogens. *IOSR J. Pharm. Biol. Sci.* **2012,** *1* (2), 38–42.
215. Mahitha, B.; Raju, B. D. P.; Dillip, G. R.; Reddy, C. M.; Mallikarjuna, K.; Manoj, L.; Priyanka, S.; Rao, K. J.; Sushma, N. J. Biosynthesis, Characterization and Antimicrobial Studies of AgNPs Extract from *Bacopa Monniera* Whole Plant. *Digest J. Nanomater. Biostruct.* **2011,** *6* (1), 135–142.
216. Maiti, S.; Krishnan, D.; Barman, G.; Ghosh, S. K.; Laha, J. K. Antimicrobial Activites of Silver Nanoparticles Synthesized from *Lycopersicon esculentum. J. Anal. Sci. Technol.* **2014,** *5,* 40.
217. Malabadi, R. B.; Mulgund, G. S.; Meti, N. T.; Nataraja, K.; Kumar, S. V. Antibacterial Activity of Silver Nanoparticles Synthesized by Using Whole Plant Extracts of *Clitoria ternatea. Res. Pharm.* **2012,** *2* (4), 10–21.
218. Malik, P.; Shankar, R.; Malik, V.; Sharma. N.; Mukherjee, T. K. Green Chemistry Based Benign Routes for Nanoparticle Synthesis. *J. Nanoparticles* **2014,** 14.
219. Mallikarjuna, K.; Narasimha, G.; Dillipa, G. R.; Praveen, B.; Shreedhar, B.; Lakshmi, C. S.; Reddy, B. V. S.; Raju, B. D. P. Green Synthesis of Silver Nanoparticles Using *Ocimum* Leaf Extract and Their Characterization. *Digest J. Nanomater. Biostruct.* **2011,** *6* (1), 181–186.

220. Marambio-Jones, C.; Hoek, E. M. V. A Review of the Antibacterial Effects of Silver Nanomaterials and Potential Implications for Human Health and the Environment. *J. Nanopart. Res.* **2010,** *12,* 1531–1551.

221. Marimuthu, S.; Rahuman, A. A.; Rajakumar, G.; Santhoshkumar, T.; Kirthi, A. V.; Jayaseelan, C.; Bagavan, A.; Zahir, A. A.; Elango, G.; Kamaraj, C. Evaluation of Green Synthesized Silver Nanoparticles Against Parasites. *Parasitol Res.* **2011,** *108,* 1541–1549.

222. Marimuthu, S.; Rahuman, A. A.; Rajakumar, G.; Santhoshkumar, T.; Jayaseelan, C.; Kirthi, A. V.; Bagavan, A.; Kamaraj, C.; Elango, G.; Zahir, A. A.; Rajakumar, G.; Velayutham K. Lousicidal Activity of Synthesized Silver Nanoparticles Using *Lawsonia Inermis* Leaf Aqueous Extract Against *Pediculus Humanus Capitis* and *Bovicola Ovis.* *Parasitol. Res.* **2012,** *111,*:2023–33.

223. Martínez-Gutierrez, F.; Thi, E. P.; Silverman, J. M.; de Oliveira, C. C.; Svensson, S. L.; Hoek, A. V.; Sánchez, E. M.; Reiner, N. E.; Gaynor, E. C. Pryzdial, E. L. G., Conway, E. M., Orrantia, E., Ruiz, F., Av-Gay, Y.; Bach, H. Antibacterial Activity, Inflammatory Response, Coagulation and Cytotoxicity Effects of Silver Nanoparticles. *Nanomed. Nanotechnol. Biol. Med.* **2012,** *8,* 328–36.

224. Mary, E. J.; Inbathamizh, L. Green Synthesis and Characterization of Nano Silver Using Leaf Extract of *Morinda Pubescens. Asian J. Pharm. Clin. Res.* **2012,** *5* (1), 159–162.

225. Mason, C.; Vivekanandhan, S.; Misra, M.; Mohanty, A. K. Switchgrass (*Panicum virgatum*) Extract Mediated Green Synthesis of Silver Nanoparticles. *World J. Nano Sci. Eng.* **2012,** *2,* 47–52.

226. Masurkar, S. A.; Chaudhari, P. R.; Shidore, V. B.; Kamble, S. P. Rapid Biosynthesis of Silver Nanoparticles Using *Cymbopogan citratus* (Lemongrass) and its Antimicrobial Activity. *Nano-Micro Lett.* **2011,** *3* (3), 189–194.

227. Mata, R.; Reddy, J.; Rani, S. S. Catalytic and Biological Activities of Green Silver Nanoparticles Synthesized From Plumeria Alba (Frangipani) Flower Extract. *Mater. Sci. Eng. C* **2015a,** *51,* 216–225.

228. Mata, R.; Nakkala, J. R.; Sadras, S. R. Biogenic Silver Nanoparticles from Abutilon Indicum: Their Antioxidant, Antibacterial and Cytotoxic Effects In Vivo. *Colloids Surf. B Biointerfaces* **2015**.

229. Mathur, A.; Kushwaha, A.; Dalakoti, V.; Dalakoti, G.; Singh, D. S. Green Synthesis of Silver Nanoparticles Using Medicinal Plants and Its Characterization. *Der Pharmacia Sinica* **2014,** *5* (5), 118–122.

230. Mendoza-Reséndez, R.; Núñez, N. O.; Díaz Barriga-Castro, E.; Luna, C. **RSC Adv.** **2013,** *3,* 20765–20771.

231. Merin, D. D.; Prakash, S.; Bhimba, B. V. Antibacterial Screening of Silver Nanoparticles Synthesized by Marine Micro Algae. *Asian Pacific J. Tropical Med.* **2010,** 797–799.

232. Minaeian, S.; Shahverdi, A. R.; Nohi, A. S.; Shahverdi, H. R. Extracellular Biosynthesis of Silver Nanoparticles by Some Bacteria. *J. Sci. IAU* **2008,** *17* (66).

233. Miri, A.; Sarani, M.; Bazaz, M. R.; Darroudi, M. Plant Mediated Biosynthesis of Silver Nanoparticles Using *Prosopis Farcta* Extract and Its Anti-bacterial Properties. *Spectrochim. Acta Part A Mol. Biomol. Spectrosc.* **2015,** *141,* 287–291.

234. Mittal, A. K.; Bhaumik, J.; Kumar, S.; Banerjee, U. C. Biosynthesis of Silver Nanoparticles: Elucidation of Prospective Mechanism and Therapeutic Potential. *J. Colloid Interface Sci.* **2014**, *415*, 39–47.

235. Mittal, A. K.; Chisti, Y.; Banerjee, U. C. Synthesis of Metallic Nanoparticles Using Plant Extracts. *Biotechnol. Adv.* **2013**, *31*, 346–356.

236. Miura, N.; Shinohara, Y. Cytotoxic Effect and Apoptosis Induction by Silver Nanoparticles in Hela Cells. *Biochem. Biophys. Res. Comm.* **2009**, *390*, 733–737.

237. Mo, Y.; Tang, Y.; Wang, S.; Ling, J.; Zhang, H.; Luo, D. Green Synthesis of Silver Nanoparticles Using Eucalyptus Leaf Extract. *Mater. Lett.* **2015**. http: //dx.doi. org/10.1016/j.matlet.2015.01.004i.

238. Moaddab, S.; Ahari, H.; Shahbazzadeh, D.; Motallebi, A. A.; Anvar, A. A.; Rahman-Nya, J.; Shokrgozar, M. R. Toxicity Study of Nanosilver (Nanocid®) on Osteoblast Cancer Cell Line. *Int. Nano. Lett.* **2011**, *1* (1), 11–16.

239. Mochochoko, T.; Oluwafemi, O. S.; Jumbam, D. N.; Songca, S. P. Green Synthesis of Silver Nanoparticles Using Cellulose Extracted from an Aquatic Weed: Water Hyacinth. *Carbohydrate Polym.* **2013**, *98*, 290–294.

240. Mondal, S.; Roy, N.; Laskar, R. A.; Sk, I.; Basu, S.; Mandal, D.; Begum, N. A. Biogenic Synthesis of Ag, A and Bimetallic Au/Ag Alloy Nanoparticles Using Aqueous Extract of Mahogany (*Swietenia Mahogany*) Leaves. *Colloids Surf. B Biointerfaces* **2015**, *82* (2), 497–504.

241. Mouxing, F. U.; Qingbiao, L. I.; Daohua, S. U. N.; Yinghua, L. U.; Ning, H. E.; Xu, D. Rapid Preparation Process of Silver Nanoparticles by Bioreduction and Their Characterizations. *Chinese J. Chem. Eng.* **2006**, *14*, 114.

242. MubarakAli, D.; Thajuddin, N.; Jeganathanb, K.; Gunasekaran, M. Plant Extract Mediated Synthesis of Silver and Gold Nanoparticles and Its Antibacterial Activity Against Clinically Isolated Pathogens. *Colloids Surf. B Biointerfaces* **2011**, *85*, 360–365.

243. Mude, N.; Ingle, A.; Gade, A.; Rai, M. Synthesis of Silver Nanoparticles Using Callus Extract of *Carica Papaya*: A First Report. *J. Plant Biochem. Biotechnol.* **2009**, *18* (1), 83–86.

244. Mukherjee, P.; Ahmad, A.; Mandal, D.; Senapati, S.; Sainkar, S. R.; Khan, M. I.; Parishcha, R.; Ajaykumar, P. V.; Alam, M.; Kumar, R.; Sastry, M. Fungus-Mediated Synthesis of Silver Nanoparticles and Their Immobilization in the Mycelial Matrix: A Novel Biological Approach to Nanoparticle Synthesis. *Nano Lett.* **2001**, *1* (10), 515–519.

245. Mukherjee, P. Roy, M.; Mandal, B. P.; Dey, G. K.; Mukherjee, P. K.; Ghatak, J.; Tyagi, A. K.; Kale, S. P. Green Synthesis of Highly Stabilized Nanocrystalline Silver Particles by a Non-pathogenic and Agriculturally Important Fungus *T. asperellum*. *Nanotechnology* **2008**, *19*, 075103.

246. Mukherjee, S.; Chowdhury, D.; Kotcherlakota, R.; Patra, S.; Vinothkumar, B.; Bhadra, M. P.; Sreedhar, B.; Patra, C. R. Potential Theranostics Application of Bio-synthesized Silver Nanoparticles (4-in-1 System). *Theranostics* **2014**, *4* (3), 316–335.

247. Mukunthan, K. S.; Elumalai, E. K.; Patel, T. N.; Murty, V. R. Catharanthus Roseus: A Natural Source for the Synthesis of Silver Nanoparticles. *Asian Pacific J. Tropical Biomed.* **2011**, 270–274.

248. Mukunthan, K. S.; Balaji, S. Cashew Apple Juice (*Anacardium occidentale* L.) Speeds Up the Synthesis of Silver Nanoparticles. *Int. J. Green Nanotechnol.* **2012**, *4*, (2), 71–79.

249. Munger, M. A.; Radwanski, P.; Hadlock, G. C. *et al. In vivo* Human Time-Exposure Study of Orally Dosed Commercial Silver Nanoparticles. *Nanomedicine* **2014**, *10* (1), 1–9.

250. Murugan, K.; Benelli, G.; Ayyappan, S.; Dinesh, D. Panneerselvam, C.; Nicoletti, M.; Hwang, S.; Kumar, P. M.; Subramaniam, J.; Suresh, U. Toxicity of Seaweed-synthesized Silver Nanoparticles Against the Filariasis Vector Culex Quinquefasciatus and Its Impact on Predation Efficiency of the Cyclopoid Crustacean Mesocyclops Longisetus. *Parasitol. Res.* **2015**. DOI 10.1007/s00436-015-4417-z.

251. Muthukumaran, U.; Govindarajan, M.; Rajeswary, M.; Hoti, S. L. Synthesis and Characterization of Silver Nanoparticles Using Gmelina Asiatica Leaf Extract Against Filariasis, Dengue, and Malaria Vector Mosquitoes. *Parasitol. Res.* **2015**. DOI 10.1007/s00436-015-4368-4.

252. Nabikhan, A.; Kandasamy, K.; Raj, A.; Alikunhi, N. M. Synthesis of Antimicrobial Silver Nanoparticles by Callus and Leaf Extracts from Saltmarsh Plant, *Sesuvium portulacastrum* L. *Colloids Surf. B Biointerfaces* **2010**, *79*, 488–493.

253. Nagajyothi, P. C.; Lee, S. E.; An, M.; Lee, K. D. Green Synthesis of Silver and Gold Nanoparticles Using *Lonicera Japonica* Flower Extract. *Bull. Korean Chem. Soc.* **2012**, *33* (8), 2609–2612.

254. Nagati, V. B.; Koyyati, R.; Donda, M. R.; Alwala, J.; Padigya, K. R. K. P. R. M. Green Synthesis and Characterization of Silver Nanoparticles from *Cajanus cajan* Leaf Extract and Its Antibacterial Activity. *Int. J. Nanomater. Biostruct.* **2012**, *2* (3), 39–43.

255. Nair, B.; Pradeep, T. Coalescence of Nanoclusters and Formation of Submicron Crystallites Assisted by *Lactobacillus sp. Cryst. Growth Des.* **2002**, *2*, 293.

256. Nakkala, J. R.; Mata, R.; Bhagat, E.; Sadras, S. R. Green Synthesis of Silver and Gold Nanoparticles from Gymnema Sylvestre Leaf Extract: Study of Antioxidant and Anticancer Activities. *J. Nanopart. Res.* **2015**, *17*, 151.

257. Nalwa, H. S. *Handbook of Nanostructured Materials and Nanotechnology*. Academic Press, 1999, Cambridge, USA

258. Nanda, A.; Saravanan, M. Biosynthesis of Silver Nanoparticles from *Staphylococcus Aureus* and Its Antimicrobial Activity Against MRSA and MRSE. *Nanomedicine* **2009**, *5* (4), 452–456.

259. Narasimha, G. Virucidal Properties of Silver Nanoparticles Synthesized from White Button Mushrooms (*Agaricus bisporus*). *Int. J. Nano Dimens.* **2013**, *3* (3), 181–184.

260. Narayanan, K. B.; Sakthivel, N. Biological Synthesis of Metal Nanoparticles by Microbes. *Adv. Colloid Interface Sci.* **2010**, *156*, 1–13.

261. Narayanan, K. B.; Sakthivel, N. Extracellular Synthesis of Silver Nanoparticles Using the Leaf Extract of Coleus amboinicus Lour. *Mater. Res. Bull.* **2011**, *46*, 1708–1713.

262. Naveen, K. S. H.; Kumar, G.; Karthik, L.; Bhaskara Rao, K. V. Extracellular Biosynthesis of Silver Nanoparticles Using the Filamentous Fungus *Penicillium* sp. *Arch. Appl. Sci. Res.* **2010**, *2* (6), 161–167.

263. Nazeema, T. H.; Sugannya, P. K. Synthesis and Characterization of Silver Nanoparticles from Two Medicinal Plants and Its Anticancer Property. *Int. J. Res. Eng. Tech.* **2014,** *2* (1), 49–56.

264. Nazir, S.; Hussain, T.; Iqbal, M.; Mazhar, K.; Muazzam, A. G.; Ismail, M. Novel and Cost-Effective Green Synthesis of Silver Nano Particles and Their In-Vivo Antitumor Properties Against Human Cancer Cell Lines. *J. Biosci. Tech.* **2011,** *2* (6), 425–430.

265. Nethradevi, C.; Sivakumar, P.; Renganathan, S. Green Synthesis of Silver Nanoparticles Using *Datura Metel* Flower Extract and Evaluation of Their Antimicrobial Activity. *Int. J. Nanomater. Biostruct.* **2012,** *2* (2), 16–21.

266. Niraimathi, K. L.; Sudha, V.; Lavanya, R.; Brindha, P. Biosynthesis of Silver Nanoparticles Using Alternanthera Sessilis (Linn.) Extract and Their Antimicrobial, Antioxidant Activities. *Colloids Surf. B Biointerfaces* **2013,** *102,* 288–291.

267. Nithya, R.; Ragunathan, R. Synthesis of Silver Nanoparticle Using Pleurotus *Sajor Caju* and Its Antimicrobial Study. *Digest J. Nanomater. Biostruct.* **2009,** *4* (4), 623–629.

268. Njagi, E. C.; Huang, H.; Stafford, L.; Genuino, H.; Galindo, H. M.; Collins, J. B.; Hoag, G. E.; Suib, S. L. Biosynthesis of Iron and Silver Nanoparticles at Room Temperature Using Aqueous Sorghum Bran Extracts. *Langmuir* **2011,** *27* (1), 264–271.

269. Nutt, M. O.; Heck, K. N.; Alvarez, P.; Wong, M. S. Improved Pd-on-Au Bimetallic Nanoparticle Catalysts for Aqueous-phase Trichloroethene Hydrodechlorination. *Appl. Catalysis B Environ.* **2006,** *69,* 115–125.

270. Ojha, A. K.; Rout, J.; Behera, S.; Nayak, P. L. Green Synthesis and Charcaterization of Zero Valent Silver Nanoparticles from the Leaf Extract of Datura Metel. *Int. J. Pharm. Res. All Sci.* **2013,** *2* (1), 31–35.

271. Packialakshmi, N.; Suganya, C.; Guru, V. Antibacterial Activity and Green Synthesis of Silver Nanoparticles Using Strychnos Potatorum Seed and Bark Extract. *Asian J. Phytomed. Clin. Res.* **2014,** *2* (3), 127–138.

272. Pandian, M.; Marimuthu, R.; Natesan,G.; Rajagopal, R. E.; Justin, J. S.; Mohideen, A. J. A. H. Development of Biogenic Silver Nano Particle from Pelargonium Graveolens Leaf Extract and Their Antibacterial Activity. *Am. J. Nanosci. Nanotechnol.* **2013,** *1* (2), 57–64.

273. Panneerselvam, C.; Ponarulselvam, S.; Murugan, K. Potential Anti-plasmodial Activity of Synthesized Silver Nanoparticle Using *Andrographis paniculata* Nees (Acanthaceae). *Arch. Appl. Sci. Res.* **2011,** *3* (6), 208–217.

274. Pantidos, N.; Horsfall, L. E. Biological Synthesis of Metallic Nanoparticles by Bacteria, Fungi and Plants. *J. Nanomed. Nanotechnol.* **2014,** *5,* 5.

275. Parashar, V.; Parashar, R.; Sharma, B.; Pandey, A. C. Parthenium Leaf Extract Mediated Synthesis of Silver Nanoparticles: A Novel Approach Towards Weed Utilization. *Digest J. Nanomater. Biostruct.* **2009,** *4* (1), 45–50.

276. Parida, U. K.; Bindhani, B. K.; Nayak, P. Green Synthesis and Characterization of Gold Nanoparticles Using Onion (*Allium cepa*) Extract. *World J. NanoSci. Eng.* **2011,** *1,* 93–98.

277. Parikh, R. Y.; Singh, S.; Prasad, B. L. V.; Patole, M. S.; Sastry, M., *et al.* Extracellular Synthesis of Crystalline Silver Nanoparticles and Molecular Evidence of Silver

Resistance from Morganella sp.: Towards Understanding Biochemical Synthesis Mechanism. *Chem. Bio. Chem.* **2008**, *9*, 1415–1422.

278. Park, S. J.; Park, H. H.; Kim, S. Y.; Kim, S. J.; Woo, K.; Ko, G. P. Antiviral Properties of Silver Nanoparticles on a Magnetic Hybrid Colloid. *Appl. Environ. Microbiol.* **2014**, *80* (8), 2343–2350.

279. Parvathy, S. K.; Vidhya, V. K.; Evanjelene, B.; Venkatraman, R. Green Synthesis of Silver Nanoparticles Using *Albizia Lebbeck* (L.) *Benth Extract* and Evaluation of Its Antimicrobial Activity. *Int. J. Innov. Res. Sci. Eng.* **2014**, *2* (1), 501–505.

280. Parveen, M.; Ahmad, F.; Malla, A. M.; Azaz, S. Microwave-assisted Green Synthesis of Silver Nanoparticles from Fraxinus Excelsior Leaf Extract and Its Antioxidant Assay. *Appl. Nanosci.* **2015**. DOI 2015 10.1007/s13204-015-0433-7.

281. Patil, C. D.; Patil, S. V.; Borase, H. P.; Salunke, B. P.; Salunkhe, R. B. Larvicidal Activity of Silver Nanoparticles Synthesized Using *Plumeria rubra* Plant Latex Against *Aedes aegypti* and *Anopheles stephensi. Parasitol. Res.* **2012**, *110*, 1815–1822.

282. Patil, R. S.; Kokate, M. R.; Kolekar, S. S. Bioinspired Synthesis of Highly Stabilized Silver Nanoparticles Using *Ocimum Tenuiflorum* Leaf Extract and Their Antibacterial Activity. *Spectrochim. Acta Part A* **2012**, *91*, 234–238.

283. Patra, C. R.; Mukherjee, S.; Kotcherlakota, R. Biosynthesized Silver Nanoparticles: A Step Forward for Cancer Theranostics? *Nanomedicine* **2014**, *9* (10), 1445–1448.

284. Pavani, K. V.; Kumar, N. S.; Gayathramma, K. Plants as Ecofriendly Nanofactories. *J. Bioscience* **2012**, *6* (1), 1–6.

285. Phanjom, P.; Sultana, A.; Sarma, H.; Ramchiary, J.; Goswami, K.; Baishya, P. Plant-mediated Synthesis of Silver Nanoparticles using *Elaeagnus Latifolia* Leaf Extract. *Digest J. Nanomater. Biostruct.* **2012**, *7* (3), 1117–1123.

286. Philip, D. Biosynthesis of Au, Ag and Au–Ag Nanoparticles Using Edible Mushroom Extract. *Spectrochim.Acta Part A* **2009**, *73* (2), 374–381.

287. Philip, D.; Unni, C.; Aromal, S. A.; Vidhu, V. K. Murraya Koenigii Leaf-assisted Rapid Green Synthesis of Silver and Gold Nanoparticles. *Spectrochim. Acta Part A* **2011**, *L78*, 899–904.

288. Philip, D.; Unni, C. Extracellular Biosynthesis of Gold and Silver Nanoparticles Using Krishna Tulsi (*Ocimum Sanctum*) Leaf. *Physica. E* **2011**, *43*, 1318–1322.

289. Philip, D. *Mangifera Indica* Leaf-assisted Biosynthesis of Well-dispersed Silver Nanoparticles. *Spectrochim. Acta Part A* **2011**, *78*, 327–331.

290. Philip, D. Green Synthesis of Gold and Silver Nanoparticles Using *Hibiscus rosa sinensis. Physica E* **2010**, *42*, 1417–1424.

291. Poinern, G. E. J.; Chapman, P.; Shah, M.; Fawcett, D. Green Biosynthesis of Silver Nanocubes Using the Leaf extracts from *Eucalyptus macrocarpa. Nano Bull.* **2013**, *2* (1), 130101–131017.

292. Ponarulselvam, S.; Panneerselvam, C.; Murugan, K.; Aarthi, N.; Kalimuthu, K.; Thangamani, S. Synthesis of Silver Nanoparticles Using Leaves of Catharanthus Roseus Linn. G. Don and Their Antiplasmodial Activities. *Asian Pacific J. Tropical Biomed.* **2012**, 574–580.

293. Prabhu, D.; Arulvasua, C.; Babu, G.; Manikandan, R.; Srinivasan, P. Biologically Synthesized Green Silver Nanoparticles from Leaf Extract of *Vitex Negundo* L.

Induce Growth-inhibitory Effect on Human Colon Cancer Cell line HCT15. *Proc. Biochem.* **2013**, *48*, 317–324.

294. Pradeepa, M.; Harini, K.; Ruckmani, K.; Geetha, N. Extracellular Bio-inspired Synthesis of Silver Nanoparticles Using Raspberry Leaf Extract Against Human Pathogens. *Int. J. Pharm. Sci. Rev. Res.* **2014**, *25* (2), 160–65.

295. Prakash, P.; Gnanaprakasam, P.; Emmanuel, R.; Arokiyaraj, S.; Saravanan, M. Green Synthesis of Silver Nanoparticles from Leaf Extract of *Mimusops Elengi*, Linn. for Enhanced Antibacterial Activity Against Multi Drug Resistant Clinical Isolates. *Colloids Surf. B Biointerfaces* **2013**, *108*, 255–259.

296. Prasad, K. S.; Pathak, D.; Patel, A.; Dalwadi, P.; Prasad, R.; Patel, P.; Selvaraj, K. Biogenic Synthesis of Silver Nanoparticles Using *Nicotiana Tobaccum* Leaf Extract and Study of Their Antibacterial Effect. *African J. Biotechnol.* **2011**, *10* (41), 8122–8130.

297. Prasad, T. N. V. K. V.; Kambala, V. S. R.; Naidu, R. Phyconanotechnology: Synthesis of Silver Nanoparticles Using Brown Marine Algae *Cystophora Moniliformis* and Their Characterization. *J. Appl. Phycol.* **2013**, *25*, 177–182.

298. Prathna, T. C.; Chandrasekaran, N.; Raichur, A. M.; Mukherjee, A. Biomimetic Synthesis of Silver Nanoparticles by *Citrus Limon* (Lemon) Aqueous Extract and Theoretical Prediction of Particle Size. *Colloids Surf. B Biointerfaces* **2011**, *82*, 152–159.

299. Praveena, V. D.; Kumar, K. V. Green Synthesis of Silver Nanoparticles from *Achyranthus aspera* Plant Extract in Chitosan Matrix and Evaluation of Their Antimicrobial Activities. *Indian J. Adv. Chem. Sci.* **2014**, *2* (3), 171–177.

300. Priya, M. M.; Selvi, B. K.; Paul, J. A. J. Green Synthesis of Silver Nanoparticles from the Leaf Extracts of *Euphorbia Hirta* and *Nerium Indicum. Digest J.Nanomater. Biostruct.* **2011**, *6* (2), 869–877.

301. Priyadarshini, K. A.; Murugan, K.; Panneerselvam, C.; Ponarulselvam, S.; Hwang, J.; Nicoletti, M. Biolarvicidal and Pupicidal Potential of Silver Nanoparticles Synthesized Using *Euphorbia Hirta* Against *Anopheles stephensi* Liston (Diptera: Culicidae). *Parasitol. Res.* **2012**, *111*, 997–1006.

302. Pugazhenthiran, N.; Anandan, S.; Kathiravan, G.; Prakash, N. K. U.; Crawford, S.; Ashokkumar, M. Microbial Synthesis of Silver Nanoparticles by Bacillus sp. *J. Nanopart. Res.* **2009**, *11*, 1811–1815.

303. Raghunandan, D.; Mahesh, B. D.; Basavaraja, S.; Balaji, S. D.; Manjunath, S. Y.; Venkataraman, A. Microwave-assisted Rapid Extracellular Synthesis of Stable Bio-functionalized Silver Nanoparticles from Guava (*Psidium guajava*) Leaf Extract. *J. Nanopart. Res.* **2011**, *13*, 2021–2028.

304. Raja, K.; Saravanakumar, A.; Vijayakumar, R. Efficient Synthesis of Silver Nanoparticles from Prosopis Juliflora Leaf Extract and Its Antimicrobial Activity Using Sewage. *Spectrochim. Acta Part A Mol. Biomol. Spectrosc.* **2012**, *97*, 490–494.

305. Rajakannu, S.; Shankar, S.; Perumal, S.; Subramanian, S.; Dhakshinamoorthy, G. P. Biosynthesis of Silver Nanoparticles using *Garcinia mangostana* Fruit Extract and Their Antibacterial, Antioxidant Activity. *Int. J. Curr. Microbiol. App. Sci.* **2015**, *4* (1), 944–952.

306. Rajakumar, G.; Rahuman, A. A. Acaricidal Activity of Aqueous Extract and Synthesized Silver Nanoparticles from Manilkara Zapota Against *Rhipicephalus (Boophilus) Microplus*. *Res. Vet. Sci.* **2012**, *93*, 303–309.
307. Rajakumar, G.; Abdul Rahuman, A. Larvicidal Activity of Synthesized Silver Nanoparticles Using *Eclipta Prostrata* Leaf Extract Against Filariasis and Malaria Vectors. *Acta Tropica* **2011**, *118*, 196–203.
308. Rajaraman, K.; Aiswarya, D. C.; Sureshkumar, P. Green Synthesis of Silver Nanparticle Using *Tephrosia Tinctoria* and Its Antidiabetic Activity. *Mater. Lett.* **2014**, *138*, 251–254.
309. Rajashekharreddy, P.; Rani, P. U.; Sreedhar, B. Qualitative Assessment of Silver and Gold Nanoparticle Synthesis in Various Plants: A Photobiological Approach. *J. Nano Res.* **2010**, *12* (5), 1711–1721.
310. Rajendran, R.; Ganesan, N.; Balu, S. K.; Alagar, S.; Thandavamoorthy, P.; Thiruvengadam, D. Green Synthesis, Characterization, Antimicrobial and Cytotoxic Effects of Silver Nanoparticles Using *Origanum heracleoticum* L. Leaf Extract. *Int. J. Pharm. Pharm. Sci.* **2015**, *7* (4), p. 288–293.
311. Rajesh, S.; Raja, D. P.; Rathi, J. M.; Sahayaraj, K. Biosynthesis of Silver Nanoparticles Using Ulva Fasciata (Delile) Ethyl Acetate Extract and Its Activity Against *Xanthomonas campestris* pv. Malvacearum. *J. Biopest.* **2012**, *5*, 119–128.
312. Rajeshkumar, S.; Kannan, C.; Annadurai, G. Green Synthesis of Silver Nanoparticles Using Marine Brown Algae *Turbinaria Conoides* and Its Antibacterial Activity. *Int. J. Pharm. Bio. Sci.* **2012**, *3* (4), 502–510.
313. Rajkuberan, C.; Sudha, K.; Sathishkumar, G.; Sivaramakrishnan, S. Antibacterial and Cytotoxic Potential of Silver Nanoparticles Synthesized Using Latex of *Calotropis gigantea* L. *Spectrochim. Acta Part A Mol. Biomol. Spectrosc.* **2015**, *136*, 924–930.
314. Raliya, R.; Tarafdar, J. C. Novel Approach for Silver Nanoparticle Synthesis Using *Aspergillus Terrus* CZR-1: Mechanism Perspective. *J. Bionanosci.* **2012**, *6* (1), 1–5.
315. Raman, J.; Reddy, G. R.; Lakshmanan, H.; Selvaraj, V.; Babu, G.; Raaman, N. Mycosynthesis and Characterization of Silver Nanoparticles from *Pleurotus djamor* Var. Roseus and Their In Vitro Cytotoxicity Effect on PC3 Cells. *Proc. Biochem.* **2014**, *50* (1), p. 140–147.
316. Ramar, M.; Manikandan, B.; Marimuthu, P. N. Synthesis of Silver Nanoparticles Using *Solanum Trilobatum* Fruits Extract and Its Antibacterial, Cytotoxic Activity Against Human Breast Cancer Cell Line MCF-7. *Spectrochim. Acta Part A Mol. Biomol. Spectrosc.* **2015**, *140*, 223–228.
317. Ramesh, P. S.; Kokila, T.; Geetha, D. Plant Mediated Green Synthesis and Antibacterial Activity of Silver Nanoparticles Using *Emblica officinalis* fruit extract. *Spectrochim. Acta Part A Mol. Biomol. Spectrosc.* **2015**, *142*, 339–343.
318. Ramteke, C.; Chakrabarti, T.; Sarangi, B. K.; Pandey, R. Synthesis of Silver Nanoparticles from the Aqueous Extract of Leaves of *Ocimum sanctum* for Enhanced Antibacterial Activity. *J. Chem.* **2013**, 7.
319. Rani, P. U.; Rajasekharreddy, P. Green Synthesis of Silver-Protein (Core–Shell) Nanoparticles Using *Piper betle* L. Leaf Extract and Its Ecotoxicological Studies on Daphnia Magna. *Colloids Surf. A Physicochem. Eng. Aspects* **2011**, *389*, 188–194.
320. Rao, C. N. R.; Müller, A.; Cheetham, A. K. *Nanomaterials Chemistry: Recent Developments and New Directions*; John Wiley & Sons, 2006, Weinheim, Germany.

321. Rao, C. N. R.; Müller, A.; Cheetham, A. K. *The Chemistry of Nanomaterials: Synthesis, Properties and Applications*; John Wiley & Sons, 2006, Weinheim, Germany.

322. Rao, K. J.; Paria, S. Green Synthesis of Silver Nanoparticles from Aqueous *Aegle marmelos* Leaf Extract. *Mater. Res. Bull.* **2010.** doi:10.1016/j.materresbull.2012.11.035

323. Rao, M. L.; Savithramma, N. Antimicrobial Activity of Silver Nanoparticles Synthesized by Using Stem Extract of *Svensonia hyderobadensis* (Walp.) Mold: A Rare Medicinal Plant. *Res. Biotechnol.* **2012,** *3* (3), 41–47.

324. Rao, M. L.; Savithramma, N. Biological Synthesis of Silver Nanoparticles Using *Svensonia Hyderabadensis* Leaf Extract and Evaluation of Their Antimicrobial Efficacy. *J. Pharm. Sci. Res.* **2011,** *3* (3), 1117–1121.

325. Rao, Y. S.; Kotakadi, V. S.; Prasad, T. N. V. K. V.; Reddy, A. V.; Gopal, D. V. R. S. Green Synthesis and Spectral Characterization of Silver Nanoparticles from Lakshmi Tulasi (*Ocimum Sanctum*) Leaf Extract. *Spectrochim. Acta Part A Mol. Biomol. Spectrosc.* **2013,** *103,* 156–159.

326. Rao, P.; Chandraprasad, M. S.; Lakshmi, Y. N.; Rao, J.; Aishwarya, P.; Shetty, S. Biosynthesis of Silver Nanoparticles Using Lemon Extract and Its Antibacterial Activity. **2014,** *2,* 165–169.

327. Raut, R. W.; Kolekar, N. S.; Lakkakula, J. R.; Mendhulkar, V. D.; Kashid, S. B. Extracellular Synthesis of Silver Nanoparticles Using Dried Leaves of *Pongamia Pinnata* (L) Pierre. *Nano-Micro Lett.* **2010,** *2,* 106–113.

328. Renugadevi, K.; Inbakandan, D.; Bavanilatha, M.; Poornima, V. *Cissus Quadrangularis* Assisted Biosynthesis of Silver Nanoparticles with Antimicrobial and Anticancer Potentials. *Int. J. Pharm. Bio. Sci.* **2012,** *3* (3), 437–445.

329. Renugadevi, K.; Aswini, V.; Raji, P. Microwave Irradiation Assisted Synthesis of Silver Nanoparticle using Leaf Extract of *Baliospermum Montanum* and Evaluation of Its Antimicrobial, Anticancer Potential Activity. *Asian J. Pharm. Clin. Res.* **2012a,** *5* (4), 283–287.

330. Resmi, C. R.; Sreejamol, P.; Pillai, P. Green Synthesis of Silver Nanoparticles Using *Azadirachta Indica* Leaves Extract and Evaluation of Antibacterial Activities. *Int. J. Adv. Biol. Res.* **2014,** *4* (3), 300–303.

331. Rodríguez-León, E.; Iñiguez-Palomares, R.; Navarro, R. E.; Herrera-Urbina, R.; Tánori, J.; Iñiguez-Palomares, C.; Maldonado, A. Synthesis of Silver Nanoparticles Using Reducing Agents Obtained from Natural Sources (*Rumex hymenosepalus* extracts). *Nanoscale Res. Lett.* **2013,** *8,* 318.

332. Roopan, S. M.; Rohit, Madhumitha, G.; Rahuman, A. A.; Kamaraj, C.; Bharathi, A. Surendra, T. V. Low-cost and Eco-friendly Phyto-synthesis of Silver Nanoparticles Using *Cocos nucifera* Coir Extract and Its Larvicidal Activity. *Ind. Crops Prod.* **2013,** *43,* 631–635.

333. Rout, Y.; Behera, S.; Ojha, A. K.; Nayak, P. L. Green Synthesis of Silver Nanoparticles Using *Ocimum sanctum (Tulashi)* and Study of Their Antibacterial and Antifungal Activities. *J. Microbiol. Antimicrob.* **2012,** *4* (6), 103–109.

334. Rupaisih, N. N.; Aher, A.; Gosavi, S.; Vidyasagar, P. B. Green Synthesis of Silver Nanoparticles Using Latex Extract of *Thevetia Peruviana*: A Novel Approach Towards Poisonous Plant Utilization. *Recent Trends Phys. Mater. Sci. Technol.* **2015,** *423,* 1–10.

335. Sable, N.; Gaikwad, S.; Bonde, S.; Gade, A.; Rai, M. Phytofabrication of Silver Nanoparticles by Using Aquatic Plant *Hydrilla verticilata*. *Bioscience* **2012**, *4* (2), 45–49.

336. Sadeghi, B.; Gholamhoseinpoor, F. A Study on the Stability and Green Synthesis of Silver Nanoparticles Using *Ziziphora Tenuior* (Zt) Extract at Room Temperature. *Spectrochim. Acta Part A* **2015**, *134*, 310–315.

337. Sadowski, Z.; Maliszewska, I. H.; Grochowalska, B.; Polowczyk, I.; Koźlecki, T. Synthesis of Silver Nanoparticles Using Microorganisms. *Mater. Sci.-Poland* **2008**, *26* (2), 419–424.

338. Sahu, N.; Soni, D.; Chandrashekhar, B.; Sarangi, B. K.; Satpute, D.; Pandey, R. A. Synthesis and Characterization of Silver Nanoparticles Using Cynodon Dactylon Leaves and Assessment of Their Antibacterial Activity. *Bioproc. Biosyst. Eng.* **2013**, *36*, 999–1004.

339. Sahu, S. C.; Zheng, J.; Graham, L.; Chen, L.; Ihrie, J.; Yourick, J. J.; Sprando, R. L. Comparative Cytotoxicity of Nanosilver in Human Liver Hepg2 and Colon Caco2 Cells in Culture. *J. Appl. Toxicol.* **2014**, 34, 1155–1166.

340. Saifuddin, N.; Wong, C. W.; Yasumira, A. A. N. Rapid Biosynthesis of Silver Nanoparticles Using Culture Supernatant of Bacteria with Microwave Irradiation. *E-J, Chem.* **2009**, *6* (1), 61–70.

341. Salkia, D. Green Synthesis and Optical Characterizations of Silver Nanoparticles. *Int. J. Latest Res. Sci. Tech.* **2014**, *3* (2), 132–135.

342. Salunkhe, R. B.; Patil, S. V.; Patil, C. D.; Salunke, B. K. Larvicidal Potential of Silver Nanoparticles Synthesized Using Fungus *Cochliobolus lunatus* Against *Aedes aegypti* (Linnaeus, 1762) and *Anopheles stephensi* Liston (Diptera; Culicidae). *Parasitol. Res.* **2011**, *109*, 823–831.

343. Sana, S. S.; Badineni, V. R.; Arla, S. K.; Boya, V. K. N. Eco-friendly Synthesis of Silver Nanoparticles Using Leaf Extract of *Grewia flavescens* and Study of Their Antimicrobial Activity. *Mater. Lett.* **2015**, *145*, 347–350.

344. Sanghi, R.; Verma, P. Biomimetic Synthesis and Characterization of Protein Capped Silver Nanoparticles. *Bioresour. Technol.* **2009**, *100*, 501.

345. Sankar, R.; Karthik, A.; Prabu, A.; Karthik, S.; Shivashangari, K. S.; Ravikumar, V. *Origanum vulgare* Mediated Biosynthesis of Silver Nanoparticles for Its Antibacterial and Anticancer Activity. *Colloids Surf. B Biointerfaces* **2013**, *108*, 80–84.

346. Santhoshkumar, T.; Rahuman, A. A.; Rajakumar, G.; Marimuthu, S.; Bagavan, A.; Jayaseelan, C.; Zahir, A. A.; Elango, G.; Kamaraj, C. Synthesis of Silver Nanoparticles Using *Nelumbo nucifera* Leaf Extract and Its Larvicidal Activity Against Malaria and Filariasis Vectors. *Parasitol. Res.* **2011**, *108*, 693–702.

347. Saravanakumar, A.; Ganesh, M.; Jayaprakash, J.; Jang, H. T. Biosynthesis of Silver Nanoparticles Using *Cassia tora* Leaf Extract and Its Antioxidant and Antibacterial Activities. *J. Ind. Eng. Chem.* **2015**. http://dx.doi.org/10.1016/j.jiec.2015.03.003

348. Sarikaya, M.; Tamerler, C.; Jen, A. K. Y.; Schulten, K.; Baneyx, F. Molecular Biomimetics: Nanotechnology Through Biology. *Nat. Mater.* **2003**, *2*, 577–585.

349. Sasikala, A.; Savithramma, N. Biological Synthesis of Silver Nanoparticles from *Cochlospermum religiosum* and Their Antibacterial Efficacy. *J. Pharm. Sci. Res.* **2012**, *4* (6), 1836–1839.

350. Sathishkumar, G.; Gobinath, C.; Karpagam, K.; Hemamalini, V.; Premkumar, K.; Sivaramakrishnan, S. Phyto-synthesis of Silver Nanoscale Particles Using *Morinda citrifolia* L. and Its Inhibitory Activity Against Human Pathogens. *Colloids and Surfaces B: Biointerfaces* **2012**, *95*, 235–240.

351. Sathishkumar, M.; Sneha, K.; Won, S. W.; Cho, C.; Kim, S.; Yun, Y. *Cinnamon zeylanicum* Bark Extract and Powder Mediated Green Synthesis of Nano-crystalline Silver Particles and Its Bactericidal Activity. *Colloids Surf. B Biointerfaces* **2009**, *73*, 332–338.

352. Sathishkumar, M.; Sneha, K.; Yun, Y. Immobilization of Silver Nanoparticles Synthesized Using *Curcuma Longa* Tuber Powder and Extract on Cotton Cloth for Bactericidal Activity. *Biores. Technol.* **2010**, *101*, 7958–7965.

353. Satyavani, K.; Gurudeeban, S.; Ramanathan, T. Balasubramanian, T. Biomedical Potential of Silver Nanoparticles Synthesized from Calli cells of *Citrullus colocynthis* (L.) Schrad. *J. Nanobiotechnol.* **2011**, *9*, 43.

354. Satyavani, K.; Ramanathan, T.;, Gurudeeban, S. Green Synthesis of Silver Nanoparticles by using Stem Derived Callus Extract of Bitter Apple (*Citrullus colocynthis*). *Digest J. Nanomater. Biostruct.* **2011**, *6* (3), 1019–1024.

355. Satyavani, K.; Ramanathan, T.; Gurudeeban, S. Plant Mediated Synthesis of Biomedical Silver Nanoparticles by Using Leaf Extract of *Citrullus colocynthis*. *Res. J. Nanosci. Nanotechnol.* **2011**, *1*, 95–101.

356. Savithramma, N.; Rao, M. L.; Devi, P. S. Evaluation of Antibacterial Efficacy of Biologically Synthesized Silver Nanoparticle using Stem Barks of *Boswellia ovalifoliolata* Bal. and Henry and *Shorea tumbuggaia* Roxb. *J. Biol. Sci.* **2011**. doi 10.3923/jbs.2011

357. Savithramma, N.; Rao, M. L.; Rukmini, K.; Devi, P. S. Antimicrobial Activity of Silver Nanoparticles Synthesized by Using Medicinal Plants. *Int. J. ChemTech. Res.* **2011**, *3* (3), 1394–1402.

358. Saxena, A.; Tripathi, R. M.; Zafar, F.; Singh, P. Green Synthesis of Silver Nanoparticles Using Aqueous Solution of *Ficus benghalensis* Leaf Extract and Characterization of Their Antibacterial Activity. *Mater. Lett.* **2012**, *67*, 91–94.

359. Saxena, A.; Tripathi, R. M.; Singh, R. P. Biological Synthesis of Silver Nanoparticles by Using Onion (*Allium cepa*) Extract and Their Antibacterial Activity. *Digest J. Nanomater. Biostruct.* **2010**, *5* (2), 427–432.

360. Selvakumar, P. M.; Antonyraj, C. A.; Babu, R.; Dakhsinamurthy, A.; Manikandan, N.; Palanivel, A. Green Synthesis and Anti-microbial Activity of Mono-dispersed Silver Nanoparticles Synthesized Using Lemon Extract. *Synthesis Reactivity Inorg. Metal-Org. Nano-Metal Chem.* **2015**. DOI: 10.1080/15533174.2014.971810

361. Shahverdi, A. R.; Minaeian, S.; Shahverdi, H. R.; Jamalifar, H.; Nohi, A. Rapid Synthesis of Silver Nanoparticles Using Culture Supernatants of Enterobacteria: A Novel Biological Approach. *Proc. Biochem.* **2007**, *42*, 919–923.

362. Shaligram, N. S.; Bule, M.; Bhambure, R.; Singhal, R. S.; Singh, S. K.; Szakacs, G.; Pandey, A. Biosynthesis of Silver Nanoparticles Using Aqueous Extract from the Compactin Producing Fungal Strain. *Proc. Biochem.* **2009**, *44*, 939–943.

363. Shameli, K.; Ahmad, M. B.; Zamanian, A.; Sangpour, P.; Shabanzadeh, P.; Abdollahi, Y.; Zargar, M. Green Biosynthesis of Silver Nanoparticles Using *Curcuma Longa* Tuber Powder. *Int. J. Nanomed.* **2012**, *7*, 5603–5610.

364. Shameli, K.; Ahmad, M. B.; Al-Mulla, E. A. J.; Ibrahim, N. A.; Shabanzadeh, P.; Rustaiyan, A.; Abdollahi, Y.; Bagheri, S.; Abdolmohammadi, S. Usman. M. S.; Zidan, M. Green Biosynthesis of Silver Nanoparticles Using *Callicarpa maingayi* Stem Bark Extraction. *Molecules* **2012,** *17,* 8506–8517.

365. Shankar, S. S.; Ahmad, A.; Sastry. M. Geranium Leaf Assisted Biosynthesis of Silver Nanoparticles. *Biotechnol. Progr.* 2003, *19* (6), 1627–1631.

366. Shankar, S. S.; Rai, A.; Ahmad, A.; Sastry, M. Rapid synthesis of Au, Ag, and Bimetallic Au Core–Ag Shell Nanoparticles Using Neem (*Azadirachta indica*) Leaf Broth. *J. Colloid Interface Sci.* **2004,** *275,* 496–502.

367. Sheeney-Haj-Ichia, L.; Pogorelova, S.; Gofer, Y.; Willner, I. Enhanced Photoelectrochemistry in CdS/Au Nanoparticle Bilayers. *Adv. Funct. Mater. Vol.* **2004,** *14* (5), 416–424.

368. Shrivastava, S.; Bera, T.; Singh, S. K.; Singh, G.; Ramachandrarao, P.; Dash, D. Characterization of Antiplatelet Properties of Silver Nanoparticles. *ACS Nano* **2012,** *3* (6), 1357–1364.

369. Singh, R.; Lillard Jr., J. W. Nanoparticle-based Targeted Drug Delivery. *Exp. Mol. Pathol.* **2009,** *86* (3), 215–223.

370. Singh, A.; Jain, D.; Upadhyay, M. K.; Khandelwal, N.; Verma, H. N. Green Synthesis of Silver Nanoparticles using *Argemone Mexicana* Leaf Extract and Evaluation of Their Antimicrobial Activities. *Digest J. Nanomater. Biostruct.* 2010, *5* (2), 483–489.

371. Singh, A. K.; Talat, M.; Singh, D. P.; Srivastava, O. N. Biosynthesis of Gold and Silver Nanoparticles by Natural Precursor Clove and Their Functionalization with Amine Group. *J. Nanopart. Res.* 2010, *12,* 1667–1675.

372. Singh, C.; Baboota, R. K.; Naik, P. K.; Singh, H. Biocompatible Synthesis of Silver and Gold Nanoparticles Using Leaf Extract of *Dalbergia sissoo. Adv. Mat. Lett.* **2012,** *3* (4), 279–285.

373. Singh, C.; Sharma, V.; Naik, P. K.; Khandelwal, V.; Singh, H. A Green Biogenic Approach for Synthesis of Gold and Silver Nanoparticles Using *Zingiber Officinale.* *Digest J. Nanomater. Biostruct.* **2011,** *6* (2), 535–542.

374. Singh, R.; Nalwa, H. S. Medical Applications of Nanoparticles in Biological Imaging, Cell Labeling, Antimicrobial Agents, and Anticancer Nanodrugs. *J. Biomed. Nanotechnol.* **2011,** *7* (4), 489503.

375. Singh, P.; Kim, Y. J.; Wang, C.; Mathiyalagan, R.; El-Agamy Farh, M.; Yang, D. C. Biogenic Silver and Gold Nanoparticles Synthesized Using Red Ginseng Root Extract, and Their Applications. *Artif. Cells Nanomed. Biotechnol.* **2015,** 1–6.

376. Singhal, G.; Bhavesh, R.; Kasariy, K.; Sharma, A. R.; Singh, R. P. Biosynthesis of Silver Nanoparticles Using *Ocimum sanctum* (Tulsi) Leaf Extract and Screening Its Antimicrobial Activity. *J. Nanopart. Res.* **2011,** *13,* 2981–2988.

377. Sinha, S. N.; Paul, D. Phytosynthesis of Silver Nanoparticles Using Andrographis paniculata Leaf Extract and Evaluation of Their Antibacterial Activities. *Spectrosc. Lett. Int. J. Rapid Comm.* **2015,** *48* (8) 600–604.

378. Sivakumar, J.; Premkumar, C.; Santhanam, P.; Saraswathi, N. Biosynthesis of Silver Nanoparticles Using *Calotropis gigantean* Leaf. *African J. Basic Appl. Sci.* **2011,** *3* (6), 265–270.

379. Sivakumar, P.; Nethradevi, C.; Renganathan, S. Synthesis of Silver Nanoparticles Using *Lantana camara* Fruit Extract and Its Effect on Pathogens. *Asian J. Pharm. Clin. Res.* **2012**, *5* (3), 97–101.

380. Solgi, M.; Taghizadeh, M. Silver Nanoparticles Ecofriendly Synthesis by Two Medicinal Plants. *Int. J. Nanomater. Biostruct.* **2012**, *2* (4), 60–64.

381. Song, J. Y.; Kim, B. S. Rapid Biological Synthesis of Silver Nanoparticles Using Plant Leaf Extracts. *Bioproc. Biosyst. Eng.* **2009**, *32*, 79–84.

382. Soni, N.; Prakash, S. Efficacy of Fungus Mediated Silver and Gold Nanoparticles Against *Aedes aegypti* Larvae. *Parasitol. Res.* **2012**, *110*, 175–184.

383. Sre, P. R. R.; Reka, M.; Poovazhagi, R.; Kumar, M. A.; Murugesan, K. Antibacterial and Cytotoxic Effect of Biologically Synthesized Silver Nanoparticles Using Aqueous Root Extract of *Erythrina Indica* Lam. *Spectrochim. Acta Part A Mol. Biomol. Spectrosc.* **2015**, *135*, 1137–1144.

384. Suganya, A.; Murugan, K.; Kovendan, K.; Kumar, P. M.; Hwang, J. Green Synthesis of Silver Nanoparticles Using Murraya Koenigii Leaf Extract Against *Anopheles Stephensi* and *Aedes aegypti*. *Parasitol. Res.* **2013**, *112*, 1385–1397.

385. Sukirtha, R.; Priyanka, K. M.; Antony, J. J.; Kamalakkannan, S.; Thangam, R.; Gunasekaran, P.; Krishnan, M.; Achiraman, S. Cytotoxic Effect of Green Synthesized Silver Nanoparticles Using Melia Azedarach Against In Vitro Hela Cell Lines and Lymphoma Mice Model. *Proc. Biochem.* **2012**, *47*, 273–279.

386. Sulaiman, G. M.; Mohammed, W. H.; Marzoog, T. R.; Al-Amiery, A. A. A.; Kadhum, A. A. H.; Mohamad, A. B. Green Synthesis, Antimicrobial and Cytotoxic Effects of Silver Nanoparticles Using *Eucalyptus chapmaniana* Leaves Extract. *Asian Pac. J. Trop. Biomed.* **2013**, *3* (1), 58–63.

387. Suman, T. Y.; Rajasree, S. R. R.; Kanchana, A.; Elizabeth, S. B. Biosynthesis, Characterization and Cytotoxic Effect of Plant Mediated Silver Nanoparticles Using *Morinda citrifolia* Root Extract. *Colloids Surf. B Biointerfaces* **2013**, *106*, 74–78.

388. Sundararajan, B.; Kumari, B. D. R. Biosynthesis of Silver Nanoparticles in *Lagerstroemia Speciosa* (L.) Pers and Their Antimicrobial Activities. *Int. J. Pharm. Pharm. Sci. 6* (3), 30–34.

389. Sundaravadivelan, C.; Padmanabhan, M. N.; Sivaprasath, P.; Kishmu, L. Biosynthesized Silver Nanoparticles from Pedilanthus Tithymaloides Leaf Extract with Anti-developmental Activity Against Larval Instars of *Aedes aegypti* L. (Diptera; Culicidae). *Parasitol. Res.* **2013**, *112*, 303–311.

390. Suresh, A. K.; Pelletier, D. A.; Wang, W.; Morrell-Falvey, J. L.; Gu, B.; Doktycz, M. J. Cytotoxicity Induced by Engineered Silver Nanocrystallites is Dependent on Surface Coatings and Cell Types. *Langmuir* **2012**, *28*, 2727–2735.

391. Suresh, U.; Murugan, K.; Benelli, G.; Nicoletti, M.; Barnard, D. R.; Panneerselvam, C.; Kumar, P. M.; Subramaniam, J.; Dinesh, D.; Chandramohan, B. Tackling the Growing Threat of Dengue: Phyllanthus Niruri-mediated Synthesis of Silver Nanoparticles and Their Mosquitocidal Properties Against the Dengue Vector *Aedes aegypti* (Diptera: Culicidae). *Parasitol. Res.* **2015**, *114*, 1551–1562.

392. Suriyakalaa, U.; Antony, J. J.; Suganya, S.; Siva, D.; Sukirtha, R.; Kamalakkannan, S.; Pichiah, P. B. T.; Achiraman, S. Hepatocurative Activity of Biosynthesized Silver Nanoparticles Fabricated Using *Andrographis Paniculata*. *Colloids Surf. B Biointerfaces* **2013**, *102*, 189–194.

393. Swamy, M. K.; Sudipta, K. M.; Jayanta, K.; Balasubramanya, S. The Green Synthesis, Characterization, and Evaluation of the Biological Activities of Silver Nanoparticles Synthesized From *Leptadenia Reticulata* Leaf Extract. *Appl. Nanosci.* **2015,** *5*, 73–81.

394. Swamy, Vs.; Prasad, R. Green Synthesis of Silver Nanoparticles from the Leaf Extract of *Santalum Album* and Its Antimicrobial Activity. *J. Optoelectronics Biomed. Mater.* **2012,** *4* (3), 53–59.

395. Syed, A.; Saraswati, S.; Kundu, G. C.; Ahmad, A. Biological Synthesis of Silver Nanoparticles Using the Fungus Humicola Sp. and Evaluation of Their Cytoxicity Using Normal and Cancer Cell Lines. *Spectrochim. Acta Part A Mol. Biomol. Spectrosc.* **2013,** *114*, 144–147.

396. Thakkar, K. N.; Mhatre, S. S.; Parikh, R. Y. Biological Synthesis of Metallic Nanoparticles. *Nanomedicine* **2010,** *6* (2), 257–262.

397. Telrandhe, R.; Mahapatra, D. K.; Kamble, M. A. Bombax Ceiba Thorn Extract Mediated Synthesis of Silver Nanoparticles: Evaluation of Anti-*staphylococcus aureus* Activity. *Int. J. Pharm. Drug Analysis* **2017,** *5* (9), 376–379.

398. Thenmozhi, B.; Suryakiran, S.; Sudha, R.; Revathy, B. Green Synthesis and Comparative Study of Silver and Iron Nanoparticle from Leaf Extract. *Int. J. Institutional Pharm. Life Sci.* **2014,** *4* (2), 5–19.

399. Thirumurugan, A.; Tomy, N. A.; Ganesh, R. J.; Gobikrishnan, S. Biological Reduction of Silver Nanoparticles Using Plant Leaf Extracts and Its Effect on Increased Antimicrobial Activity Against Clinically Isolated Organism. *Der Pharma. Chemica* **2010,** *2* (6), 279–284.

400. Thirunavoukkarasu, M.; Balaji, U.; Behera, S.; Panda, P. K.; Mishra, B. K. Biosynthesis of Silver Nanoparticle from Leaf Extract of *Desmodium Gangeticum* (L.) Dc. and Its Biomedical Potential. *Spectrochim. Acta Part A Mol. Biomol. Spectrosc.* **2013,** *116*, 424–427.

401. Thombre, R.; Parekh, F.; Patil, N. Green Synthesis of Silver Nanoparticles Using Seed Extract of *Argyreia nervosa. Int. J. Pharm. Bio. Sci.* **2014,** *5* (1), 114–9.

402. Tran, H. V.; Tran, L. D.; Ba, C. T.; Vu, H. D.; Nguyen, T. N.; Pham, D. G.; Nguyen, P. X. Synthesis, Characterization, Antibacterial and Antiproliferative Activities of Monodisperse Chitosan-based Silver Nanoparticles. *Colloids Surf. A Physicochem. Eng. Aspects* **2010,** 360, 32–40.

403. Tripathi, R. M.; Kumar, N.; Shrivastava, A.; Singh, P.; Shrivastav, B. R. Catalytic Activity of Biogenic Silver Nanoparticles Synthesized by *Ficus panda* Leaf Extract. *J. Mol. Catal. B Enzymatic* **2013,** *96*, 75– 80.

404. Tripathy, A.; Raichur, A. M.; Chandrasekaran, N.; Prathna, T. C.; Mukherjee, A. Process Variables in Biomimetic Synthesis of Silver Nanoparticles by Aqueous Extract of *Azadirachta indica* (Neem) Leaves. *J. Nanopart. Res.* **2010,** *12*, 237–246.

405. Tsakalakos, T.; Ovid'ko, I. A.; Vasudevan, A. K. Springer Science and Business Media, 2003.

406. Udayasoorian, C.; Vinoth, K. K.; Jayabalakrishnan, R. M. Extracellular Synthesis of Silver Nanoparticles Using Leaf Extract of *Cassia auriculata. Digest J. Nanomater. Biostruct.* **2011,** *6* (1), 279–283.

407. Umadevi, M.; Bindhu, M. R.; Sathe, V. A Novel Synthesis of Malic Acid Capped Silver Nanoparticles Using *Solanum lycopersicums* Fruit Extract. *J. Mater. Sci. Technol.* **2013,** *29* (4), 317–322.

408. Ulug, B.; Turkdemir, M. H.; Cicek, A.; Mete, A. Role of Irradiation in the Green Synthesis of Silver Nanoparticles Mediated by Fig (*Ficus carica*) Leaf Extract. *Spectrochim. Acta Part A Mol. Biomol. Spectrosc.* **2015**, *135*, 153–161.

409. Vahabi, K.; Mansoori, G. A.; Karimi, S. Biosynthesis of Silver Nanoparticles by Fungus *Trichoderma Reesei*. *Insci. J.* **2011**, *1* (1), 65–79.

410. Vaishnavi, B.; Rameshkumar, G.; Rajagopal, T.; Ponmanickam, P. Evaluation of Bactericidal and Fungicidal Properties of Silver Nanoparticles Fabricated Using *Jasminum Sambac* (L.). *Global J. Biotechnol. Biochem.* **2015**, *10* (1), 22–31.

411. Valli, J. S.; Vaseeharan, B. Biosynthesis of Silver Nanoparticles by *Cissus Quadrangularis* Extracts. *Mater. Lett.* **2012**, *82*, 171–173.

412. Valodkar, M.; Jadeja, R. N.; Thounaojam, M. C.; Devkar, R. V.; Thakore, S. In Vitro Toxicity Study of Plant Latex Capped Silver Nanoparticles in Human Lung Carcinoma Cells. *Mater. Sci. Eng. C* **2011**, *31* (8), 1723–1728.

413. Van Aerle, R.; Lange, A.; Moorhouse, A. Molecular Mechanisms of Toxicity of Silver Nanoparticles in Zebrafish Embryos. *Environ. Sci. Technol.* **2013**, *47* (14), 8005–8014.

414. Vanaja, M.; Annadurai, G. *Coleus Aromaticus* Leaf Extract Mediated Synthesis of Silver Nanoparticles and Its Bactericidal Activity. *Appl. Nanosci.* **2013**, *3*, 217–223.

415. Vanaja, M.; Rajeshkumar, S.; Paulkumar, K.; Gnanajothiba, G.; Malarkodi, C.; Annadurai, G. Phytosynthesis and Characterization of Silver Nanoparticles Using Stem Extract of *Coleus aromaticus*. *Int. J. Nanomater. Biostruct.* **2013**, *3* (1), 1–4.

416. Varshney, R.; Mishra, A. N.; Bhadauria, S.; Gaur, M. S.. A Novel Microbial Route to Synthesize Silver Nanoparticles Using Fungus *Hormoconis Resinae*. *Digest J. Nanomater. Biostruct.* **2009**, *4* (2), 349–355.

417. Varshney, R.; Bhadauria, S.; Gaur, M. S. Biogenic Synthesis of Silver Nanocubes and Nanorods Using Sundried *Stevia Rebaudiana* Leaves. *Adv. Mat. Lett.* **2010**, *1* (3), 232–237.

418. Veerakumar, K.; Govindarajan, M.; Rajeswary, M. Green Synthesis of Silver Nanoparticles Using *Sida acuta* (Malvaceae) Leaf Extract Against Culex Quinquefasciatus, Anopheles Stephensi, and *Aedes Aegypti* (Diptera: Culicidae). *Parasitol. Res.* **2013**, *112*, 4073–4085.

419. Veerasamy, R.; Xin, T. Z.; Gunasagaran, S.; Xiang, T. F. W.; Yang, E. F. C.; Jeyakumar, N.; Dhanaraj, S. A. Biosynthesis of Silver Nanoparticles Using Mangosteen Leaf Extract and Evaluation of Their Antimicrobial Activities. *J. Saudi Chem. Soci.* **2011**, *15*, 113–120.

420. Velayutham, K.; Rahuman, A.A.; Rajakumar, G.; Roopan, S. M.; Elango, G.; Kamaraj, C.; Marimuthu, S.; Santhoshkumar, T.; Iyappan, M.; Siva, C. Larvicidal Activity of Green Synthesized Silver Nanoparticles Using Bark Aqueous Extract of *Ficus Racemosa* Against Culex Quinquefasciatus and Culex Gelidus. *Asian Pacific J. Trop. Med.* **2013**, 95–101.

421. Velmurugan, P.; Cho, M.; Lim, S.; Seo, S.; Myung, H.; Bang, K.; Sivakumar, S.; Cho, K.; Oh, B. Phytosynthesis of Silver Nanoparticles by Prunus Yedoensis Leaf Extract and Their Antimicrobial Activity. *Mater. Lett.* **2015**, *138*, 272–275.

422. Verma, V. C. Kharwar, R. N.; Gange, A. C. Biosynthesis of Antimicrobial Silver Nanoparticles by the Endophytic Fungus *Aspergillus clavatus*. *Nanomedicine* **2010**, *5* (1), 33–40.

423. Vidhu, V. K.; Philip, D. Spectroscopic, Microscopic and Catalytic Properties of Silver Nanoparticles Synthesized Using Saraca Indica Flower. *Spectrochim. Acta Part A Mol. Biomol. Spectrosc.* **2014,** *117,* 102–108.

424. Vidhu, V. K.; Aromal, S. A.; Philip, D. Green Synthesis of Silver Nanoparticles Using Macrotyloma Uniflorum. *Spectrochim. Acta Part A* **2011,** *83,* 392–397.

425. Vigneshwaran, N.; Kathe, A. A.; Varadarajan, P. V.; Nachne, R. P.; Balasubramanya, R. H. Biomimetics of Silver Nanoparticles by White Rot Fungus, *Phaenerochaete chrysosporium. Colloids Surf. B Biointerfaces* **2006,** *53,* 55–59.

426. Vijayakumar, M.; Priya, K.; Nancy, F. T.; Noorlidah, A.; Ahmed, A. B. A. Biosynthesis, Characterisation and Anti-bacterial Effect of Plant-mediated Silver Nanoparticles Using *Artemisia nilagirica. Ind. Crops Prod.* **2013,** *41,* 235–240.

427. Vijayaraghavan, K.; Nalini, S. P. K.; Prakash, N. U.; Madhankumar, D. Biomimetic Synthesis of Silver Nanoparticles by Aqueous Extract of *Syzygium Aromaticum. Mater. Lett.* **2012,** *75,* 33–35.

428. Vilchis-Nestor, A. R.; Sánchez-Mendieta, V.; Camacho-López, M. A.; Gómez-Espinosa, R. M.; Camacho-López, M. A.; Arenas-Alatorre, J. A. Solventless Synthesis and Optical Properties of Au and Ag Nanoparticles Using *Camellia sinensis* Extract. *Mater. Lett.* 2008, *62,* 3103–3105.

429. Vimala, R. T. V.; Sathishkumar, G.; Sivaramakrishnan, S. Optimization of Reaction Conditions to Fabricate Nano-silver Using *Couroupita Guianensis* Aubl. (Leaf and Fruit) and Its Enhanced Larvicidal Effect. *Spectrochim. Acta Part A Mol. Biomol. Spectrosc.* **2015,** *135,* 110–115.

430. Vishnudas, D.; Mitra, B.; Sant, S. B.; Annamalai, A. Green-synthesis and Characterization of Silver Nanoparticles by Aqueous Leaf Extracts of *Cardiospermum helicacabum* Leaves. *Drug Invention Today* **2012,** *4* (2), 340–344.

431. Vivek, M.; Kumar, P. S.; Steffi, S.; Sudha, S. Biogenic Silver Nanoparticles by *Gelidiella acerosa* Extract and Their Antifungal Effects. *Avicenna J. Med. Biotech.* **2011,** *3* (3), 143–148.

432. Vivek, R.; Thangam, R.; Muthuchelian, K.; Gunasekaran, P.; Kaveri; K.; Kannan, S. Green Biosynthesis of Silver Nanoparticles from *Annona squamosa* Leaf Extract and Its In Vitro Cytotoxic Effect on Mcf-7 Cells. *Proc. Biochem.* **2012,** *47,* 2405–2410.

433. Wang, A. Z.;, Gu, F.; Zhang, L.; Chan, J. M.; Radovic-Moreno, A.; Shaikh, M. R.; Langer, R. S.; Farokhzad, O. C. Biofunctionalized Targeted Nanoparticles for Therapeutic Applications. *Expert Opin. Biol. Ther.* **2008,** *8* (8), 1063–1070.

434. Wang, Z. L. *Characterization of Nanophase Materials.* John Wiley & Sons, 2001, Weinheim, Germany.

435. Wei. L., *et al.* Silver Nanoparticles: Synthesis, Properties, and Therapeutic Applications. *Drug Discov. Today* **2015.** http://Dx.Doi.Org/10.1016/ J.Drudis.2014.11.014.

436. White II, G.V.; Kerscher, P.; Brown, R.M.; Morella, J.D.; McAllister, W.; Dean, D.; Kitchens, C. L. Green Synthesis of Robust, Biocompatible Silver Nanoparticles Using Garlic Extract. *J. Nanomater.* **2012,** *2012,* 730746.

437. Who Reports. http://Www.Who.Int/Mediacentre/Factsheets/Fs094/En/ (accessed April 16, 2015).

438. Wong, K. K.Y.; Cheung, S. O. F.; Huang, L.; Niu, J.; Tao, C.; Ho, C.; Che, C.; Tam, P. K. H. Further Evidence of the Anti-inflammatory Effects of Silver Nanoparticles. *Chemmedchem.* **2009,** *4,* 1129–1135.

439. Xiang, D.; Zheng, Y.; Duan, W.; Li, X.; Yin, J.; Shigdar, S.; O'connor. M. L.; Marappan, M.; Zhao, X.; Miao, Y.; Xiang, B.; Zheng. C. Inhibition of A/Human/ Hubei/3/2005 (H3n2) Influenza Virus Infection by Silver Nanoparticles *In Vitro* and *In Vivo*. *Int. J. Nanomed.* **2013,** *8,* 4103–4114.

440. Yilmaz, M.; Turkdemir, H.; Kilic, M. A.; Bayram, E.; Cicek, A.; Mete, A.; Ulug, B. Biosynthesis of Silver Nanoparticles Using Leaves of *Stevia rebaudiana*. *Mater. Chem. Phys.* **2011,** *130,* 1195–1202.

441. Yudha, S. S.; Notriawan, D.; Angasa, E.; Suharto, T. E.; Hendri, J.; Nishina, Y. Green Synthesis of Silver Nanoparticles Using Aqueous Rinds Extract of *Brucea Javanica* (L.) Merr at Ambient Temperature. *Mater. Lett.* **2013,** *97,* 181–183.

442. Zahir, A. A.; Rahuman, A. A. Evaluation of Different Extracts and Synthesised Silver Nanoparticles from Leaves of Euphorbia Prostrata Against *Haemaphysalis bispinosa* and *Hippobosca maculate*. *Vet. Parasitol.* **2012,** *187,* 511–520.

443. Zahir, A. A.; Bagavan, A.; Kamaraj, C.; Elango, G.; Rahuman, A. A. Efficacy of Plant-mediated Synthesized Silver Nanoparticles Against *Sitophilus oryzae*. *J. Biopest.* **2012,** *5,* 95–102.

444. Zanette, C.; Pelin, M.; Crosera, M.; Adami, G.; Bovenzi, M.; Larese, F. F.; Florio, C. Silver Nanoparticles Exert a Long-lasting Antiproliferative Effect on Human Keratinocyte Hacat Cell Line. *Toxicol. in Vitro* **2011,** *25,* 1053–1060.

445. Zarchi, A. A. K.; Mokhtari, N.; Arfan, M.; Rehman, T.; Ali, M.; Amini, M.; Majidi, R. F.; Shahverdi, A. R. A Sunlight-induced Method for Rapid Biosynthesis of Silver Nanoparticles Using an *Andrachnea chordifolia* Ethanol Extract. *Appl. Phys. A* **2011,** *103,* 349–353.

446. Zargar, M.; Hamid, A. A.; Bakar, F. A.; Shamsudin, M. N.; Shameli, K.; Jahanshiri, F.; Farahani, F. Green Synthesis and Antibacterial Effect of Silver Nanoparticles Using *Vitex Negundo* L. *Molecules* **2011,** *16,* 6667–6676.

447. Zayed, M. F.; Eisa, W. H.; Shabaka, A. A. Malva Parviflora Extract Assisted Green Synthesis of Silver Nanoparticles. *Spectrochim. Acta Part A Mol. Biomol. Spectrosc.* **2012,** *98,* 423–428.

448. Zhang, H.; Li, Q.; Lu, Y.; Sun, D. Lin, X.; Deng, X. Biosorption and Bioreduction of Diamine Silver Complex by *Corynebacterium*. *J. Chem. Technol. Biotechnol.* **2005,** *80,* 2850.

449. Zhang, T.; Wang, L.; Chen, Q.; Chen, C. Cytotoxic Potential of Silver Nanoparticles. *Yonsei Med. J.* **2014,** *55* (2), 283–291.

450. Zhou, C.; Long, M.; Qin, Y.; Sun, X.; Zheng, J. Luminescent Gold Nanoparticles with Efficient Renal Clearance. *Angew. Chem. Int. Ed. Engl.* **2011,** *50* (14), 3168–3172.

CHAPTER 12

RESEARCH PROGRESS TO IMPROVE SOLUBILITY AND DRUG-RELEASING CURCUMIN PROPERTIES WITH NANOENCAPSULATION

RAJPREET KAUR and POONAM KHULLAR*

Department of Chemistry, BBK DAV College for Women, Krishna Nagar, Dayanand Nagar, Amritsar 143001, Punjab, India

Corresponding author. E-mail: virgo16sep2005@gmail.com

ABSTRACT

Curcumin is widely used as a spice as well as coloring agent. It has also shown some clinical activities. Major drawback of curcumin is its poor solubility and less bioavailability. Encapsulation method of hydrophobic drug, curcumin, is found to be very effective to increase the solubility and bioavailability of curcumin. Water-soluble polymer biomaterials combined with curcumin in form of micelles, BSA nanoparticles, silver hydrogels, MMT formulation, polymer block have increased the loading capacity of drug and efficiency of drug-release encapsulation. They can be used as nanocarriers for hydrophobic drug in target drug delivery. These nanocarriers also increased the stability on surface as well as interior of nanoparticles. Release rate of drug is determined by the properties of polymer matrix of nanocarriers.

12.1 INTRODUCTION

Curcumin is a natural polyphenol obtained from rhizome of *Curcuma longa*,[1] a perennial herb, also widely known as turmeric. Curcumin (diferuloylmethane) with low-molecular-weight polyphenol (formula weight = 368.37) is found in 31 species of curcuma plants, a member of the ginger family *Zingiberaceae*. Curcumin has medicinal properties like anti-inflammatory,[2] antibacterial,[3] antifungal,[4] anticancer,[5] antispasmodic,[6] antioxidant,[7] antiamoebic,[8] anti-HIV,[8] antidiabetic,[9] and antifertility.[10] Apart from these properties, the use of curcumin is very restricted because of its poor solubility and low bioavailability, fast metabolic rate, rapid degradation, and fast systemic elimination.[11] Various necessary systems have been designed to increase the solubility in aqueous medium and for drug delivery to the targeted site with increased efficiency and reduced side effects on healthy cells and organs. Solid dispersion technique is most preferred to improve the solubility of curcumin in aqueous medium. It fastens with inert transporter or matrix in solid state by melting, melting solvent method, or dissolution in solvent. Various water-soluble polymers such as PEG, PVP, and PVA have been used as dispersion matrix. Curcumin, a hydrophobic drug, could enhance its use clinically by encapsulation with hydrogels, microparticles and nanoparticles, polymer drug conjugates, cyclodextrins, and micelles. Their diameter ranges from micrometers to nanometers. These nanocarriers have large bioapplicability due to their large specific surface area. Also the release rate of drug in intravenous fluid depends upon the properties of matrix prepared by encapsulation method. When the encapsulated drug is inserted in the fluid media, polymer dissolves and the drug is released as colloidal particles.[12] Curcumin drug, with medicinal properties,[13] conjugated with micelles can locate itself into the tumor without effecting natural tissues. Keto-enol tautomerism is basically observed in curcumin .In neutral and acidic medium, keto form of curcumin is prominent and shows influential hydrogen patron.[15-16] Some micelles like PDEA blocks are amphoteric in nature by protonation and deprotonation. These pH-sensitive micelles are attracted by tumor, inflammatory tissue, etc., due to difference between tumor tissues and normal tissues. So, they are also used in pH-triggered drug release in tumor tissues. Reduction in particles size from micronization to nanonization improves the dissolution rate and oral bioavailability. The stabilized and controlled size of particles distributes themselves finely in the network of polymer matrix, hydrogels. Hydrogels with stabilized Ag nanoparticles with reduced size have wide application in burn dressings, skin donor, and recipient sites health products, in purification of water system. Silver nanoparticles, size-dependent interaction with HIV-I virus via suitable binding with proteins, inhibits virus from interaction with host cells.

12.2 DISCUSSION

12.2.1 SOLUBILITY OF CURCUMIN

Curcumin has poor solubility in water that result in poor bioavailability. To improve the solubility of curcumin in water, solid dispersion method like hot melt method and solvent evaporation were used. PEG-4000, PEG-6000, Tween 80, and PVP K 30 were chosen as carriers for solvent evaporation method.[14] Solubility of curcumin in solid dispersion increased about 100 folds with PEG-6000 by hot melt method. In comparison with hot melt method, solubility of curcumin decreases with PVP K 30 by solvent evaporation method. Further evaluation of studies like SEM, TLC, IR, and X-ray analysis[17] have been done. Pure curcumin, curcumin solid dispersions by hot melt method (SDHM) and by solvent evaporation method in vitro dissolution profile, and release of drug in medium were 2.6%, 10.03%, and 8.5%, respectively. Three spots were identified with TLC studies, carried out for pure dry and solid dispersion method which showed curcumin (Rf = 0.96), demethoxy curcumin (Rf = 0.94), and bismethoxy curcumin (Rf = 0.88–0.9) and there was no interaction between drug and carrier.[17–18]

X-ray diffraction studies revealed the decrease in intensities of curcumin in solid dispersion in comparison with pure curcumin. IR spectrum studies showed curcumin did not interact with carriers as the peaks shown in pure curcumin were also observed in solid dispersions IR spectrum.

Solubility of curcumin with reduced side effects could be enhanced by its physically encapsulation into the hydrophobic cores of polymeric micelles polycaprolactone (PCL)–PDEA–PSBMA, inulin-d-α-tocopherol succinate (INVITE) micelles, MMT formulation, etc.

12.2.2 DRUG LOAD AND RELEASE

12.2.2.1 pH EFFECT ON DRUG LOADING AND RELEASING WITH PDEA BLOCK

The pH plays a major role in loading and releasing of drug release process. Rate of drug release was increased when pH value changed from 7.4 to 5.0. Protonation of curcumin loaded micelles, that is, PDEA block at lower pH at 5.0 formed hydrophilic segments which resulted in faster release of curcumin.[19]

Hindrance of drug release in the cores medium at pH 7.4 was due to no protonation of PDEA block and remained in compact form, whereas at pH 5.0, PDEA block swelled and ruptured, attributed to protonation of amine groups from PDEA blocks. DLS studies showed that the average size of curcumin-loaded micelles was 182.2 nm, which was larger than empty micelles. The calculated values of DLC and DLE in PCL–PDEA–PSBMA micelles were 8.06% and 64.46%, respectively.

12.2.3 IN VITRO CURCUMIN RELEASE STUDY WITH INVITE MICELLES

A most acceptable method, that is, fusion of hydrophobic drug into miceller body like INVITE are valuable for intravenous application in drug delivery systems.[20] A hypothesis, INVITE micelle cores are formed by pi–pi interaction of aromatic portion of VITE rather than hydrophobic interaction between the aliphatic chains.[20–21] Loading of hydrophobic drug curcumin having aromatic ring in the INVITE micelles should not form permanent stable system. So, without bursting effect, encapsulated drug could be released in a controlled manner.

The releasing behavior of the INVITE at pH 7.4 or 5.5, in a physiologic fluids, or in a conditions stimulating the endosome or the tumor environments had been studied as shown in Figure 12.1. Curves are showing about the release of free drug, curcumin, and the controlled release of curcumin in INVITE micelles system without bursting effect

at both pH 7.4 or pH 5.5 in PBS solution in the presence of 0.5% polysorbate 80 up to 48 h.

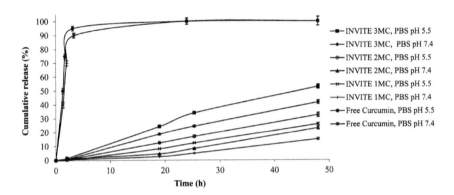

FIGURE 12.1   Release of curcumin drug by INVITE micelles at pH 7.4 and pH 5.5.
Source: Reprinted with permission from Ref. [20]. © 2013 American Chemical Society.

A comparison was figured out that 42% from INVITE 3MC, 23% from
INVITE 2MC, and 15% from INVITE 1MC after 48 h in PBS at pH had
been released. At pH 5.5, there was 10% increased in value, that is, 53%
for INVITE 3MC, 23% for INVITE 2MC, and 25% for INVITE 1MC.
The order of release trend was
INVITE 3MC > INVITE 2MC > INVITE 1MC.

This was because "bridge-like structure" of VITE in INVITE 3M
with curcumin hydrophobic portion of the micelles resulted into larger
particles formation with respect to INVITE 2M and INVITE 1MC. TEM
analysis as shown in Figure 12.2 depicts that there was no size difference
between drug loaded and empty micelles. Furthermore, it was noted that
release rate of drug almost constant at no time fluctuations confirmed
the higher physical stability.[21,22] Similarly , a detectable burst effects of
curcumin had been reported by Abouzeid et al. and by Sun et al.[23,24]
Hence, rate of release at pH 7.4 or 5.5 may be useful for tumor delivery
in acidic environment.

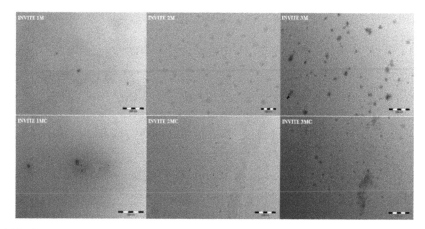

FIGURE 12.2   TEM images of empty INVITE M's and curcumin-loaded INVITE MC's of equal size of 200 nm.
Source: Reprinted with permission from Ref. [20]. © 2013 American Chemical Society.

12.2.4   CURCUMIN-LOADED BSA NPs

To remove the instability, cross-linked treatment of BSA with glutaral-dehyde and curcumin-loaded BSA NPs with equal molar ratio of 1 were used. With this treatment, size of nanoparticles and stability with increased polydispersity index had been observed 1 month at room temperature under stirring. It neither increased the stability on surface of NPs, but also increased the stability in the interior of NPs. These stable NPs have wide potential as biocompatible carriers for hydrophobic curcumin.[25]

12.2.5   IN VITRO CURCUMIN RELEASE STUDY WITH BSA NPs

To evaluate in vitro curcumin release (Figure 12.3), the dialysis tubing method was used from BSA NPs. The lyophilized curcumin-loaded NPs showed that during 72 h, about 70% of curcumin was released, which showed that curcumin molecules which were attached on surface could be released easily, while curcumin molecules which were loaded inside BSA NPs were not released with amount of 30%. This suggests the ability of BSA NPs for controlled release of curcumin.[25]

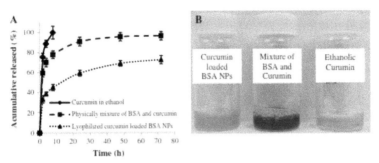

FIGURE 12.3 In vitro release profile of curcumin.
Source: Reprinted with permission from Ref. [25]. © 2014 Springer Science Media.

As shown in Figure 12.4 the redispersion study of curcumin-loaded BSA NPs showed that with lyophilization treatment, free curcumin molecules

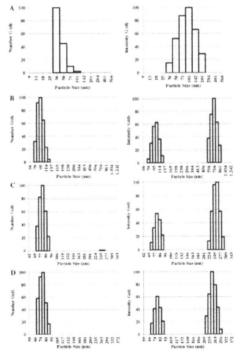

FIGURE 12.4 (A) Showing distribution of size of curcumin-loaded BSA NPs after glutaraldehyde (B) and (C) showing after redispersion in water for 24 h and sonication for 10 s and 20 s, respectively, and (D) showing number mode and intensity.
Source: Reprinted with permission from Ref. [25]. © 2014 Springer Science Media.

after coprecipitation with BSA on the surface of BSA NPs could be loaded. After lyophilization, NPs could be dispersed in water, in spite of larger size and polydispersity index of redispersed NPs as compared to original NPs size before lyophilization. Both number and intensity distribution profile of DLS showed that the average size of most nanoparticles were smaller than 100 nm. In redispersion method, sonication is important and effective treatment for dispersion of aggregated NPs, which got swollen in redispersion in water.

12.2.6 IN VITRO CURCUMIN RELEASE AND LOADING WITH % MMT

The kinetic parameters had been studied regarding release of curcumin from nanoformulations with PCL/organomodified montmorillonite. By applying regression coefficient analysis ($R^2 > 0.95$), power law reported the mechanism regarding curcumin release from nanoformulation from stimulated intestinal fluid, with faster release of curcumin at initial stage than later stage. Besides this, percent composition of MMT encapsulation also effected the release of curcumin drug, that is, 15% MMT formulation revealed faster drug release than 30% MMT and without MMT nanoformulation encapsulation.[26]

Table 1   Properties of curcumin nanoparticles immediately after production

Formulation	Size (nm)	PdI	Zeta potential (mV)	Encapsulation Efficiency (%)	Drug Loading (%)
0 % MMT in PCL	155.0 ± 5.2	0.168 ± 0.05	−5.49 ± 5.30	54 ± 3.5	0.15 ± 0.02
15 % MMT in PCL	180.3 ± 6.1	0.333 ± 0.04	−15.60 ± 6.23	71 ± 2.8	0.18 ± 0.01
30 % MMT in PCL	206.5 ± 8.5	0.369 ± 0.07	−16.80 ± 11.7	60 ± 3.2	0.19 ± 0.01

FIGURE 12.5   Effect of encapsulation on curcumin properties.
Source: Reprinted with permission from Ref. [26]. © 2016 Springer Nature.

As recorded in Figure 12.5 (table), the drug-loading percentage and encapsulation efficiency percentage of curcumin nanoparticles appeared to be high for 15% MMT with formulation than formulation containing 0% and 30% MMT.

For drug release, size of nanoparticle is the important factor that plays model role in drug delivery system. Curcumin nanoformulation with MMT had larger sizes than zero MMT nanoformulation, as nanoparticles size is directly related to diffusion rate of organic solvent to outer aqueous medium. Formulation containing MMT had high viscosity that resulted to larger size nanoparticles. This high encapsulation efficiency leads to availability of drug at target site with increased residence time of drug. Release of drug with MMT formulation due to presence of nanoclay inhibited the diffusion of curcumin and slows down their aggregation, which further leads to their dissolution. About 15% of MMT formulation had faster release of drug than 30% because higher concentration of MMT decreases the penetration of water into the PCL nanocomposite and inhibits the erosion of the polymer matrix.

Figure 12.6 showed drug release with 0% , 15%, and 30% MMT in PCL with respect to time.

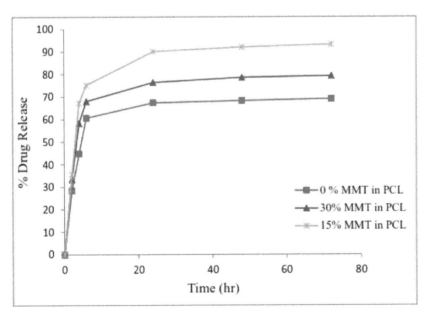

FIGURE 12.6   Drug release from curcumin formulation with respect to time.
Source: Reprinted with permission from Ref. [26]. © 2016 Springer Nature.

12.2.7 CURCUMIN LOADING AND RELEASE STUDIES WITH Ag NANOCOMPOSITE HYDROGELS

Figure 12.7 resulted in the encapsulation efficiency of loading and releasing drug curcumin from which it was observed that AgNPs-loaded hydrogels had higher efficiency than simple hydrogels like plain poly(AM-co-AMPs) and Ag+ loaded hydrogel.[27]

Table 2 Results of % encapsulation efficiency and % of cumulative releases of placebo hydrogels, silver ions loaded hydrogels and silver nanocomposite hydrogels

Hydrogels code	% of encapsulation efficiency	n	k (10^2)	R^2	% of cumulative releases at 37 °C
P (AM-AMPS) based hydrogels					
PAM-A1	31.14	0.6829	0.9629	0.9334	84.95
PAM-A5	59.47	0.4611	1.3631	0.9386	100
PAM-M1	43.42	0.5323	1.2398	0.9428	96.95
PAM-M5	35.42	0.6436	1.0381	0.9228	84.84
Silver ions loaded hydrogels					
PAM-A1+Ag⁺	36.4	0.4998	1.292	0.9285	81.45
PAM-A5+Ag⁺	60.2	0.7811	0.817	0.9138	95.8
PAM-M1+Ag⁺	45.6	0.5826	1.1558	0.9346	93.15
PAM-M5+Ag⁺	37.14	0.75	0.8785	0.898	80.54
Silver nanocomposite hydrogels					
PAM-A1+Ag	61.9	1.0857	0.3918	0.8367	71.57
PAM-A5+Ag	79.2	0.9555	0.619	0.8172	77.95
PAM-M1+Ag	63.42	0.7841	0.8395	0.8886	81.95
PAM-M5+Ag	56.42	1.0407	0.4813	0.8259	74.84

FIGURE 12.7   Percent efficiency of encapsulation of curcumin drug and drug release at 37°C.

Source: Reprinted with permission from Ref. [27]. © 2012 Springer Nature.

Order of loading capacity of curcumin into hydrogels was:

AgNPs-loaded hydrogels > hydrogels > Ag+ ions-loaded hydrogels.

Reason behind least loading efficiency of Ag+ ions-loaded hydrogels was AMP chains which were interacted by Ag+ ions and therefore inhibit the loading capacity of drug into the hydrogels.

For drug release, experiments were conducted at pH 7.4 of buffer solution at 37°C on plain poly(AM-co-AMPs) hydrogel, silver nanocomposite hydrogels, and curcumin-loaded silver nanocomposite hydrogel. It was observed that with interval of time, curcumin-loaded silver nanocomposite had slower release of drug than plain poly(AM-co-AMPs) hydrogel and AgNPs-loaded hydrogels because of presence of number of curcumin molecules adsorbed on silver NPs in addition to entrapment in the hydrogels.

Further antibacterial activity was checked on plain poly(AM-co-AMPs) hydrogel, HSNC, and curcumin-loaded HSNC's against *E. coli* bacteria (Fig. 12.8). It was observed that plain hydrogel did not show any effect on bacterial growth. Both HSNC's and curcumin-loaded HSNC's suppressed the growth of bacteria.

Order of antibacterial activity

Curcumin-loaded HSNC > HSNC's > plain poly(AM-co-AMPs) hydrogel

This was due to curcumin and silver NPs the suppressed the growth of bacteria and act as superior antibacterial materials and wounds cure.

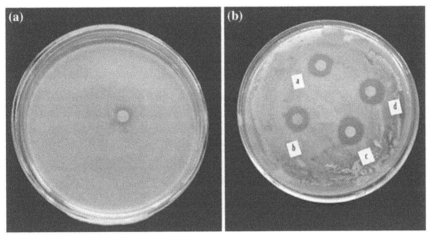

FIGURE 12.8   Antibacterial activity of curcumin-loaded HSNC.
Source: Reprinted with permission from Ref. [27]. © 2012 Springer Nature.

12.2.8   DRUG RELEASE FROM CCM/SiO₂-FO

Figure 12.9 shows steps for formation of silica-loaded folate-functionalized silica nanoparticles. Silica particles structure had microporous and mesoporous compartments. CCM molecules merged into the pore space of elementary silica particles. Here, folate-functionalized silica nanoparticles were suitable for killing PC3 prostate cancer cells than normal prostate epithelial cells.[28]

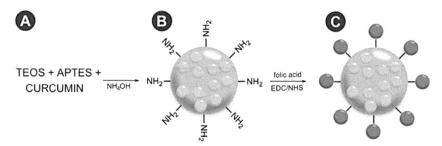

FIGURE 12.9 Systematic synthesis steps of folate-functionalized silica-loaded nanoparticles with curcumin. Orange sphere represents curcumin and green spheres showing folate.

Source: Reprinted with permission from Ref. [28]. © 2016 American Chemical Society.

Experiment was conducted that showed dispersion of CCM/SiO$_2$-FO sample in a RPMI medium at 37°C, attributed to release of CCM from silica particles which was about 35% after 48 h incubation, suitable for drug delivery platform. Figure 12.10 shows the release of curcumin drug with respect to time. To judge the accurate toxicity of curcumin drug on tumor cell, experiment was conducted with DMSO. In this case, cytotoxicity profile was totally changed. It was observed that DMSO itself had killed 50% population of cell, which was not viable for calculating the exact curcumin toxicity against tumor cells also on normal cell. On contrary, SiO$_2$-FO and CCM/SiO$_2$-FO was observed less cytotoxicity upto 80% after 48 h on normal cell and highly selective and toxic on tumor cells.

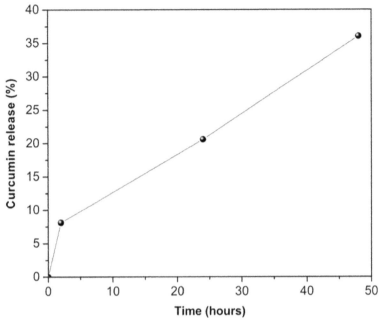

FIGURE 12.10   Release profile of curcumin from CCM/SiO$_2$-FO.
Source: Reprinted with permission from Ref. [28]. © 2016 American Chemical Society.

12.3   CONCLUSION

Simple and progressive encapsulation method is adopted to enhance the bioavailability of curcumin clinically. Solid dispersion method is used to increase the solubility of curcumin in aqueous medium. Various encapsulation method have been introduced nowadays, like encapsulation with INVITE micelles, AgNP hydrogels network, functionalized silica nanoparticles, MMT formulation, etc. Apart from this, factors like pH, curcumin concentration, and % formulation also affect the efficiency of in vitro release of drug. With nanoencapsulation, controlled environment loading and release of curcumin drug can be maintained. Thus, these nanoencapsulated curcumin NPs can be directly used for antibacterial treatment and cure of wound.

KEYWORDS

- curcumin
- hydrophobic
- encapsulation
- nanocarrier
- bioavailability

REFERENCES

1. Bharath, L. D.; Anushree, K.; Agarwal, M. S.; Shishodia, S. Curcumin Derived from Turmeric *Curcuma Longa*: Phytochemicals in Cancer Chemoprevention. CRC Press LLC, 2005; pp 349–387.
2. Negi, P. S.; Jayaprakash, G. K.; Jagan, M. R. L.; Sakariah, K. K. Antibacterial Activity of Turmeric Oil: A Byproduct From Curcumin Manufactuer. *J. Agric.Food Chem.* **1999**, *47*, 4297–4300.
3. Apisariyakul, A.; Vanittanakomm, N.; Buddhasukh, D. Antifungal Activity of Turmeric Oil Extracted From *Curcuma Longa* (Zingiberaceae). *J. Ethnopharmacol.***1995**, *49*, 163–169.
4. Ghatak, N.; Basu, N. Sodium Curcuminate as an Effective Anti-inflammatory Agent. *Indian J. Exp. Biol.***1972**, *10*, 235–236.
5. Itthipanichpong, C.; Ruangrungsi, N.; Kemsri, W.; Sawasdipanich, A. Antispasmodic Effects of Curcuminioids on Isolated Guinea Pig Ileum and Rat Uterus. *J. Med. Assoc. Thai.* **2003**, *86*, 299–309.
6. Ruby, A. J.; Kuttan, G.; Dinesh, B. K.; Rajasekharan, K. N.; Kuttan, R. Antitumor and Antioxidant Activity of Natural Curcuminoids. *Cancer Lett.* **1995**, *94*, 79–83.
7. Dhar, M. L.; Dhar, M. M.; Dhawan, B. N.; Mehrotra, B. N.; Ray, C. Screening of Indian Plants for Biological Activity. *Indian J. Exp. Biol.* **1968, 6**, 232–247.
8. Mazumdar, A.; Raghavan, K.; Weinstein, J.; Kohn, K. W.; Pommer, Y. Inhibition of Human Immunodeficiency Virus Type-1 Integrase by Curcumin. *Biochem. Pharmacol.* **1995**, *49*, 1165–1170.
9. Halim, E. M.; Ali, H. Hypoglycemic, Hypolipidemic and Antioxidant Properties of Combination of Curcumin from *Curcuma Longa* Lin and Partially Purified Product from *Abroma Augusta* Lin. in Streptozotocin Induced Diabetes. *Ind. J. Clinic. Biochem.* **2002**, *17* (12), 33–43.
10. Garg, S. K. Effect of *Curcuma Longa* (Rhizomes) on Fertility in Experimental Animals. *Planta Med.* **1974**, *26*, 225–227.
11. Ling, G.; Wei, Y.; Ding, X. Transcriptional Regulation of Human CYP2A13 Expression in the Respiratory Tract by CCAAT/Enhancer Binding Protein and Epigenetic Modulation. *Mol. Pharmacol.* **2006**, *71* (3), 807–816.

12. Okonogi, S.; Oguchi, T.; Yonemochi, E.; Puttipipatkhachorn, S.; Yamamoto, K. Physicochemical Properties of Ursodeoxycholic Acid Dispersed in Controlled Pore Glass. *J Colloid Interface Sci.*1999, *216,* 276–284.

13. Kapoor, L. D. *Handbook of Ayurvedic Medicinal Plants;* CRC Press: Boca Raton, Florida, 1990, p 185.

14. Modasiya, M. K.; Patel, V. M. Studies of Solubility of Curcumin. *Int. J. of Pharm. Sci.* **2012,** *3*, 1490–1497.

15. Kiuchi, F.; Goto, Y.; Sugimoto, N.; Akaon, N.; Kondo,K.; Tsuda, Y. Nemotodial Activity of Turmeric. *Chem. Pharm. Bull Tokyo* **1993,** *41* (a), 1640–1643.

16. Ravindranath, V.; Chandrasekhara, N. In Vitro Studies on the Intestinal Absorption of Curcumin in Rats. *Toxicol.* **1981,** *20,* 251–257.

17. Wang, Y. J.; Pan, M. H.; Chang, A. L.; Hsieh, C. Y.; Lin, J. K. Stability of Curcumin in Buffer Solutions and Characterization of its Degradation Products. *J. Pharmaceut. Biomed. Anal.* **1997,** *15*, 1867–1876.

18. Newa, M.; Bhandari, K.; Jong, Oh, K.; Seob, J.; Jung, Ae. K., Bong, Kyu, Y.; Jong, Soo,W.; Han Gon, C.; Chul, Soon, Y. Enhancement of Solubility, Dissolution and Bioavailability of Ibuprofen in Solid Dispersion Systems. *Chem. Pharm. Bull.* **2008,** *56,* (4) 569–574.

19. Zhai, S.; Ma, Y.; Chen, Y.; Li, D.; Cao, J.; Liu, Y.; Cai, M.; Xie, X.; Chen, Y.; Luo, X. Synthesis of an Amphiphilic Block Copolymer Containing Zwitterionic Sulfobetaine as a novel Ph-sensitive Drug Carrier. *Poly. Chem.* **2014,** *5*, 1285–1297.

20. Tripodo, G.; Pasut, G.; Trapani, A.; Mero, A.; Lasorsa, M. F.; Chlapanidas, T.; Trapani, G.; Mandracchia, D. Inulin-D-a-Tocopherol Succinate (INVITE) Nanomicelles as a Platform for Effective Intravenous Administration of Curcumin. *Biomacromolecules* **2015,** *16*, 550–557.

21. Tripodo, G.; Mandracchia, D.; Dorati, R.; Latrofa, A.; Genta, I.; Conti, B. Nanostructured Polymeric Functional Micelles for Drug Delivery Applications. *Macromol. Symp.* **2013,** *334* (1), 17–23.

22. Catenacci, L.; Mandracchia, D.; Sorrenti, M.; Colombo, L.; Serra, M.; Tripodo, G. In-Solution Structural Considerations by ^{1}H NMR and Solid-State Thermal Properties of Inulin-d-α-Tocopherol Succinate (INVITE) Micelles as Drug Delivery Systems for Hydrophobic Drugs. *Macromol. Chem. Phys.* **2014,** *215*, 2084–2096.

23. Mandracchia, D.; Tripodo, G.; Latrofa, A.; Dorati, R. Amphiphilic Inulin-D-A-Tocopherol Succinate (Invite) Bioconjugates for Biomedical Applications. *Carbohydr. Polym.* **2014,** *103*, 46–54.

24. Abouzeid, A. H.; Patel, N. R.; Torchilin, V. P. Polyethylene Glycol Phosphatidylethanolamine (Peg-Pe)/Vitamin E Micelles for Co-delivery of Paclitaxel and Curcumin to Overcome Multi-drug Resistance in Ovarian Cancer. *Int. J. Pharm.* **2014,** *464* (1–2), 178–184.

25. Sadeghi, R.; Moosavi-Movahedi, A. A.; Emam-jomeh, Z.; Kalbasi, A.; Razavi, S. H.; Karimi, M.; Kokini, J. The Effect of Different Desolvating Agents on BSA Nanoparticle Properties and Encapsulation of Curcumin. *J. Nanopart. Res.* **2014,** *16,* 2565.

26. Bakre, G. L.; Sarvaiya, I. J.; Agarwal, K. Y. Synthesis, Characterization, and Study of Drug Release Properties of Curcumin from Polycaprolactone/Organomodified Montmorillonite Nanocomposite. *J. Pharma. Innov.* **2016,** *11*, 4, 300–307.

27. Ravindra, S.; Mulaba-Bafubiandi, F. A.; Rajinikanth, V.; Varaprasad, K.; Reddy, N. N.; Mohana Raju, K. Development and Characterization of Curcumin Loaded Silver Nanoparticle Hydrogels for Antibacterial and Drug Delivery Applications. *J. Inorg. Organomet. Polym.* **2012,** *22*, 1254–1262.

28. Oliveira, F. L.; Bouchmella, K.; Goncalves, A. D. K.; Bettini, J.; Kobarg, J.; Cardoso, M. Funtionalized Silica Nanoparticles As an Alternative Platform for Targeted Drug-Delivery of Water Insoluble Drugs. *Langmuir* **2016,** *32*, 3217–3225.

CHAPTER 13

INSIGHTS INTO THE THERANOSTIC POTENTIAL OF CORE–SHELL NANOPARTICLES: A COMPREHENSIVE APPROACH TO CANCER

P. L. RESHMA, B. S. UNNIKRISHNAN, H. P. SYAMA, and T. T. SREELEKHA*

Laboratory of Biopharmaceuticals and Nanomedicine, Division of Cancer Research, Regional Cancer Centre, Thiruvananthapuram 695011, Kerala, India

Corresponding author. E-mail: ttsreelekha@gmail.com; ttsreelekha@rcc.gov.in

ABSTRACT

Cancer nanomedicine is a rapidly progressing field dealing with the application of nanotechnology to solve the limitations of conservative cancer treatment. Over the past decade, there has been a rapid advancement in the field of nanotechnology. Nanotechnology can increase the selectivity and effectiveness of cancer treatment with the least damage to normal cells. They are used to target cancer cells either by active targeting or by passive targeting through enhanced permeation retention effect. Core–shell nanoparticles are of a novel type, typically consisting of a core of one material covered by a shell of another material. Coating with an inert biocompatible material can overcome the toxicity and hydrophobicity of the core. The shell, in turn, protects the core from degrading agents and delivers the cargo safely to the region of interest. Nanoparticle chemotherapeutics are widely studied, especially of the core–shell type. Gene silencing by small-interfering RNA is a promising therapy, but progress is hindered because it is incompetent to enter the cell due to increased degradation. Coating with a biocompatible, inert substance overcomes the drawbacks of small-interfering RNA delivery and surface functionalization leads to targeted delivery to tumor site. The use of lasers for cancer management has highly progressed with heat-producing nanoparticles for photothermal therapy, and photosensitizers that produce oxygen radicals for reactive oxygen species-induced damage, and core–shell nanoparticles, especially upconversion nanoparticles, have great potential due to the ability to use near infrared rays with high penetration. Core–shell nanoparticles have great potential as radiosensitizers because the targeted delivery of nanoparticles improves the efficiency of radiotherapy. Core–shell fluorescent nanoparticles with good biocompatibility are efficient for cellular imaging. This chapter highlights the advancements, problems, and prospects in cancer nanomedicine and discusses novel core–shell nanoparticles to develop innovative and effective nanotherapeutics for cancer management.

13.1   INTRODUCTION

Cancer is a biological disorder that requires multidisciplinary care, and essential treatment modalities comprise surgery, radiotherapy, and chemotherapy. Chemotherapy forms mainstay of cancer care. Chemotherapy is the use of chemical agents, biological agents, and natural derivatives, which have a destructive effect on cancer cells.[1] The chemotherapy drugs hinder the capability of the cell to grow and proliferate. Though chemotherapeutics affect all cells in the body, they act against rapidly growing cells.[2] Cancer cells proliferate quickly than normal cells. As a result, the most common side-effects of chemotherapy are linked to their effects on other fast-growing cells. Systemic cancer chemotherapy agents play a valuable role in cancer treatment; however, there are undesirable effects both to normal replicating and quiescent cells. Rapidly dividing normal cells that are vulnerable to damage include cells of the bone marrow, hair follicles, and mucous membrane and reproductive system. The side-effects of cancer chemotherapy agents may be acute, self-limited, and mild.[3] Although much progress has been made, the management of side effects continues to be of utmost importance or the tolerability of therapy and impact on overall quality of life.

Nanomedicine is an emerging field of science and has several applications in the diagnosis and treatment of cancer. Nanoparticles show remarkable properties because of their size and increased surface-to-volume ratio.[4] Among various nanoparticles, core–shell nanoparticles, in particular, are heterogeneous with a core of one material and a coating of another material (Scheme 13.1) and have increased attention due to its application in biology and material science.[5]

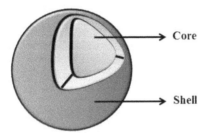

SCHEME 13.1   Schematic representation of a nanoparticle with core–shell structure.

By coating, the surface of a nanoparticle with a different material changes its properties. Core–shell nanoparticles with different shapes and morphologies can be synthesized (Scheme 13.2) and are advantageous due to increased biocompatibility, less cytotoxicity, improved stability, and better conjugation with other molecules.[6]

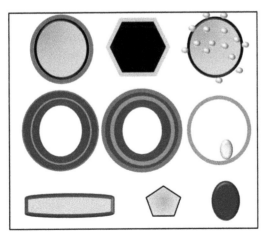

SCHEME 13.2   Different types of core–shell nanoparticles based on shape structure and composition.
Source: Adapted with permission from Ref. [7], copyright (2018) MDPI. https://creativecommons.org/licenses/by/4.0/

The properties of the core–shell nanoparticle can be changed by changing the material of the core or shell. For instance, if a therapeutic or drug for cancer is hydrophobic, it cannot be applied to a biological system. It will be rejected and it cannot reach its target site, and several toxic materials have to be used to deliver the therapeutic to the tumor. Coating the therapeutic with a hydrophilic material forming a core–shell nanoparticle, the therapeutic can reach its target site by enhanced permeation retention (EPR) effect (Scheme 13.3).

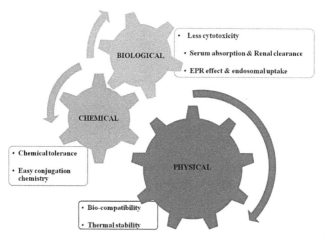

SCHEME 13.3   Advantages of core–shell nanoparticles.

There are several smart nanomaterials with unique quality; their properties can be changed in different environments inside the cell, like a change in pH, redox, and by external stimuli. This property can be employed to deliver drugs or therapeutics to the target site specifically and by protecting it from degradation. The toxicity of a material to be delivered can be changed by coating it with a nontoxic material to form a core–shell nanoparticle for making it more biocompatible. Theranostics is a combination of imaging and therapy into one program for use in cancer management. It involves using image-guided surgery and delivery of therapeutic agents like drugs, antibodies, small-interfering RNA (siRNA), microRNA (miRNA), and nanotherapeutic agents. The therapeutic agents may be light activated to produce reactive oxygen species (ROS) (photo-dynamic therapy [PDT]) or heat (photothermal therapy) to destroy the tumor. The applications of core–shell nanoparticles in different aspects of tumor management are explained in this chapter.

13.2   CORE–SHELL NANOPARTICLES FOR siRNA DELIVERY

The mechanism of siRNA-mediated gene silencing has two main stages: direct sequence-specific cleavage leading to translation repression and consequent degradation and transcriptional gene silencing.[8] siRNA delivery to the tumor has to face several obstacles like they are charged and large to enter the cell, degradation by RNAses, rapid renal filtration, entrapment by phagocytes, and extravasation from blood to tumor tissues.[9] Upon reaching the tumor, it has to overcome the vascular barrier and be internalized in cancer cells by cellular uptake, escape from the endosome into the cytoplasm, and finally be released from the siRNA payload to RNA-induced silencing complex that recognizes the mRNA of interest and cleaves by Argonaute.[10] Using nanoparticle to deliver siRNA is advantageous as they have desirable particle size to overcome barriers; they are inert and nonimmunogenic and, by EPR effect, they can penetrate and accumulate in tumor cells more efficiently.[11] Efficient siRNA delivery to the target site requires several prerequisites. The siRNA is highly susceptible to nucleases and efficient delivery needs transfection agents. The core–shell structure can efficiently protect the siRNA from degradation, and in order to improve the delivery of siRNA to the tumor site, core–shell nanoparticles with stimuli-responsive motifs are used extensively.

The pH at the tumor site (pH 7.2–6.0) is different from normal tissue and when inside the tumor cell particles have to pass through the endosome, lysosome, and then enter the cytoplasm, which is the site of action (Scheme 13.4). Several pH-responsive core–shell nanocarriers utilize this property. In a new study, a medium pore mesoporous core–shell silica nanoparticle with a multifunctional polymer cap was used for siRNA delivery to Human HeLa contaminant carcinoma cell line (KB) cells based on pore size, pore morphology, and pH.[12] In another promising study, a poly(ethylene glycol) (PEG)– reduction-sensitive poly(N-(2,2'-dithiobis(ethylamine)) aspartamide) PAsp(AED)–copolymer consisting of pH-sensitive poly(2-(diisopropyl amino)ethyl methacrylate) (PDPA), nanoparticle was synthesized, Dox was encapsulated in a pH-sensitive core, and Bcl-2 siRNA in a reduction sensitive inner layer and trigger caused rapid release of Dox and silencing of antiapoptotic protein Bcl-2, which lead to enhanced apoptosis of SKOV3 cell line. Further, PEG helps the particle to remain in the tumor site by passive accumulation and escape from the reticuloendothelial system.[13] Xiong et al. developed a magnetic core–shell silica nanoparticle with an acid labile tannic acid coating which increased the dispersion stability and acted as a pH-responsive release switch for the siRNA.[14]

The ROS level in the tumor cell is higher than the normal cell. The glutathione (GSH) level is high in tumor cell to maintain the redox balance. Numerous core–shell nanoparticles utilize this property for siRNA targeting.[15] A core–shell nanoparticle was developed with a bovine serum albumin arginylglycylaspartic acid (RGD) peptide outer shell that detached in the acidic environment of lysosome and a positively charged chitosan nona-arginine redox-sensitive inner core that escaped the lysosomal environment for the efficient delivery of vascular endothelial growth factor (siVEGF) to cytoplasm of hepatocellular carcinoma cells and inhibited cell proliferation.[16] In a promising work, Chen et al. developed a pH/redox dual-sensitive nanoparticle with eminent endosomal/lysosomal escape mechanism with a core with disulphide bonds which cleaved in the presence of GSH and pH (pH 5.3)-sensitive bonds that undergo siGFP gene silencing in epidermal growth factor receptor overexpressing cancer cells.[17]

There is an increased expression of certain enzymes in the tumor microenvironment for metastasis of tumor-like matrix metalloproteinase 2/9, alkaline phosphates, hyaluronidase, cathepsin, etc. which can cleave certain peptide linkers in core–shell nanocarriers to release the cargo. Xi et al. developed a PEG–PLA core–shell nanoparticle with a matrix metalloproteinase 2/9 cleavable peptide linker Pro-Val-GLy-Leu-Ile-Gly linker (PVGLIG) for delivery of drug paclitaxel (PTX) and siVEGF. These nanoparticles had enhanced accumulation and deeper penetration in tumor with moderate toxicity to normal cells.[18]

External stimuli like near infrared (NIR) penetrate cells but cannot start a reaction as the energy is very low. In upconversion nanoparticles, two or more photons are absorbed giving out emission with higher energy. External stimuli like NIR-triggered siRNA release have been employed in core–shell nanoparticles. Zhang et al. developed a core–shell upconversion nanoparticle coated with mesoporous silica when irradiated at 980-nm produce blue emissions that activate hypocrellin A to produce ROS causing the endosomal escape of loaded cargo and a photocleavable linker for siRNA release causing suppression of proliferation and inhibition of tumor growth.[19,20] In a promising study, a 78-nm 1,2-distearoyl-sn-glycero-3-phosphoethanolamine-Polyethyleneglycol (DSPE–PEG) nanoparticle contains photosensitizer (PS) chlorin e6, docetaxel (DTX), and antitwist siRNA which utilized PDT for the release of siRNA and DTX for the destruction of tumor by trimodality therapy, PDT, chemotherapy, and gene therapy.[21]

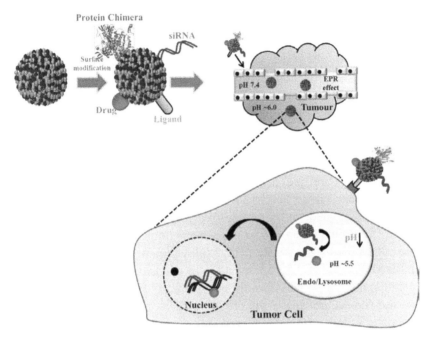

SCHEME 13.4   (See color insert.) Core–shell nanoparticle for siRNA and drug delivery based on pH; targeted core–shell nanoparticles deliver siRNA and drug into the tumor site via EPR effect followed by pH-sensitive behavior of delivery.

13.3   CORE–SHELL NANOPARTICLES FOR DRUG AND ANTIGEN DELIVERY

A handful of core–shell nanoparticles are used as drug or antigen delivery agents. In a study, a multifunctional superparamagnetic iron oxide nanoparticle with RGD peptide and PTX-conjugated beta-cyclodextrin immobilized on the surface was prepared. These core–shell magnetic nanoclusters exhibited enhanced magnetic resonance, improved cellular uptake, and efficient PTX loading and sustained release at the desired time point.[22] In a study, a functional platinum core gold platinum shell bimetallic nanoparticle with chitosan and doxorubicin was synthesized. The particles induce a controlled and pH-triggered release causing cancer cell death by apoptosis.[23] Core–shell nanoparticles were prepared by coating superparamagnetic Fe_3O_4 core and reduction/pH dual-responsive poly(methacrylic acid) as shells and thermal-responsive poly(*N*-isopropylacrylamide) as a "gatekeeper." These nanoparticles showed increased release of Dox at pH 5 and 10 mM GSH accelerating triggered release (Figs. 13.1 and 13.2).[24] In a study, several Au/Ag nanoparticles of core–shell type were tested for antitumor activity, and metastatic inhibition and NPs with Ag core covered by Au shell were the most effective.[25] Ag@Pd core–shell nanoparticles synthesized by using extracts of almond nuts and blackberry fruits showed increased cytotoxicity against breast cancer cell lines human breast adenocarcinoma cell line (MCF7) and Human hepato-carcinoma cell line (HepG2) than normal cell line Human HeLa contaminant cell line (WISH).[26] Yolk–shell nanoparticles with pH-responsive cross-linked polyacrylic acid core and temperature-responsive cross-linked Polyhydroxyethylmethacrylate (PHEMA) shell were successfully synthesized with template-assisted and template-free methods. These nanoparticles exhibited controlled release of Dox when compared to an immediate release of free Dox.[27] Iron-oxide–zinc-oxide core–shell nanoparticles were synthesized as antigen delivery agents into dendritic cells for DC-based immunotherapy. These particles did not show any toxicity up to 200 mg/kg but showed accumulation and granulomatous inflammation at the site of injection.[28]

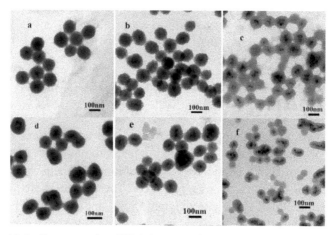

FIGURE 13.1 Representative TEM images of Fe_3O_4@PMAA nanoparticles with different core contents of Fe_3O_4@MPS [(a) 1 wt%, (b) 2 wt%, (c) 3 wt%, (d) 4 wt%, (e) 5 wt%, and (f) 10 wt%].

Source: Reprinted with permission from Ref. [24], copyright (2015) American Chemical Society.

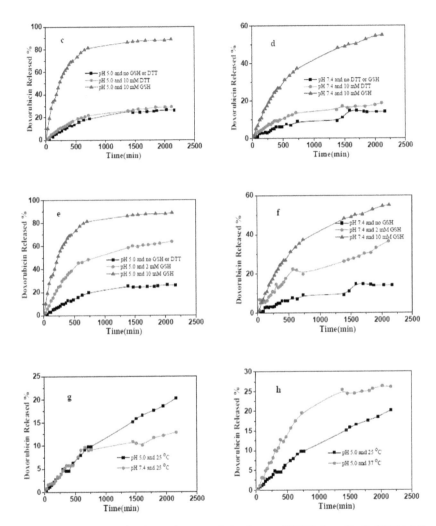

FIGURE 13.2   (See color insert.) Controlled release of DOX from Fe_3O_4@PMAA@ PNIPAM nanoparticles under different reductive agent DTT/GSH concentrations and pH values at 37°C: (a) 10 mM DTT and (b) 10 mM GSH; different pH values and reductive agent DTT/GSH concentrations at 37°C: (c) pH 5.0 and (d) pH 7.4; different pH values and GSH concentrations at 37°C: (e) pH 5.0 and (f) pH 7.4; different pH and temperatures (g and h).

Source: Reprinted with permission from Ref. [24], copyright (2015) American Chemical Society.

13.4 PHOTOTHERMAL THERAPY

Phototherapy is the introduction of light of a specific wavelength to activate photosensitive molecules to produce therapeutic effects.[29] Phototherapy can be classified as photothermal therapy and PDT. Photothermal therapy is a phenomenon by which when light of a specific wavelength is illuminated on a nanoparticle, some of the incident photons will be absorbed while others will be scattered.[30] The energy absorbed by the nanoparticle will be given out as luminescence or as heat (Scheme 13.5). Heating properties of nanoparticles depend upon size, shape, location of the nanoparticle, concentration and incubation time and in in-vivo targeting, biodistribution, and tracking of the nanoparticle.[31] Nanoparticles with more complex geometries, such as core–shell nanoparticles, nanocages, nanorods, or nanostars, are better photothermal agents.[32]

Several core–shell theranostic nanoparticles with image-guided photothermal therapy have been synthesized. The complex core–shell structure is advantageous as several therapeutic agents can be incorporated into the complex structures. Xu et al. synthesized a yolk–shell mesoporous silica nanoparticle doped with Fluorescein isothiocyanate (FITC) and patchy gold and loaded with Dox with applications in chemo and photothermal therapy and luminescence-based tracking.[33] Li et al. developed a nanoparticle with a gold nanostar core for photothermal therapy (PTT) and two-photon photoluminescence, and a shell of gadolinium and gemcitabine monophosphate for chemotherapy and magnetic resonance imaging (MRI) imaging, and inhibited tumor growth in vivo.[34] In another promising study, Au was coated on the surface of DTX-loaded Poly (lactic-co-glycolic acid) (PLGA) nanoparticle and angiopep-2 was used for the targeted therapy. This NP had the theranostic property of antitumor efficiency due to IR-induced DTX release and showed X-ray imaging ability.[35] Parchur et al. developed an Au nanorod core/Gd shell nanoparticle (Au/Gd$_2$O$_3$) for site-specific delivery as imaging contrast agent and for PTT.[36] In another work, a metal-shell (Au) dielectric-core (BaTiO$_3$) nanoparticle (BT-Au-NP) was used for in-vivo real-time tracking and photothermal therapy.[37] Li et al. synthesized Tam functionalized bismuth sulphide@mesoporous silica nanoparticle specific to Her2-positive breast cancer, which can be used as a contrast agent for computed tomography (CT) imaging and tumor reduction by photothermal chemotherapy.[38]

NIR excitable upconversion nanoparticles that produce photons of higher energy when excited by low-energy NIR rays are widely used in photothermal therapy. In a promising work, upconversion core–shell–shell nanoparticle with mesoporous silica (mSiO$_2$) and polyallylamine for IR808 loading and polyethylene glycol–folic acid (PEG–FA) for tumor targeting was used for PTT and luminescence imaging using a single wavelength NIR light.[39] In another study, an upconversion nanoparticle with Cu$_2$S shell plays the role of bio-imaging and photothermal conversion to generate heat to kill *Escherichia coli* and HepG-2 cells.[40] In a study, nanographene oxide-upconversion nanoparticles (UCNP)-Ce6 nanocomposite was used for upconversion luminescence image-guided PDT/PTT therapy.[41]

Some core–shell nanoparticles were used for combined chemophotothermal therapy. Nanospheres with Dox Au-hybridized matrix as the shell and Dox as the core caused enhanced super plasmon resonance in the NIR region causing enhanced photothermal conversion and strong tumor-killing ability.[42] Cai et al. developed EBSUCNPs@SiO$_2$@Au 3D nanoclusters that exhibit near I/R on/off switch for Dox release exhibiting combined chemotherapy with photothermal therapy.[43] Mesoporous silica nanoparticle coated with DSPE–PEG2000–COOH and DSPE–PEG2000–FA modified lecithin for the targeted delivery of Dox and PS indocyanine green for synergistic chemo-photothermal therapy.[44] In a separate study (Pd@Au–PEG–Pt), nanoplates showed better tumor accumulation due to 2D structure than spherical and was used for combined photothermal chemotherapy.[45]

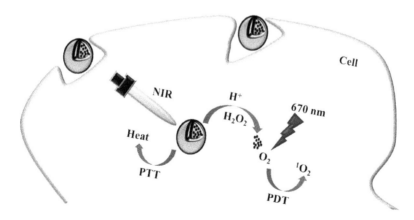

SCHEME 13.5   Core–shell nanoparticles for photothermal and photodynamic therapy.

13.5 PHOTODYNAMIC THERAPY

PDT is a phenomenon by which when light of a specific wavelength is illuminated on a PS, it produced ROS that is harmful to the surrounding tissue (Scheme 13.5). PDT requires a PS light source and oxygen. A PS absorbs ultraviolet or light in the visible region and transfers it to adjacent molecules, thereby producing singlet oxygen. During irradiation of light at a specific wavelength, the PS absorbs light, gets transformed to an excited state, and undergoes intersystem crossing with oxygen to form ROS, predominantly singlet oxygen. The ROS kills tumor cells by apoptosis or necrosis or by disrupting tumor vasculature. Fastidious accumulation of PS in the tumor is essential to prevent damage to normal tissue. But most PSs are hydrophobic and thus have less tumor accumulation. Nanoparticles, especially of the core–shell type, can solve the problem of solubility and cause selective accumulation in tumor by enhanced permeability and retention effect.[46]

Core–shell nanoparticles are used to increase the efficacy of PSs. The PSs used in core–shell nanoparticles are rose bengal,[47–52] methylene blue,[51,53–55] indocyanine green,[56] squaraine, boron-dipyrromethene, and phenalenones. Natural products such as hypericin (Hy), hypocrellin,[19,57–59] riboflavin,[60] curcumin etc.; tetrapyrrole-related molecules such as heme, chlorophyll[46] and cytochromes, porphyrin,[61–64] chlorin e6,[65,66] and phthalocyanines[67,68]; and inorganic nanoparticles like titanium oxide[69,70] are also employed for PDT.

In upconversion, two or more incident photons of lower energy are absorbed, and one photon of higher energy is emitted. Absorption usually occurs in the infrared region and emission in the ultraviolet or visible region. Lanthanide-doped upconversion nanoparticles are typically used where lanthanide ions are dispersed in a dielectric host. Upconversion nanoparticles can convert NIR to short IR, visible (blue, green, red), or ultraviolet. NIR light has deeper tissue penetration, lower phototoxicity, reduced auto-fluorescence, and scattering and no harmful effect on the tissue. Upconversion nanoparticles are usually applied in core–shell structure. Several upconversion nanoparticles of core–shell structure used in PDT in cancer are given in Table 13.1.

TABLE 13.1 Upconversion Nanoparticles Used in Photodynamic Therapy.

Laser	Upconversion nanoparticle	Coating	Photosensitizer	Emission	In vitro	In vivo	References
980 nm	Yb/Tm/GZO	SiO$_2$	Host sensitizers	—	HeLa cells	—	[71]
980 nm	LiLuF4:Yb,Er@ nLiGdF4	SiO$_2$	Merocyanine 549 (MC540)	510–530 530–570 630–680 nm	MCF-7 cells	Female Bal/c nude mice	[72]
808 nm	β-Sodium gadolinium fluoride nanoparticle (NaGdF4):Yb^{3+}/Er^{3+}	PEGylated nanographene oxide	Chlorin e6	540–560 640–680 nm	L929, human cervical adenocarcinoma cell line (HeLa) cells	Balb/c mice	[41]
980 nm	NaYF$_4$:Yb,Er UCNPs	SiO$_2$	Rose bengal	410 and 543 nm	—	—	[73]
980 nm	NaYF$_4$:Yb,Tm@ NaYF$_4$	Carbon dot embodied silica sandwich shell and mesoporous silica outer shell	Hypocrellin A	450–480 nm	HePG2 cells	—	[57]
975 nm	NaYF$_4$:Yb^{3+}Tm^{3+}/ NaYF$_4$	PMAO DARPin-mCherry	Endogenous	345, 360, and 410 nm	Human metastatic breast adenocarcinoma cell line (SK-BR-3) and Chinese hamster ovarian cell line (CHO) cells	Lewis lung cancer mouse model	[74]
808 nm	NaYF$_4$:Yb,Er,Tm UCNPs	SiO$_2$	Methylene blue	540, 655, and 800 nm	MCF-7, MCF-10A, HeLa, and HepG2 cells	—	[75]

TABLE 13.1 (Continued)

Laser	Upconversion nanoparticle	Coating	Photosensitizer	Emission	In vitro	In vivo	References
808 nm	NaGdF4:Yb,Er	Molybdenum sulphate (MoS$_2$) nanosheet	IR-808	750–1050 nm	mouse sub-cutaneous connective tissue cell line (L929) cells	Balb/c	[66]
980 nm	Fe$_3$O$_4$/g-C$_3$N$_4$	NaGdF$_4$:Yb,Tm@ NaGdF$_4$	mesoporous graphitic phase Carbon nitride g-C$_3$N$_4$	475 nm	L929 and HeLa cells	Balb/c mice	[76]
980 nm	NaYF$_4$:Yb^{3+}:Tm^{3+}/ NaYF$_4$		Riboflavin Rf	375 and 450 nm	SK-BR-3 and CHO cells	Induced Lewis lung cancer in BDF1 Balb/c nu/nu mice	[60]
980 nm	NaFY4: ErYb(PEI–FA)	Mesoporous silica SiO$_2$@mSiO$_2$	Methylene blue	540 and 660 nm	–	–	[53]
980 nm	Upconversion core NaYF4:Yb3$^+$,Er3$^+$	Graphene oxide quantum dot shell	Hypocrellin A	560 and 660 nm	HeLa cells	–	[58]
808 nm	[NaGdF4:Yb/Nd@ NaGdF4:Yb/Er@ NaGdF4]	PAA	Rose bengal	550 and 660 nm	Human ovarian carcinoma cell line (A2780) and human ovarian carcinoma cell line cisplatin resistant (A2780cisR) cells	–	[47]

TABLE 13.1 *(Continued)*

Laser	Upconversion nanoparticle	Coating	Photosensitizer	Emission	In vitro	In vivo	References
808 nm	NaYbF4:Nd@ NaGdF4:Yb/Er@ NaGdF4	poly(maleic anhydride-alt-1-octadecene) (C18PMH) — PEG	Ppa	540 and 660 nm	HeLa cells	—	[77]
980 nm	NaYF4:Yb,Er Upconversion core	silica sandwich shell, β-cyclodextrin mesoporous silica outermost shell	Methylene blue	540 and 660 nm	A549 cells	—	[78]
980 and 796 nm	NaGdF4:Yb,Er@ NaYF4:Yb@ NaGdF4:Yb,Nd@ NaYF4@ NaGdF4:Yb,Tm@ NaYF4	mSiO$_2$	Rose bengal	520–570 440–490 620–650 nm	HeLa cells	—	[79]
980 nm	NaYbF4:Nd@ NaGdF4:Yb/Er@ NaGdF4	FA–PEG	Chlorin e6	550 and 660 nm	KB, A549 cells	—	[80]

FA–PEG, folic acid– polyethylene glycol; mSiO$_2$, mesoporous silica; PAA, polyacrylic acid; PEG, polyethylene glycol; Ppa, pyropheophorbide a.

13.6 IMAGING

The optical properties of the core–shell nanomaterials are used as bio-imaging agents in different biological and cell imaging techniques. A novel radionuclide bio-imaging agent was synthesized by coating radioactive iodine gold nanoballs with PEGylated crushed gold shells.[81] A dysfunctional Y_2O_3:Nd^{3+}/Yb^{3+}@SiO_2@Cu_2S core–shell structure is used for NIR bio-imaging and photothermal therapy in *E. coli* and hepatoma cells.[40] Piperine micelles loaded into trimethyl chitosan to form a core–shell structure were able to effectively kill glioma cells.[82] In another work, a gold shell and dielectric $BaTiO_3$ core nanoparticle with photothermal and bimodal imaging application were studied.[37] Poly-L-lysine was coated on silver nanoparticle with various thicknesses, and folic acid was functionalized along with fluorphore (5-(and-6)-carboxyfluorescein-succinimidyl ester). These nanoparticles are used for cellular imaging and targeting folic acid-positive cancer cells.[83] An X-ray excited core–shell–shell ScNP–PAH–RB–PEG–FA nanocomposite has been constructed for dual PDT and luminescence diagnostic imaging to image and kill MDA–MB-231 and MCF-7 cancer cells with low dark cytotoxicity and efficient photocytotoxicity.

Surface-enhanced Raman scattering (SERS) is a technique where a molecule was absorbed on a rough metal surface such as silver or gold and when excited by a laser resulting in an increase in an electric field that is proportional to the Raman intensities. This property is utilized by several core–shell nanoparticles for bio-imaging. In a study, liposome@Au/Ag nanoparticles were prepared for controlled Dox release triggered by NIR laser irradiation. It can also monitor drug molecules by SERS and fluorescence signal during the release process.[84] A sensitive biorthogonal SERS tag gold silver core–shell nanoparticle was prepared with Raman reporter dye (*E*)-2-((4-(phenylethynyl)benzylidene) amino) ethanethiol for liver cancer cell imaging and future biomedical diagnostic applications.[85] A pH-sensitive nanocarrier of multiwalled carbon nanotubes decorated with gold/silver core–shell nanoparticles for tracking intracellular drug-release process. Drug release from the nanoparticle is triggered by pH change and the endocytosis pathway and drug release can be measured by fluorescence and SERS imaging.

13.7   RADIOSENSITIZATION

Radiation therapy is an effective treatment modality for cancer. Radiation interacts with DNA and produces a series of free electrons such as hydroxyl radicals, which destroy DNA and destroy the atomic structure of biomolecules. The radiation delivery techniques need to be improved so that it can effectively kill tumor cells sparing normal cells so as to avoid secondary effects. Radiosensitizers make tumor cells more susceptible to radiation. Elements of high Z energy such as iodine, gadolinium, and gold have the ability to radiosensitize tumors, of which gold has better biocompatibility.[86] The mechanism by which radiation interacts with nanoparticle is by Compton effect, photoelectric effect, and Auger effect (Scheme 13.6).[87] Core–shell nanoparticles are advantageous because core improves physical properties and shell improves stability as well as biocompatibility. Thus, these particles can be used for dual purposes.

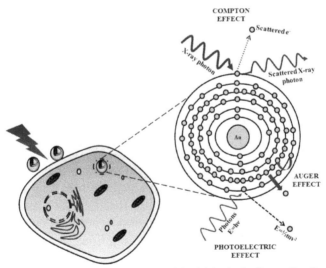

SCHEME 13.6   Schematic representation of the biological effects of radiosensitization causing oxidative stress, DNA damage, and cell death. In the inset schematic representation of photoelectric effect, Compton effect, and Auger effect, the main mechanisms how radiosensitization interacts with nanoparticles. When an incident photon is absorbed by an electron leading to its ejection, it causes the Compton effect. In photoelectric effect, electrons are ejected from an inner orbital and the vacancy is filled by an outer shell electron causing a release of low energy photons starting a cascade release of secondary electrons (Auger effect) causing ionizing radiation.
Source: Adapted with changes with permission from Springer from Ref. [87].

In a study, Au@MnS@ZnS–PEG core/shell/shell nanoparticles were used for imaging-guided enhanced RT treatment for cancer. The presence of paramagnetic Mn^{2+} ions makes it a contrast agent for T1-weighted magnetic resonance (MR) imaging.[88] Gold nanoclusters with fluorescent and radiosensitizing gold and RGD peptide (c(RGDyC)) shell are used to target $\alpha_v\beta_3$ integrin-positive cancer cells.[89] Bimetallic hybrid Zn@Au NPs were advantageous as they can be used for in vivo tracking through positron emission tomography due to the zinc core as well as radiosensitization through the gold shell.[90] Gold–silicon oxide core–shell NPs ($AuN@SiO_2$ and $AuS@SiO_2$) radiated with X-ray energies of 6 and 18 MV are effective in radiotherapy of MCF-7 breast cancer cell lines.[91] Core–shell nanoparticles containing gold nanoparticles and doxorubicin are used for combination therapy of radiation and chemotherapy.[92] Fe_3O_4/TaO_x (core/shell) nanoparticles used as contrast agents for CT and MRI are used as effective cancer cell radiosensitizers with proton therapy.[93] Tungsten oxide-coated Fe/Pt core–shell nanoparticle, where the tungsten was converted to radioisotope W-187 and showed enhanced heating under magnetic field, can be used for advanced radiopharmaceutical applications.94 The ZnO/SiO_2 core–shell nanoparticles exhibited fluorescence and radiosensitizing properties due to X-ray-induced radiocatalysis by the particle.[95]

13.8   BIOSENSORS

Biosensors are devices that contain biological materials that interact with an analyte to produce a physical, chemical, and electrical signal that can be measured.[96] Usually, certain biochemical reactions such as receptor–ligand, enzyme–substrate, and antigen–antibody are employed and the signal is converted to an electrical signal, amplified and converted to sensible output.[97] Biosensors are classified as electrochemical, amperometric, potentiometric, thermometric, optical, and piezoelectric biosensors. Core–shell nanomaterials are important components in bio-analytical devices since they clearly enhance the performances regarding sensitivity and detection limits down to single molecules detection.[98] Based on the property measured by the transducer of the biosensor, the nanomaterials are classified as amperometric and optical.

Amperometry is the most common electrochemical technique used in biosensors. Amperometry measures the amplitude of a reduction or oxidation current at a given specific potential over a fixed period.[99] Hybrid nanomaterials, especially in the core–shell structure, are greatly employed as electrochemical/amperometric biosensors since they exhibit unique characteristics[100] and are used as electrochemical signal-generating probes, signal amplifiers, and as immobilization support. Several nanoparticles, especially gold (Au), platinum (Pt),[101,102] and cobalt oxide (Co_3O_4),[103] have catalytic activity and are able to reduce H_2O_2 and are used in electrochemical amperometric biosensors. Trimetallic nanoparticles of any one of Au/Pt/Pd/Ni showed great ability to reduce H_2O_2 and thus can be used in amperometric biosensors.[104,105] Cobalt oxide (CO_3O_4) can enhance the catalytic activity of the biosensor by reduction of H_2O_2.[100] $Cu_2O@$ CeO_2 core–shell loaded with Au nanoparticles showed synergistic ability toward reduction of H_2O_2 than individually and is used for the detection of prostate-specific antigen (PSA).[106] Au/Ag/Au core double shell nanoparticles were developed to enhance the electrocatalytic activity of Au and to increase the chemical stability of silver and thus reduce H_2O_2 for the detection of squamous cell carcinoma antigen.[107] Hemispherical platinum silver nanoparticles regioselectively functionalized with a probe for miR-132 of neuroblastoma for electrochemical reduction of H_2O_2.[108] Platinum-coated gold (Pt@Au) nanoparticles with core@shell structure cause reduction of hydrogen peroxide and are used in electrochemical biosensors.[109]

Some nanomaterials cannot reduce H_2O_2 by itself, and so, other reducing agents are used for the reduction of H_2O_2. In Au@Pd core–shell nanoparticle-modified magnetic Fe_3O_4/MnO_2 beads $(Fe_3O_4/MnO_2/Au@$ Pd), horseradish peroxidase was added for the oxidation of hydroquinine to H_2O_2, thus amplifying the electrochemical signal.[110] In some others, other reducing agents are used. A highly ordered ZnO/ZnS core/shell nanotube arrays exhibit a superior electrochemical response toward ferrocyanide/ferricyanide and in glucose sensing.[111] $Fe_3O_4@Au/$Glucose oxidase-modified SPCE was used in a flow-injection analysis for the detection of glucose.

Metal nanoparticles have optical and electronic properties and display localized surface plasmon resonance due to collective oscillation of conduction bands.[112] Semiconductor materials can generate electron–hole pairs upon photoexcitation, and the generated electrons could be utilized to reduce the nearby metal cations to prepare hybridized metal–semiconductor nanostructures.[113] Core–shell Au/Ag/SnO nanoparticle uses this process for the detection of GSH.[114] A silica core–gold shell nanoparticle SiO_2:Au NPs were used for the development of long period grating sensor which is used as a ligand for streptavidin detection.[115] A core–shell composite fluorescent nanoparticle ($Ag@SiO_2@SiO_2$–RuBpy) was fabricated for the detection of PSA, which provides a photoluminescence enhancement of up to ~threefold when the separation distance between the surface of silver core and the center of the third RuBpy doped silica shell is about 10 nm.[116] One of the SERS-based biosensors consists of a nickel iron core and a gold shell, which gets magnetically focused on a microfluidic platform for SERS and detection of carcinoembryonic antigen. It utilizes a nanocomposite probe for the detection of cancer biomarkers by magnetic focusing and SERS detection in a microfluidic platform (Fig. 13.3).[117]

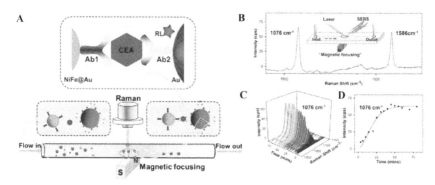

FIGURE 13.3   (See color insert.) (A) Illustration of SERS detection of cancer biomarker CEA using functional nanoprobes consisting of Au-coated NiFe magnetic nanoparticle (NiFe@Au), Ab1, capture antibody; Ab2, detection antibody; and RL, Raman label B; (B) illustration of magnetic focusing of NiFe@Au NPs in amicrofluidic channel (~0.5 mm); (C) SERS spectra in 1000–1150 cm^{-1} for an aqueous solution of MBA-labeled NiFe@Au NPs in the channel as a function of time when a magnet was applied to the focal point; (D) kinetic plot of peak intensity at 1076 cm^{-1} (dots). The fitting curve was based on Avrami theoretical model with a rate constant of 0.0075 $min^{-1.6}$.
Source: Reprinted with permission from Ref. [117], copyright (2015) American Chemical Society.

13.9   CONCLUSION

Nanoparticles with core–shell structures have a wide range of applications in different aspects of cancer starting from diagnosis to chemotherapy, radiotherapy, imaging, and surgery. It was found that gold, silver, iron, and mesoporous silica are the most widely used materials in core–shell nanoparticles with gold being the most widely used because of its unique properties and biocompatibility. Nanoparticles of different core–shell structures are discussed in this paper from a simple core–shell to multilayered core–shell. But the theranostic application of a nanoparticle will not be possible if not for the core–shell structure. Nanoparticles that are toxic are usually coated with an inert substance to form a core–shell structure. The coating increases the biocompatibility and reduces cytotoxicity of the nanoparticle. Although extensive research in this field of nanotechnology has been done in the past decade, there are very few that have come from "the bench to the bedside." There is vast scope in the near future for nanomaterials with core–shell structure to be used in real-life applications.

KEYWORDS

- cancer
- core–shell nanoparticles
- photothermal therapy
- drug delivery
- imaging

REFERENCES

1. Perez-Herrero, E.; Fernandez-Medarde, A. Advanced Targeted Therapies in Cancer: Drug Nanocarriers, the Future of Chemotherapy. *Eur. J. Pharm. Biopharm.* **2015,** *93,* 52–79. PubMed PMID: 25813885. Epub 2015/03/31.eng.
2. Albertsson, P.; Lennernas, B.; Norrby, K. Chemotherapy and Antiangiogenesis: Drug-Specific Effects on Microvessel Sprouting. *Acta Pathol. Microbiol. Immunol. Scand.* **2003,** *111* (11), 995–1003. PubMed PMID: 14629265. Epub 2003/11/25.eng.
3. Schulze, C.; McGowan, M.; Jordt, S. E.; Ehrlich, B. E. Prolonged Oxaliplatin Exposure Alters Intracellular Calcium Signaling: A New Mechanism to Explain Oxaliplatin-Associated Peripheral Neuropathy. *Clin. Colorect. Cancer* **2011,** *10* (2), 126–133. PubMed PMID: 21859566. Pubmed Central PMCID: PMC3388801. Epub 2011/08/24.eng.

4. Baetke, S. C.; Lammers, T.; Kiessling, F. Applications of Nanoparticles for Diagnosis and Therapy of Cancer. *Br. J. Radiol.* **2015,** *88* (1054), 20150207. PubMed PMID: 25969868. Pubmed Central PMCID: PMC4630860. Epub 2015/05/15.eng.

5. Ghosh Chaudhuri, R.; Paria, S. Core/Shell Nanoparticles: Classes, Properties, Synthesis Mechanisms, Characterization, and Applications. *Chem. Rev.* **2012,** *112* (4), 2373–2433.

6. Chatterjee, K.; Sarkar, S.; Jagajjanani Rao, K.; Paria, S. Core/Shell Nanoparticles in Biomedical Applications. *Adv. Colloid Interface Sci.* **2014,** *209,* 8–39. PubMed PMID: 24491963. Epub 2014/02/05.eng.

7. Khatami, M.; Alijani, H. Q.; Nejad, M. S.; Varma, R. S. Core@Shell Nanoparticles: Greener Synthesis Using Natural Plant Products. *Appl. Sci.* **2018,** *8* (3), 411.

8. Tatiparti, K.; Sau, S.; Kashaw, S. K.; Iyer, A. K. siRNA Delivery Strategies: A Comprehensive Review of Recent Developments. Nanomaterials (Basel, Switzerland). 2017, 7(4), pii: E77.

9. Kanasty, R. L.; Whitehead, K. A.; Vegas, A. J.; Anderson, D. G. Action and Reaction: The Biological Response to siRNA and Its Delivery Vehicles. *Mol. Ther.: J. Am. Soc. Gene Therapy* **2012,** *20* (3), 513–524. PubMed PMID: 22252451. Pubmed Central PMCID: PMC3293611. Epub 2012/01/19.eng.

10. Johannes, L.; Lucchino, M. Current Challenges in Delivery and Cytosolic Translocation of Therapeutic RNAs. *Nucl. Acid Ther.* **2018,** *28* (3), 178–93. PubMed PMID: 29883296. Pubmed Central PMCID: PMC6000866. Epub 2018/06/09.eng.

11. Luo, D.; Carter, K. A.; Lovell, J. F. Nanomedical Engineering: Shaping Future Nanomedicines. *Wiley Interdiscipl. Rev. Nanomed. Nanobiotechnol.* **2015,** *7* (2), 169–188. PubMed PMID: 25377691. Pubmed Central PMCID: PMC4308429. Epub 2014/11/08.eng.

12. Moller, K.; Muller, K.; Engelke, H.; Brauchle, C.; Wagner, E.; Bein, T. Highly Efficient siRNA Delivery from Core–Shell Mesoporous Silica Nanoparticles with Multifunctional Polymer Caps. *Nanoscale* **2016,** *8* (7), 4007–4019. PubMed PMID: 26819069. Epub 2016/01/29.eng.

13. Chen, W.; Yuan, Y.; Cheng, D.; Chen, J.; Wang, L.; Shuai, X. Co-delivery of Doxorubicin and siRNA with Reduction and pH Dually Sensitive Nanocarrier for Synergistic Cancer Therapy. *Small (Weinheim Bergstrasse, Germany)* **2014,** *10* (13), 2678–2687. PubMed PMID: 24668891. Epub 2014/03/29.eng.

14. Xiong, L.; Bi, J.; Tang, Y.; Qiao, S. Z. Magnetic Core–Shell Silica Nanoparticles with Large Radial Mesopores for siRNA Delivery. *Small (Weinheim Bergstrasse, Germany)* **2016,** *12* (34), 4735–4742. PubMed PMID: 27199216. Epub 2016/05/21. eng.

15. Saravanakumar, G.; Kim, J.; Kim, W. J. Reactive-Oxygen-Species-Responsive Drug Delivery Systems: Promises and Challenges. *Adv. Sci. (Weinheim, Baden-Wurttemberg, Germany)* **2017,** *4* (1), 1600124. PubMed PMID: 28105390. Pubmed Central PMCID: PMC5238745. Epub 2017/01/21.eng.

16. Xu, B.; Xu, Y.; Su, G.; Zhu, H.; Zong, L. A Multifunctional Nanoparticle Constructed with a Detachable Albumin Outer Shell and a Redox-Sensitive Inner Core for Efficient siRNA Delivery to Hepatocellular Carcinoma Cells. *J. Drug Target.* **2018,** *26* (10), 941–954.

17. Chen, G.; Wang, Y.; Xie, R.; Gong, S. Tumor-Targeted pH/Redox Dual-Sensitive Unimolecular Nanoparticles for Efficient siRNA Delivery. *J. Controlled Release* **2017**, *259*, 105–114. PubMed PMID: 28159516. Pubmed Central PMCID: PMC5538929. Epub 2017/02/06.eng.

18. Li, X.; Sun, A.; Liu, Y.-J.; Zhang, W.-J.; Pang. N.; Cheng S-x, et al. Amphiphilic Dendrimer Engineered Nanocarrier Systems for Co-delivery of siRNA and Paclitaxel to Matrix Metalloproteinase-Rich Tumors for Synergistic Therapy. *NPG Asia Mater.* **2018**, *10* (4), 238–254.

19. Zhang, Y.; Ren, K.; Zhang, X.; Chao, Z.; Yang, Y.; Ye, D.; et al. Photo-Tearable Tape Close-Wrapped Upconversion Nanocapsules for Near-Infrared Modulated Efficient siRNA Delivery and Therapy. *Biomaterials* **2018**, *163*, 55–66. PubMed PMID: 29452948. Epub 2018/02/18.eng.

20. Zhang, Z.; Suo, H.; Zhao, X.; Sun, D.; Fan, L.; Guo, C. NIR-to-NIR Deep Penetrating Nanoplatforms $Y_2O_3:Nd^{3+}/Yb^{3+}@SiO_2@Cu_2S$ toward Highly Efficient Photothermal Ablation. *ACS Appl. Mater. Interfaces* **2018**, *10* (17), 14570–14576.

21. Meng, Q.; Meng, J.; Ran, W.; Wang, J.; Zhai, Y.; Zhang, P.; et al. Light-Activated Core–Shell Nanoparticles for Spatiotemporally Specific Treatment of Metastatic Triple-Negative Breast Cancer. *ACS Nano* **2018**, *12* (3), 2789–2802. PubMed PMID: 29462553. Epub 2018/02/21.eng.

22. Nguyen, D. H.; Lee, J. S.; Choi, J. H.; Park, K. M.; Lee, Y.; Park, K. D. Hierarchical Self-Assembly of Magnetic Nanoclusters for Theranostics: Tunable Size, Enhanced Magnetic Resonance Imagability, and Controlled and Targeted Drug Delivery. *Acta Biomater.* **2016**, *35*, 109–117. PubMed PMID: 26884278. Epub 2016/02/18.eng.

23. Maney, V.; Singh, M. An In Vitro Assessment of Novel Chitosan/Bimetallic PtAu Nanocomposites as Delivery Vehicles for Doxorubicin. *Nanomedicine (London, Engl.).* **2017**, *12* (21), 2625–2640. PubMed PMID: 28965478. Epub 2017/10/03.eng.

24. Zeng, J.; Du, P.; Liu, L.; Li, J.; Tian, K.; Jia, X.; et al. Superparamagnetic Reduction/ pH/Temperature Multistimuli-Responsive Nanoparticles for Targeted and Controlled Antitumor Drug Delivery. *Mol. Pharm.* **2015**, *12* (12), 4188–4199. PubMed PMID: 26554495. Epub 2015/11/12.eng.

25. Shmarakov, I.; Mukha, I.; Vityuk, N.; Borschovetska, V.; Zhyshchynska, N.; Grodzyuk, G.; et al. Antitumor Activity of Alloy and Core–Shell-Type Bimetallic AgAu Nanoparticles. *Nanoscale Res Lett.* **2017**, *12* (1), 333. PubMed PMID: 28476089. Pubmed Central PMCID: PMC5418356. Epub 2017/05/10.eng.

26. Abdel-Fattah, W. I.; Eid, M. M.; Abd El-Moez, S. I.; Mohamed, E.; Ali, G. W. Synthesis of Biogenic Ag@Pd Core–Shell Nanoparticles Having Anti-cancer/Anti-microbial Functions. *Life Sci.* **2017**, *183*, 28–36. PubMed PMID: 28642073. Epub 2017/06/24.eng.

27. Nikravan, G.; Haddadi-Asl, V.; Salami-Kalajahi, M. Synthesis of Dual Temperature- and pH-Responsive Yolk–Shell Nanoparticles by Conventional Etching and New Deswelling Approaches: DOX Release Behavior. *Colloids Surf., B, Biointerfaces* **2018**, *165*, 1–8. PubMed PMID: 29448215. Epub 2018/02/16.eng.

28. Yun, J. W.; Yoon, J. H.; Kang, B. C.; Cho, N. H.; Seok, S. H.; Min, S. K.; et al. The Toxicity and Distribution of Iron Oxide-Zinc Oxide Core–Shell Nanoparticles in C57BL/6 Mice After Repeated Subcutaneous Administration. *J Appl Toxicol.* **2015**, *35* (6), 593–602. PubMed PMID: 25572658. Epub 2015/01/13.eng.

29. Chitgupi, U.; Qin, Y.; Lovell, J. F. Targeted Nanomaterials for Phototherapy. *Nanotheranostics* **2017,** *1* (1), 38–58. PubMed PMID: 29071178. Pubmed Central PMCID: PMC5646723. Epub 2017/10/27.eng.

30. Huang, X.; El-Sayed, M. A. Gold Nanoparticles: Optical Properties and Implementations in Cancer Diagnosis and Photothermal Therapy. *J. Adv. Res.* **2010,** *1* (1), 13–28.

31. Jaque, D.; Martinez Maestro, L.; del Rosal, B.; Haro-Gonzalez, P.; Benayas, A.; Plaza, J. L.; et al. Nanoparticles for Photothermal Therapies. *Nanoscale* **2014,** *6* (16), 9494–9530. PubMed PMID: 25030381. Epub 2014/07/18.eng.

32. Wang, Y.; Black, K. C. L.; Luehmann, H.; Li, W.; Zhang, Y.; Cai, X.; et al. Comparison Study of Gold Nanohexapods, Nanorods, and Nanocages for Photothermal Cancer Treatment. *ACS Nano* **2013,** *7* (3), 2068–2077.

33. Xu, J.; Wang, X.; Teng, Z.; Lu, G.; He, N.; Wang, Z. F. Multifunctional Yolk–Shell Mesoporous Silica Obtained via Selectively Etching the Shell: a Therapeutic Nanoplatform for Cancer Therapy. *ACS Appl. Mater. Interfaces.* **2018,** *10* (29), 24440–24449.

34. Li, M.; Li, L.; Zhan, C.; Kohane, D. S. Core–Shell Nanostars for Multimodal Therapy and Imaging. *Theranostics* **2016,** *6* (13), 2306–2313. PubMed PMID: 27877236. Pubmed Central PMCID: PMC5118596. Epub 2016/11/24.eng.

35. Hao, Y.; Zhang, B.; Zheng, C.; Ji, R.; Ren, X.; Guo, F.; et al. The Tumor-Targeting Core–Shell Structured DTX-Loaded PLGA@Au Nanoparticles for Chemo-Photothermal Therapy and X-Ray Imaging. *J. Controlled Release* **2015,** *220* (Pt. A), 545–555. PubMed PMID: 26590021. Epub 2015/11/22.eng.

36. Parchur, A. K.; Sharma, G.; Jagtap, J. M.; Gogineni, V. R.; LaViolette, P. S.; Flister, M. J.; et al. Vascular Interventional Radiology Guided Photothermal Therapy of Colorectal Cancer Liver Metastasis with Theranostic Gold Nanorods. *ACS Nano.* **2018,** *12* (7), 6597–6611.

37. Wang, Y.; Barhoumi, A.; Tong, R.; Wang, W.; Ji, T.; Deng, X.; et al. BaTiO$_3$–Core Au–Shell Nanoparticles for Photothermal Therapy and Bimodal Imaging. *Acta Biomater.* **2018,** *72,* 287–294. PubMed PMID: 29578086. Pubmed Central PMCID: PMC5938150. Epub 2018/03/27.eng.

38. Li, L.; Lu, Y.; Jiang, C.; Zhu, Y.; Yang, X.; Hu, X.; et al. Actively Targeted Deep Tissue Imaging and Photothermal-Chemo Therapy of Breast Cancer by AntibodyFunctionalized Drug-Loaded X-Ray-Responsive Bismuth Sulfide@ Mesoporous Silica Core–Shell Nanoparticles. Adv. Funct. Mater. **2018,** *28* (5), pii: 1704623.

39. Li, J.; Jiang, H.; Yu, Z.; Xia, H.; Zou, G.; Zhang, Q.; et al. Multifunctional Uniform Core–Shell Fe$_3$O$_4$@mSiO$_2$ Mesoporous Nanoparticles for Bimodal Imaging and Photothermal Therapy. *Chem., Asian J.* **2013,** *8* (2), 385–391. PubMed PMID: 23225542. Epub 2012/12/12.eng.

40. Zhang, Z.; Suo, H.; Zhao, X.; Sun, D.; Fan, L.; Guo, C. NIR-to-NIR Deep Penetrating Nanoplatforms Y$_2$O$_3$:Nd(3+)/Yb(3+)@SiO$_2$@Cu$_2$S Toward Highly Efficient Photothermal Ablation. *ACS Appl. Mater. Interfaces* **2018,** *10* (17), 14570–14576. PubMed PMID: 29637783. Epub 2018/04/12.eng.

41. Gulzar, A.; Xu, J.; Yang, D.; Xu, L.; He, F.; Gai, S.; et al. Nano-Graphene Oxide-UCNP-Ce6 Covalently Constructed Nanocomposites for NIR-Mediated Bioimaging

and PTT/PDT Combinatorial Therapy. *Dalton Trans. (Cambridge, Engl.: 2003)* **2018,** *47* (11), 3931–3939. PubMed PMID: 29459928. Epub 2018/02/21.eng.

42. Zhou, J.; Wang, Z.; Li, Q.; Liu, F.; Du, Y.; Yuan, H.; et al. Hybridized Doxorubicin–Au Nanospheres Exhibit Enhanced Near-Infrared Surface Plasmon Absorption for Photothermal Therapy Applications. *Nanoscale* **2015,** *7* (13), 5869–5883. PubMed PMID: 25757809. Epub 2015/03/12.eng.

43. Cai, H.; Shen, T.; Kirillov, A. M.; Zhang, Y.; Shan, C.; Li, X.; et al. Self-Assembled Upconversion Nanoparticle Clusters for NIR-Controlled Drug Release and Synergistic Therapy after Conjugation with Gold Nanoparticles. *Inorg. Chem.* **2017,** *56* (9), 5295–5304. PubMed PMID: 28402112. Epub 2017/04/13.eng.

44. Sun, K.; You, C.; Wang, S.; Gao, Z.; Wu, H.; Tao, W. A.; et al. NIR Stimulus-Responsive Core–Shell Type Nanoparticles Based on Photothermal Conversion for Enhanced Antitumor Efficacy Through Chemo-Photothermal Therapy. *Nanotechnology* **2018,** *29* (28), 285302.

45. Shi, S.; Chen, X.; Wei, J.; Huang, Y.; Weng, J.; Zheng, N. Platinum(IV) Prodrug Conjugated Pd@Au Nanoplates for Chemotherapy and Photothermal Therapy. *Nanoscale* **2016,** *8* (10), 5706–5713. PubMed PMID: 26900670. Epub 2016/02/24. eng.

46. Duan, X.; Chan, C.; Guo, N.; Han, W.; Weichselbaum, R. R.; Lin, W. Photodynamic Therapy Mediated by Nontoxic Core–Shell Nanoparticles Synergizes with Immune Checkpoint Blockade to Elicit Antitumor Immunity and Antimetastatic Effect on Breast Cancer. *J. Am. Chem. Soc.* **2016,** *138* (51), 16686–16695. PubMed PMID: 27976881. Pubmed Central PMCID: PMC5667903. Epub 2016/12/16.eng.

47. Ai, F.; Sun, T.; Xu, Z.; Wang, Z.; Kong, W.; To, M. W.; et al. An Upconversion Nanoplatform for Simultaneous Photodynamic Therapy and Pt Chemotherapy to Combat Cisplatin Resistance. *Dalton Trans. (Cambridge, Engl.: 2003)* **2016,** *45* (33), 13052–13060. PubMed PMID: 27430044. Epub 2016/07/19.eng.

48. Hou, B.; Yang, W.; Dong, C.; Zheng, B.; Zhang, Y.; Wu, J.; et al. Controlled Co-release of Doxorubicin and Reactive Oxygen Species for Synergistic Therapy by NIR Remote-Triggered Nanoimpellers. *Mater. Sci. Eng. C Mater. Biol. Appl.* **2017,** *74,* 94–102. PubMed PMID: 28254338. Epub 2017/03/04.eng.

49. Hsu, C. C.; Lin, S. L.; Chang, C. A. Lanthanide-Doped Core–Shell–Shell Nanocomposite for Dual Photodynamic Therapy and Luminescence Imaging by a Single X-ray Excitation Source. *ACS Appl. Mater. Interfaces* **2018,** *10* (9), 7859–7870. PubMed PMID: 29405703. Epub 2018/02/07.eng.

50. Lu, S.; Tu, D.; Hu, P.; Xu, J.; Li, R.; Wang, M.; et al. Multifunctional Nano-bioprobes Based on Rattle-Structured Upconverting Luminescent Nanoparticles. *Angew. Chem. (Int. Ed. Engl.).* **2015,** *54* (27), 7915–7919. PubMed PMID: 26013002. Epub 2015/05/28.eng.

51. Pena Luengas, S. L.; Marin, G. H.; Aviles, K.; Cruz Acuna, R.; Roque, G.; Rodriguez Nieto, F.; et al. Enhanced Singlet Oxygen Production by Photodynamic Therapy and a Novel Method for Its Intracellular Measurement. *Cancer Biother. Radiopharm.* **2014,** *29* (10), 435–443. PubMed PMID: 25490599. Pubmed Central PMCID: PMC4267548. Epub 2014/12/10.eng.

52. Zhou, F.; Zheng, B.; Zhang, Y.; Wu, Y.; Wang, H.; Chang, J. Construction of Near-Infrared Light-Triggered Reactive Oxygen Species-Sensitive (UCN/SiO$_2$-RB +

DOX)@PPADT Nanoparticles for Simultaneous Chemotherapy and Photodynamic Therapy. *Nanotechnology* **2016**, *27* (23), 235601. PubMed PMID: 27139178. Epub 2016/05/04.eng.

53. Han, R.; Yi, H.; Shi, J.; Liu, Z.; Wang, H.; Hou, Y.; et al. pH-Responsive Drug Release and NIR-Triggered Singlet Oxygen Generation Based on a Multifunctional Core–Shell–Shell Structure. *Phys. Chem. Chem. Phys.* **2016**, *18* (36), 25497–25503. PubMed PMID: 27711590. Epub 2016/10/07.eng.

54. Patel, K.; Raj, B. S.; Chen, Y.; Lou, X. Novel Folic Acid Conjugated Fe_3O_4–ZnO Hybrid Nanoparticles for Targeted Photodynamic Therapy. *Colloids Surf. B, Biointerfaces* **2017**, *150*, 317–325. PubMed PMID: 27810128. Epub 2016/11/05. eng.

55. Seo, S. H.; Kim, B. M.; Joe, A.; Han, H. W.; Chen, X.; Cheng, Z.; et al. NIR-Light-Induced Surface-Enhanced Raman Scattering for Detection and Photothermal/Photodynamic Therapy of Cancer Cells Using Methylene Blue-Embedded Gold Nanorod@SiO_2 Nanocomposites. *Biomaterials* **2014**, *35* (10), 3309–3318. PubMed PMID: 24424205. Pubmed Central PMCID: PMC4576838. Epub 2014/01/16.eng.

56. Tan, X.; Wang, J.; Pang, X.; Liu, L.; Sun, Q.; You, Q.; et al. Indocyanine Green-Loaded Silver Nanoparticle@Polyaniline Core/Shell Theranostic Nanocomposites for Photoacoustic/Near-Infrared Fluorescence Imaging-Guided and Single-Light-Triggered Photothermal and Photodynamic Therapy. *ACS Appl. Mater. Interfaces* **2016**, *8* (51), 34991–35003. PubMed PMID: 27957854. Epub 2016/12/14.eng.

57. Chen, Y.; Zhang, F.; Wang, Q.; Tong, R.; Lin, H.; Qu, F. Near-Infrared Light-Mediated LA–UCNPs@SiO_2–C/HA@mSiO_2–DOX@NB Nanocomposite for Chemotherapy/PDT/PTT and Imaging. *Dalton Trans. (Cambridge, Engl.: 2003)* **2017**, *46* (41), 14293–14300. PubMed PMID: 29019363. Epub 2017/10/12.eng.

58. Choi, S. Y.; Baek, S. H.; Chang, S. J.; Song, Y.; Rafique, R.; Lee, K. T.; et al. Synthesis of Upconversion Nanoparticles Conjugated with Graphene Oxide Quantum Dots and Their Use against Cancer Cell Imaging and Photodynamic Therapy. *Biosens. Bioelectron.* **2017**, *93*, 267–273. PubMed PMID: 27590213. Epub 2016/09/04.eng.

59. Qin, C.; Fei, J.; Wang, A.; Yang, Y.; Li, J. Rational Assembly of a Biointerfaced Core@Shell Nanocomplex towards Selective and Highly Efficient Synergistic Photothermal/Photodynamic Therapy. *Nanoscale* **2015**, *7* (47), 20197–20210. PubMed PMID: 26574662. Epub 2015/11/18.eng.

60. Khaydukov, E. V.; Mironova, K. E.; Semchishen, V. A.; Generalova, A. N.; Nechaev, A. V.; Khochenkov, D. A.; et al. Riboflavin Photoactivation by Upconversion Nanoparticles for Cancer Treatment. *Sci Rep.* **2016**, *6*, 35103. PubMed PMID: 27731350. Pubmed Central PMCID: PMC5059683. Epub 2016/10/13.eng.

61. Jia, H. R.; Jiang, Y. W.; Zhu, Y. X.; Li, Y. H.; Wang, H. Y.; Han, X.; et al. Plasma Membrane Activatable Polymeric Nanotheranostics with Self-Enhanced Light-Triggered Photosensitizer Cellular Influx for Photodynamic Cancer Therapy. *J. Controlled Release* **2017**, *255*, 231–241. PubMed PMID: 28442408. Epub 2017/04/27.eng.

62. Pramual, S.; Lirdprapamongkol, K.; Svasti, J.; Bergkvist, M.; Jouan-Hureaux, V.; Arnoux, P.; et al. Polymer–Lipid–PEG Hybrid Nanoparticles as Photosensitizer Carrier for Photodynamic Therapy. *J. Photochem. Photobiol. B, Biol.* **2017**, *173*, 12–22. PubMed PMID: 28554072. Epub 2017/05/30.eng.

63. Tan, J.; Sun, C.; Xu, K.; Wang, C.; Guo, J. Immobilization of ALA–Zn(II) Coordination Polymer Pro-photosensitizers on Magnetite Colloidal Supraparticles for Target Photodynamic Therapy of Bladder Cancer. *Small (Weinheim Bergstrasse, Germany)* **2015,** *11* (47), 6338–6346. PubMed PMID: 26514273. Epub 2015/10/31. eng.

64. Wang, J.; Zhong, Y.; Wang, X.; Yang, W.; Bai, F.; Zhang, B.; et al. pH-Dependent Assembly of Porphyrin–Silica Nanocomposites and Their Application in Targeted Photodynamic Therapy. *Nano Lett.* **2017,** *17* (11), 6916–6921. PubMed PMID: 29019240. Epub 2017/10/12.eng.

65. Chen, Q.; Wang, X.; Wang, C.; Feng, L.; Li, Y.; Liu, Z. Drug-Induced Self-Assembly of Modified Albumins as Nano-theranostics for Tumor-Targeted Combination Therapy. *ACS Nano* **2015,** *9* (5), 5223–5233. PubMed PMID: 25950506. Epub 2015/05/08.eng.

66. Xu, J.; Gulzar, A.; Liu, Y.; Bi, H.; Gai, S.; Liu, B.; et al. Integration of IR-808 Sensitized Upconversion Nanostructure and MoS2 Nanosheet for 808 nm NIR Light Triggered Phototherapy and Bioimaging. Small (Weinheim Bergstrasse, Germany) **2017,** *13*(36), 1701841.

67. Duchi, S.; Ramos-Romero, S.; Dozza, B.; Guerra-Rebollo, M.; Cattini, L.; Ballestri, M.; et al. Development of Near-Infrared Photoactivable Phthalocyanine-Loaded Nanoparticles to Kill Tumor Cells: An Improved Tool for Photodynamic Therapy of Solid Cancers. *Nanomed.: Nanotechnol., Biol., Med.* **2016,** *12* (7), 1885–1897. PubMed PMID: 27133189. Epub 2016/05/03.eng.

68. Kim, Y. K.; Na, H. K.; Kim, S.; Jang, H.; Chang, S. J.; Min, D. H. One-Pot Synthesis of Multifunctional Au@Graphene Oxide Nanocolloid Core@Shell Nanoparticles for Raman Bioimaging, Photothermal, and Photodynamic Therapy. *Small (Weinheim Bergstrasse, Germany)* **2015,** *11* (21), 2527–2535. PubMed PMID: 25626859. Epub 2015/01/30.eng.

69. Hou, Z.; Zhang, Y.; Deng, K.; Chen, Y.; Li, X.; Deng, X.; et al. UV-Emitting Upconversion-Based TiO_2 Photosensitizing Nanoplatform: Near-Infrared Light Mediated In Vivo Photodynamic Therapy via Mitochondria-Involved Apoptosis Pathway. *ACS Nano* **2015,** *9* (3), 2584–2599. PubMed PMID: 25692960. Epub 2015/02/19.eng.

70. Lucky, S. S.; Muhammad Idris, N.; Li, Z.; Huang, K.; Soo, K. C.; Zhang, Y. Titania Coated Upconversion Nanoparticles for Near-Infrared Light Triggered Photodynamic Therapy. *ACS Nano* **2015,** *9* (1), 191–205. PubMed PMID: 25564723. Epub 2015/01/08.eng.

71. Li, Y.; Wang, R.; Xu, Y.; Zheng, W.; Li, Y. Influence of Silica Surface Coating on Operated Photodynamic Therapy Property of Yb(3+)–Tm(3+):Ga(III)-Doped ZnO Upconversion Nanoparticles. *Inorg. Chem.* **2018,** *57* (13), 8012–8018.

72. Yu, Z.; Xia, Y.; Xing, J.; Li, Z.; Zhen, J.; Jin, Y.; et al. Y1-Receptor-Ligand-Functionalized Ultrasmall Upconversion Nanoparticles for Tumor-Targeted Trimodality Imaging and Photodynamic Therapy with Low Toxicity. *Nanoscale* **2018.** PubMed PMID: 29850734. Epub 2018/06/01.eng.

73. Huang, K.; Liu, H.; Kraft, M.; Shikha, S.; Zheng, X.; Agren, H.; et al. A Protected Excitation-Energy Reservoir for Efficient Upconversion Luminescence. *Nanoscale* **2017,** *10* (1), 250–259. PubMed PMID: 29210408. Epub 2017/12/07.eng.

74. Mironova, K. E.; Khochenkov, D. A.; Generalova, A. N.; Rocheva, V. V.; Sholina, N. V.; Nechaev, A. V.; et al. Ultraviolet Phototoxicity of Upconversion Nanoparticles Illuminated with Near-Infrared Light. *Nanoscale* **2017**, *9* (39), 14921–14928. PubMed PMID: 28952637. Epub 2017/09/28.eng.

75. Cen, Y.; Deng, W. J.; Yang, Y.; Yu, R. Q.; Chu, X. Core–Shell–Shell Multifunctional Nanoplatform for Intracellular Tumor-Related mRNAs Imaging and Near-Infrared Light Triggered Photodynamic–Photothermal Synergistic Therapy. *Anal. Chem.* **2017**, *89* (19), 10321–10328. PubMed PMID: 28872842. Epub 2017/09/06.eng.

76. Feng, L.; Yang, D.; He, F.; Gai, S.; Li, C.; Dai, Y.; et al. A Core–Shell–Satellite Structured Fe3O4@g–C3N4–UCNPs–PEG for T1/T2-Weighted Dual-Modal MRI-Guided Photodynamic Therapy. Adv. Healthcare Mater. **2017**, *6* (18), 1700502

77. Zhang, X.; Ai, F.; Sun, T.; Wang, F.; Zhu, G. Multimodal Upconversion Nanoplatform with a Mitochondria-Targeted Property for Improved Photodynamic Therapy of Cancer Cells. *Inorg. Chem.* **2016**, *55* (8), 3872–3880. PubMed PMID: 27049165. Epub 2016/04/07.eng.

78. Wang, H.; Han, R. L.; Yang, L. M.; Shi, J. H.; Liu, Z. J.; Hu, Y.; et al. Design and Synthesis of Core–Shell–Shell Upconversion Nanoparticles for NIR-Induced Drug Release, Photodynamic Therapy, and Cell Imaging. *ACS Appl. Mater. Interfaces* **2016**, *8* (7), 4416–4423. PubMed PMID: 26816249. Epub 2016/01/28.eng.

79. Li, X.; Guo, Z.; Zhao, T.; Lu, Y.; Zhou, L.; Zhao, D.; et al. Filtration Shell Mediated Power Density Independent Orthogonal Excitations–Emissions Upconversion Luminescence. *Angew. Chem. (Int. Ed. Engl.)* **2016**, *55* (7), 2464–2469. PubMed PMID: 26762564. Epub 2016/01/15.eng.

80. Ai, F.; Ju, Q.; Zhang, X.; Chen, X.; Wang, F.; Zhu, G. A core–Shell–Shell Nanoplatform Upconverting Near-Infrared Light at 808 nm for Luminescence Imaging and Photodynamic Therapy of Cancer. *Sci. Rep.* **2015**, *5*, 10785. PubMed PMID: 26035527. Pubmed Central PMCID: PMC4451683. Epub 2015/06/04.eng.

81. Lee, S. B.; Kumar, D.; Li, Y.; Lee, I. K.; Cho, S. J.; Kim, S. K.; et al. PEGylated Crushed Gold Shell–Radiolabeled Core Nanoballs for In Vivo Tumor Imaging with Dual Positron Emission Tomography and Cerenkov Luminescent Imaging. *J. Nanobiotechnol.* **2018**, *16* (1), 41. PubMed PMID: 29669544. Pubmed Central PMCID: PMC5907375. Epub 2018/04/20.eng.

82. Sedeky, A. S.; Khalil, I. A.; Hefnawy, A.; El-Sherbiny, I. M. Development of Core–Shell Nanocarrier System for Augmenting Piperine Cytotoxic Activity against Human Brain Cancer Cell Line. *Eur. J. Pharm. Sci.* **2018**, *118*, 103–112. PubMed PMID: 29597041. Epub 2018/03/30.eng.

83. Tang, F.; Wang, C.; Wang, X.; Li, L. Facile Synthesis of Biocompatible Fluorescent Nanoparticles for Cellular Imaging and Targeted Detection of Cancer Cells. *ACS Appl. Mater. Interfaces* **2015**, *7* (45), 25077–25083. PubMed PMID: 26544019. Epub 2015/11/07.eng.

84. Zhao, Y.; Zhao, J.; Shan, G.; Yan, D.; Chen, Y.; Liu, Y. SERS-Active Liposome@ Ag/Au Nanocomposite for NIR Light-Driven Drug Release. *Colloids Surf., B, Biointerfaces* **2017**, *154*, 150–159. PubMed PMID: 28334692. Epub 2017/03/24.eng.

85. Chen, M.; Zhang, L.; Gao, M.; Zhang, X. High-Sensitive Bioorthogonal SERS tag for Live Cancer Cell Imaging by Self-Assembling Core–Satellites Structure Gold–Silver

Nanocomposite. *Talanta* **2017,** *172*, 176–181. PubMed PMID: 28602292. Epub 2017/06/13.eng.

86. Dorsey, J. F.; Sun, L.; Joh, D. Y.; Witztum, A.; Kao, G. D.; Alonso-Basanta, M.; et al. Gold Nanoparticles in Radiation Research: Potential Applications for Imaging and Radiosensitization. *Transl. Cancer Res.* **2013,** *2* (4), 280–291. PubMed PMID: 25429358. Pubmed Central PMCID: PMC4241969. Epub 2013/08/01.eng.

87. Rosa, S.; Connolly, C.; Schettino, G.; Butterworth, K. T.; Prise, K. M. Biological Mechanisms of Gold Nanoparticle Radiosensitization. *Cancer Nanotechnol.* **2017,** *8* (1), 2. PubMed PMID: 28217176. Pubmed Central PMCID: PMC5288470. Epub 2017/02/22.eng.

88. Li, M.; Zhao, Q.; Yi, X.; Zhong, X.; Song, G.; Chai, Z.; et al. Au@MnS@ZnS Core/Shell/Shell Nanoparticles for Magnetic Resonance Imaging and Enhanced Cancer Radiation Therapy. *ACS Appl. Mater. Interfaces* **2016,** *8* (15), 9557–9564. PubMed PMID: 27039932. Epub 2016/04/05.eng.

89. Liang, G.; Jin, X.; Zhang, S.; Xing, D. RGD Peptide-Modified Fluorescent Gold Nanoclusters as Highly Efficient Tumor-Targeted Radiotherapy Sensitizers. *Biomaterials* **2017,** *144*, 95–104. PubMed PMID: 28834765. Epub 2017/08/24.eng.

90. Cho, J.; Wang, M.; Gonzalez-Lepera, C.; Mawlawi, O.; Cho, S. H. Development of Bimetallic (Zn@Au) Nanoparticles as Potential PET-Imageable Radiosensitizers. *Med. Phys.* **2016,** *43* (8), 4775. PubMed PMID: 27487895. Pubmed Central PMCID: PMC4967079. Epub 2016/08/05.eng.

91. Darfarin, G.; Salehi, R.; Alizadeh, E.; Nasiri Motlagh, B.; Akbarzadeh, A.; Farajollahi, A. The effect of SiO2/Au Core–Shell Nanoparticles on Breast Cancer Cell's Radiotherapy. *Artif. Cells Nanomed. Biotechnol.* **2018,** *46*(sup2) 836–846.

92. Kim, K.; Oh, K. S.; Park, D. Y.; Lee, J. Y.; Lee, B. S.; Kim, I. S.; et al. Doxorubicin/Gold-Loaded Core/Shell Nanoparticles for Combination Therapy to Treat Cancer Through the Enhanced Tumor Targeting. *J. Controlled Release* **2016,** *228*, 141–149. PubMed PMID: 26970205. Epub 2016/03/13.eng.

93. Ahn, S. H.; Lee, N.; Choi, C.; Shin, S. W.; Han, Y.; Park, H. C. Feasibility Study of Fe_3O_4/TaO_x Nanoparticles as a Radiosensitizer for Proton Therapy. *Phys. Med. Biol.* **2018,** *63* (11), 114001. PubMed PMID: 29726404. Epub 2018/05/05.eng.

94. Seemann, K. M.; Luysberg, M.; Revay, Z.; Kudejova, P.; Sanz, B.; Cassinelli, N.; et al. Magnetic Heating Properties and Neutron Activation of Tungsten-Oxide Coated Biocompatible FePt Core–Shell Nanoparticles. *J. Controlled Release* **2015,** *197*, 131–137. PubMed PMID: 25445697. Epub 2014/12/03.eng.

95. Generalov, R.; Kuan, W. B.; Chen, W.; Kristensen, S.; Juzenas, P. Radiosensitizing Effect of Zinc Oxide and Silica Nanocomposites on Cancer Cells. *Colloids Surf., B, Biointerfaces* **2015,** *129*, 79–86. PubMed PMID: 25829130. Epub 2015/04/02.eng.

96. Bhalla, N.; Jolly, P.; Formisano, N.; Estrela, P. Introduction to Biosensors. *Essays Biochem.* **2016,** *60* (1), 1–8. PubMed PMID: 27365030. Pubmed Central PMCID: PMC4986445. Epub 2016/07/02.eng.

97. Grieshaber, D.; MacKenzie, R.; Voros, J.; Reimhult, E. Electrochemical Biosensors—Sensor Principles and Architectures. *Sensors (Basel)* **2008,** *8* (3), 1400–1458. PubMed PMID: 27879772. Pubmed Central PMCID: PMC3663003. Epub 2008/03/07.eng.

98. Wang, E. C.; Wang, A. Z. Nanoparticles and Their Applications in Cell and Molecular Biology. *Integr. Biol.: Quant. Biosci. Nano Macro* **2014,** *6* (1), 9–26. PubMed PMID: 24104563. Pubmed Central PMCID: PMC3865110. Epub 2013/10/10.eng.

99. Hayat, A.; Catanante, G.; Marty, J. L. Current Trends in Nanomaterial-Based Amperometric Biosensors. *Sensors (Basel, Switzerland)* **2014,** *14* (12), 23439–23461. PubMed PMID: PMC4299072.

100. Li, Y.; Zhang, Y.; Li, F.; Feng, J.; Li, M.; Chen, L.; et al. Ultrasensitive Electrochemical Immunosensor for Quantitative Detection of SCCA Using $Co_3O_4@CeO_2$–$Au@Pt$ Nanocomposite as Enzyme-Mimetic Labels. *Biosens. Bioelectron.* **2017,** *92,* 33–39. PubMed PMID: 28182976. Epub 2017/02/10.eng.

101. Wu, D.; Ma, H.; Zhang, Y.; Jia, H.; Yan, T.; Wei, Q. Corallite-Like Magnetic $Fe_3O_4@$ $MnO_2@Pt$ Nanocomposites as Multiple Signal Amplifiers for the Detection of Carcinoembryonic Antigen. *ACS Appl. Mater. Interfaces* **2015,** *7* (33), 18786–18793. PubMed PMID: 26244448. Epub 2015/08/06.eng.

102. Li, F.; Han, J.; Jiang, L.; Wang, Y.; Li, Y.; Dong, Y.; et al. An Ultrasensitive Sandwich-Type Electrochemical Immunosensor Based on Signal Amplification Strategy of Gold Nanoparticles Functionalized Magnetic Multi-Walled Carbon Nanotubes Loaded with Lead Ions. *Biosens. Bioelectron.* **2015,** *68,* 626–632. PubMed PMID: 25656779. Epub 2015/02/07.eng.

103. Gao, Z.; Zhang, L.; Ma, C.; Zhou, Q.; Tang, Y.; Tu, Z.; et al. TiO_2 Decorated Co_3O_4 Acicular Nanotube Arrays and Its Application as a Non-Enzymatic Glucose Sensor. *Biosens. Bioelectron.* **2016,** *80,* 511–518. PubMed PMID: 26890826. Epub 2016/02/19.eng.

104. Zhang, X.; Du, B.; Wu, D.; Ma, H.; Zhang, Y.; Li, H.; et al. Signal Amplification Strategy of Triple-Layered Core-Shell $Au@Pd@Pt$ Nanoparticles for Ultrasensitive Immunoassay Detection of Squamous Cell Carcinoma Antigen. *J. Biomed. Nanotechnol.* **2015,** *11* (2), 245–252. PubMed PMID: 26349300. Epub 2015/09/10. eng.

105. Tian, L.; Liu, L.; Li, Y.; Wei, Q.; Cao, W. Ultrasensitive Sandwich-Type Electrochemical Immunosensor Based on Trimetallic Nanocomposite Signal Amplification Strategy for the Ultrasensitive Detection of CEA. *Sci. Rep.* **2016,** *6,* 30849. PubMed PMID: 27488806. Pubmed Central PMCID: PMC4973229. Epub 2016/08/05.eng.

106. Li, F.; Li, Y.; Feng, J.; Dong, Y.; Wang, P.; Chen, L.; et al. Ultrasensitive Amperometric Immunosensor for PSA Detection Based on $Cu_2O@CeO_2$–Au Nanocomposites as Integrated Triple Signal Amplification Strategy. *Biosens. Bioelectron.* **2017,** *87,* 630–637. PubMed PMID: 27619526. Epub 2016/09/14.eng.

107. Wang, Y.; Zhang, Y.; Su, Y.; Li, F.; Ma, H.; Li, H.; et al. Ultrasensitive Non-Mediator Electrochemical Immunosensors Using Au/Ag/Au Core/Double Shell Nanoparticles as Enzyme-Mimetic Labels. *Talanta* **2014,** *124,* 60–66. PubMed PMID: 24767446. Epub 2014/04/29.eng.

108. Spain, E.; Adamson, K.; Elshahawy, M.; Bray, I.; Keyes, T. E.; Stallings, R. L.; et al. Hemispherical Platinum:Silver Core:Shell Nanoparticles for miRNA Detection. *Analyst* **2017,** *142* (5), 752–762. PubMed PMID: 28091676. Epub 2017/01/17.eng.

109. Li, Y.; Lu, Q.; Wu, S.; Wang, L.; Shi, X. Hydrogen Peroxide Sensing Using Ultrathin Platinum-Coated Gold Nanoparticles with Core@Shell Structure. *Biosens. Bioelectron.* **2013,** *41,* 576–581. PubMed PMID: 23062554. Epub 2012/10/16.eng.

110. Seo, S. J.; Han, S. M.; Cho, J. H.; Hyodo, K.; Zaboronok, A.; You, H.; et al. Enhanced Production of Reactive Oxygen Species by Gadolinium Oxide Nanoparticles Under Core–Inner–Shell Excitation by Proton or Monochromatic X-Ray Irradiation: Implication of the Contribution from the Interatomic de-Excitation-Mediated Nanoradiator Effect to Dose Enhancement. *Radiat. Environ. Biophys.* **2015,** *54* (4), 423–431. PubMed PMID: 26242374. Epub 2015/08/06.eng.

111. Tarish, S.; Xu, Y.; Wang, Z.; Mate, F.; Al-Haddad, A.; Wang, W.; et al. Highly Efficient Biosensors by Using Well-Ordered ZnO/ZnS Core/Shell Nanotube Arrays. *Nanotechnology* **2017,** *28* (40), 405501. PubMed PMID: 28749787. Epub 2017/07/28.eng.

112. Grzelczak, M.; Liz-Marzan, L. M. The Relevance of Light in the Formation of Colloidal Metal Nanoparticles. *Chem. Soc. Rev.* **2014,** *43* (7), 2089–2097.

113. Dinh, C.-T.; Nguyen, T.-D.; Kleitz, F.; Do, T.-O. A New Route to Size and Population Control of Silver Clusters on Colloidal TiO_2 Nanocrystals. *ACS Appl. Mater. Interfaces* **2011,** *3* (7), 2228–2234.

114. Zhou, N.; Ye, C.; Polavarapu, L.; Xu, Q. H. Controlled Preparation of $Au/Ag/SnO_2$ core–Shell Nanoparticles Using a Photochemical Method and Applications in LSPR Based Sensing. *Nanoscale* **2015,** *7* (19), 9025–9032. PubMed PMID: 25921493. Epub 2015/04/30.eng.

115. Marques, L.; Hernandez, F. U.; James, S. W.; Morgan, S. P.; Clark, M.; Tatam, R. P.; et al. Highly Sensitive Optical Fibre Long Period Grating Biosensor Anchored with Silica Core Gold Shell Nanoparticles. *Biosens. Bioelectron.* **2016,** *75*, 222–231. PubMed PMID: 26319165. Epub 2015/09/01.eng.

116. Xu, D. D.; Deng, Y. L.; Li, C. Y.; Lin, Y.; Tang, H. W. Metal-Enhanced Fluorescent Dye-Doped Silica Nanoparticles and Magnetic Separation: A Sensitive Platform for One-Step Fluorescence Detection of Prostate Specific Antigen. *Biosens. Bioelectron.* **2017,** *87,* 881–887. PubMed PMID: 27662582. Epub 2016/09/24.eng.

117. Li, J.; Skeete, Z.; Shan, S.; Yan, S.; Kurzatkowska, K.; Zhao, W.; et al. Surface Enhanced Raman Scattering Detection of Cancer Biomarkers with Bifunctional Nanocomposite Probes. *Anal. Chem.* **2015,** *87* (21), 10698–10702. PubMed PMID: 26479337. Epub 2015/10/20.eng.

INDEX

Milton Keynes UK
Ingram Content Group UK Ltd.
UKHW050257161024
449569UK00042B/1746

9 781774 634486